Sustainable Development Goals Series

The **Sustainable Development Goals Series** is Springer Nature's inaugural cross-imprint book series that addresses and supports the United Nations' seventeen Sustainable Development Goals. The series fosters comprehensive research focused on these global targets and endeavours to address some of society's greatest grand challenges. The SDGs are inherently multidisciplinary, and they bring people working across different fields together and working towards a common goal. In this spirit, the Sustainable Development Goals series is the first at Springer Nature to publish books under both the Springer and Palgrave Macmillan imprints, bringing the strengths of our imprints together.

The Sustainable Development Goals Series is organized into eighteen subseries: one subseries based around each of the seventeen respective Sustainable Development Goals, and an eighteenth subseries, "Connecting the Goals", which serves as a home for volumes addressing multiple goals or studying the SDGs as a whole. Each subseries is guided by an expert Subseries Advisor with years or decades of experience studying and addressing core components of their respective Goal.

The SDG Series has a remit as broad as the SDGs themselves, and contributions are welcome from scientists, academics, policymakers, and researchers working in fields related to any of the seventeen goals. If you are interested in contributing a monograph or curated volume to the series, please contact the Publishers: Zachary Romano [Springer; zachary.romano@springer.com] and Rachael Ballard [Palgrave Macmillan; rachael.ballard@palgrave.com].

Jin Sun

Sustainable Financialization

Leveraging FinTech, Social Innovation and Inclusive Partnership for Sustainable Development Goals

palgrave
macmillan

Jin Sun
Chinese University of Hong Kong
Hong Kong, Hong Kong

ISSN 2523-3084 ISSN 2523-3092 (electronic)
Sustainable Development Goals Series
ISBN 978-981-96-3579-5 ISBN 978-981-96-3580-1 (eBook)
https://doi.org/10.1007/978-981-96-3580-1

Cover illustration: Barbara Boensch Alamy Stock Photo

This Palgrave Macmillan imprint is published by the registered company Springer Nature Singapore Pte Ltd.
The registered company address is: 152 Beach Road, #21-01/04 Gateway East, Singapore 189721, Singapore

Preface

Learning Global Governance in Harvard, Geneva, and Europe

Ten years ago, while sitting on the campus of Harvard's John F. Kennedy School of Government, I could barely have imagined that I would one day participate so much to the Sustainable Development Goals (SDGs). Back then, the only hint was that I excelled in Professor John Gerard Ruggie's Global Governance course. In 2013, global governance was a fresh concept to most Chinese individuals including me. The Belfer Center for Science and International Affairs at Harvard has been the world's best think tank in international affairs. In 2013, the Center introduced an international and global affairs concentration in the Master of Public Policy program at the Kennedy School. I was the sole Chinese student in the inaugural class. My outstanding performance in Professor Ruggie's course in this new track led me to "mistakenly" believe I have a gift for global governance.

Before that, I graduated from Hong Kong's best university, the Chinese University of Hong Kong. In Hong Kong, Asia's premier international financial hub, whether in business, law, or economics, the prevailing education emphasized maximizing shareholder interests. The focus was on ensuring listed companies adhered to corporate governance principles, avoiding actions that would harm shareholders' interests. It was also stressed that listed companies should embrace corporate social responsibility (CSR), such as encouraging employee participation in voluntary services or supporting social welfare organizations. However, environmental responsibility, climate justice, and social impact were rarely considered in investment decisions or business models in Asia one to two decades ago.

When Professor Ruggie shared his United Nations (UN) experiences with us, it seemed that global sustainable development issues were quite distant from Asian business practices. Professor Ruggie, a former Assistant Secretary-General for Strategic Planning at the UN, played a key role in planning the Millennium Development Goals (the precursor of the Sustainable Development Goals) and the UN Global Compact along with Secretary-General Kofi Annan. His early advocacy for the Principles for Responsible Investment (PRI) laid the groundwork for today's principle of Environmental, Social, and Governance (ESG). However, convincing bankers worldwide to prioritize these sustainable principles in investment decisions proved challenging from Wall Street to London, Hong Kong, and Shanghai. By the time I graduated from Harvard in 2014 as part of the first international and global affairs class, global governance faced major setbacks, from North Korean and Iranian nuclear to climate change negotiations.

The tide began to turn between 2015 and 2017. The 2015 United Nations Climate Change Conference in Paris saw significant progress, with China, the US, and the EU reaching a consensus. The Paris Agreement came into effect in 2016, coinciding with successful negotiations and implementation of the Iran nuclear deal, respectively, in 2015 and 2016. In 2017, all 15 members of the UN Security Council unanimously adopted sanctions resolutions against North Korea, especially the Resolution 2371, showcasing global cooperation in addressing challenges through peaceful means. The focus of global governance on these matters shifted from setting a global agenda and negotiating global goals to implementing them, spanning nuclear disarmament to climate change.

In 2017, Professor Gregoire Mallard secured a grant from the European Research Council for us to research global governance at the Global Governance Center founded by Prof Thomas Biesterker at the Geneva Graduate Institute of International Studies and Development. Secretary-General Kofi Annan is a distinguished alumnus of the Institute, which is widely considered to be the best school of international studies in Europe. From 2017 to 2022, I conducted doctoral research on global governance in Geneva. In the summer of 2019, I participated in directed studies of public international law at the Hague Academy of International Law. Traveling from Geneva to The Hague, Berlin, and Paris, I observed the rise of climate protests in Europe and found inspiration in the Dutch courts' legal rulings on state accountability for climate change and their backing of climate activism. In Geneva, I assisted the UN Children Foundation (UNICEF) in implementing and enforcing the principles of 17 SDGs into the operation and practices of private fundraising and partnerships.

Why Is It Challenging to Implement Global Governance Objectives?

Can these strategies for implementing global governance objectives be applied in Asia? This is an inquiry that has been on my mind since my return to Asia in the summer of 2022, particularly after serving my role as an assistant professor at the Chinese University of Hong Kong. Through collaborative research with Professor Chen Ling from Tsinghua University, we systematically analyzed the empirical results of the initial ten-year pilot of the emission trading system introduced from Europe to China. Our study revealed that the carbon trading market had a limited impact on reducing carbon emissions in China during its first decade. Recent academic studies also indicate that most global governance or domestic policy measures aimed at combating climate change have shown little effectiveness.

Why do global governance mechanisms or public initiatives yield minimal results in Asia (and other regions)? Let's consider the example of addressing global climate change. There are various approaches that appear to facilitate carbon emission reduction, each with its own advantages, disadvantages, and constraints. One significant approach involves controlling and decreasing carbon emissions by adjusting energy sources, increasing the use of clean energy like solar and wind power, reducing reliance on fossil fuels, enhancing energy efficiency, and promoting the adoption of electric vehicles. Another key approach is achieving decarbonization by exploring viable carbon capture and storage (CCS) solutions, which include afforestation and reforestation. In addition, the carbon market aids in establishing carbon emission targets, encouraging voluntary reductions in carbon emissions, and facilitating the attainment of regulatory objectives for phased mandatory reductions in carbon emissions. In China, despite notable progress by both the private and public sectors in advancing clean energy, reducing fossil fuel usage, enhancing energy efficiency, and popularizing electric vehicles, local governments and the power generation sector have expressed concerns that stringent mandatory targets for carbon emission reduction could impede local economic growth and employment. Instead, local governments in China often view climate change as an opportunity to nurture local technology champions in new energy industries like solar and electric vehicles and revitalize mercantilism in international trade. In Europe, although there is public consensus on reducing carbon emissions, the Ukrainian war and subsequent sanctions have posed significant challenges to Europe's energy security. Trade protectionists aspire for Europe's fiscal subsidies aimed at controlling and reducing carbon emissions to primarily benefit European employment rather than China's photovoltaic and electric vehicle export sectors. This underscores that while major powers agree on the global governance objectives associated with addressing climate change, there are substantial disparities in how these global initiatives should be implemented, as they prioritize transforming global initiatives to align with their national or commercial interests. It underlies that it is challenging to implement global governance objectives without effective public participation and social mobilization.

Are there alternative solutions? This question underpins the essence of this book. The book originated from a somewhat audacious concept: to devise a viable incentive mechanism to mobilize corporate sectors like global FinTech leaders to advance the achievement of SDGs through effective public participation and social mobilization empowered by ESG innovation, particularly sustainable financial practices. This endeavor thus combines the knowledge of corporate governance I acquired in Asia's premier international financial hub with the latest practices of global governance I gained from the top academic institutions in the United States and Europe and the UN agency over the past decade.

Alipay+ for SDGs: Leveraging Digital Payments Revolution for Public Participation to SDGs

The Alipay+ for SDGs, implemented since November 2022, is led by Ant International in partnership with my team at the Department of Sociology, The Chinese University of Hong Kong. Participants include over 30 digital payment platforms and 300 partners from nine regions including Thailand, Malaysia, Philippines, Indonesia, Singapore, South

Korea, Japan, Hong Kong, and Macau. The Alipay+ for SDGs aims to leverage the Alipay+ digital payment infrastructure to support the SDGs, drawing from China's experience in digital payments for SDG initiatives such as Ant Forest, digital poverty alleviation, and digital economy-empowered job opportunity programs. By serving global consumers and small and micro-enterprises, the project aims to lead in nine Asian and more markets in using digital payments to support the SDGs. This project is possible because many countries in Asia have entered a cashless society, in which digital payment platforms make it possible to advance the achievement of SDGs through effective public participation and social mobilization empowered by ESG innovation, particularly sustainable financial practices

What is Alipay+? Alipay+ is a comprehensive suite of innovative cross-border digital payment and marketing solutions. Since 2022, Alipay+ has experienced a quick growth in global marketplaces, for which it now covers over 66 markets globally with more than 90 million merchants, seamlessly connecting with over 1.5 billion Asian consumers. Alipay+ enables Asian consumers to use their home country's digital wallets for overseas payments without opening local digital wallets or bank accounts while traveling abroad. Alipay+ offers a range of merchant solutions including Alipay+ Rewards, a new digital marketing platform developed in partnership with e-wallets. Alipay+ Rewards helps global brands better engage with consumers worldwide by providing incentives and exclusive offers that can drive new growth channels. Alipay+ also provides marketing solutions like Alipay+ D-store, a digital store solution that provides a suite of tools to digitalize a merchant's payments, operations, services, and marketing programs. It helps merchants build an "all-in-one store" across various platforms and social media, unlocking new revenue opportunities. It also enables merchants, especially brick-and-mortar businesses in the service industry, to embrace digital transformation in retail services from payments to marketing and promotion.

Serving as an innovative strategy for public participation in SDGs, Alipay+ offers a range of solutions that contribute to the SDGs and promote public participation in sustainable development. For example, initiatives like GForest in the Philippines have successfully engaged local users in planting trees, leading to a significant reduction in carbon emissions. Similarly, partnerships with tourism boards in Hong Kong and Macau have boosted tourism recovery and economic growth through digital payment solutions. The collaboration between Alipay+ and various stakeholders has resulted in tangible outcomes that align with the SDGs. For instance, the partnership with Huggs Coffee in Singapore has facilitated a digital transformation in the local coffee chain industry. In Thailand and Malaysia, Alipay+ has supported the digital capacity building of central banks to enhance cross-border digital payment facilities for international tourists, thereby promoting sustainable tourism.

The Alipay+ for SDGs serves as a catalyst for promoting sustainable development through digital payments and innovative solutions. By leveraging its global reach and partnerships, Alipay+ is driving public participation in achieving the SDGs and creating a positive impact on environmental, social, and governance aspects. The success stories and collaborations highlighted in this book demonstrate the potential of digital payments to contribute to a more sustainable and inclusive future.

Hong Kong Jin Sun

Acknowledgements First and foremost, I want to express my gratitude to Professor Tony Tam, Chairman of the Department of Sociology at the Chinese University of Hong Kong, for his unwavering support. This project owes a great deal to Professor Tam and the numerous talks and meetings we had since the summer of 2022. At that time, Professor Mallard and I would soon publish our article titled Viral Governance in a prestigious journal, the American Journal of Sociology. Our article delved into the impact of financialization, particularly financial sanctions facilitated by the global banking system, on global governance. Professor Tam was also aware of my upcoming article in another top journal, the Journal of Public Administration (4/173 in Chinese political science journals), where Professor Chen Ling and I explored the role of financialization in addressing climate change by comparing carbon markets in China and Europe.

In the Fall of 2022, I came back to the Chinese University of Hong Kong and started my work at the Department of Sociology, as one of the first batch of visitors to Hong Kong following the relaxation of COVID-19 border controls. From fall to winter, I forged a close partnership with Ant International, the overseas division of Ant Group (the operator of Alipay) based in Hong Kong. We collaborated with Hong Kong and Macau's governments and tourism boards to revitalize the tourism sector and boost the local economy using Alipay, the world's largest payment platform. We checked and upgraded the digital infrastructure in Hong Kong and Macau, ensuring that QR-code devices are available at every entrance of subway stations, on major bus routes and in landmark sites and main scenarios. We explored strategies to leverage the super-app—a mobile application offering various services to over one billion users—to engage with travelers. Through this experience with Ant International and further discussions with Professor Tam, I came to realize that in Asia, digital payment platforms could be more effective tools of financialization for global sustainable governance than traditional banking systems or emissions trading systems. Taking China as an example, the number of Alipay users (1.3 billion) is nearly equal to the Chinese population, and that of WeChat pay users (1.133 billion) is about 80% of the Chinese population. Traditional banks or emissions trading systems cannot directly reach such a vast user base. Almost half of these users regularly utilize the mini-programs offered by digital payment platforms. The

FinTech-empowered trial runs in Hong Kong and Macau proved highly successful, with the number of visitors to Macau in 2023 surging by 3.9 times compared to the previous year, reaching 71.6% of the pre-pandemic levels recorded in 2019. Ant International recognized an unprecedented strategic opportunity for global expansion as international travel is poised to rebound post-pandemic. This indicates that ESG innovation, especially sustainable financial practices facilitated by digital payment platforms, can effectively drive public engagement and social mobilization for SDGs. Inspired by the successful trial runs in Macau and Hong Kong, Ant International became interested in their broad implications to its globalization strategy, as my research suggests that incorporating SDGs into a multinational company's global strategy would strengthen its partnership with local businesses, governments, and communities. In line with this vision, Ant International decided to establish Alipay+, the world's largest cross-border digital payment infrastructure, and leverage it to implement SDGs in 66 countries, which becomes its top strategy for global expansion. As such, Chinese digital payment platforms like Ant's Alipay are actively expanding their presence in other parts of Asia by which they are establishing cross-border digital payment infrastructure with other prominent platforms in the region. This presents a valuable opportunity to advance and implement the SDGs through collaborative efforts with key FinTech players in Asia. This idea evolved into the Alipay+ for SDGs including Sustainable Travel Program detailed in this book through our partnership with Ant International. In 2023, Alipay International expanded pilots to Malaysia, Thailand, Singapore, South Korea, and Japan through the Sustainable Travel Program (STP) with 13 international e-wallets. In 2024, Alipay+ connects 90 million local merchants with 1.5 billion users across 32 popular digital wallets in Asia and beyond. In 2025, Ant International launched the Alipay+ Global Travel Gala, covering ten popular tourist destinations favored by mainland Chinese tourists, including Japan, South Korea, Hong Kong, Macau, Singapore, Malaysia, Australia, Italy, France, and the United States, to raise their awareness of SDGs. Partners in Japan include the Osaka Nagai Botanical Garden, Universal Studios Japan, Legoland Discovery Center Osaka, Osaka Solaniwa Onsen, Tokyo Tower, Tokyo Skytree, Shinjuku Gyoen National Garden, and Sanrio Puroland. Partners in South Korea include N Seoul Tower, Lotte World, Gyeongbokgung Palace, Myeongdong Theater, Busan Songdo Marine Cable Car, Busan X the Sky, Busan Spa Land, and Busan SEA LIFE Aquarium. Partners in Hong Kong include Ocean Park, Hong Kong Disneyland, Ngong Ping 360, and the Central Ferris Wheel. Partners in Macau include Studio City Water Park, Macau Tower, teamLab SuperNature Macao, and The Parisian Macao's Eiffel Tower. Partners in Singapore include Universal Studios Singapore, Gardens by the Bay, S.E.A. Aquarium, and Night Safari. Partners in Malaysia include the National Zoo of Malaysia, Sunway Lagoon, KL Hop-On Hop-Off Bus, Aquaria KLCC, Langkawi Crocodile Adventureland, Langkawi Wildlife Park, Langkawi Underwater World, and Langkawi SkyCab. Partners in Australia include Taronga Zoo, Scenic World Blue Mountains, Featherdale Wildlife Park, Luna Park Sydney, Melbourne Skydeck, Puffing Billy Railway, Melbourne Museum, and Werribee Open Range Zoo. Partners in Italy include the Milan Cathedral, Teatro alla Scala, Pinacoteca di Brera, Uffizi Gallery, Galleria dell'Accademia, Palazzo Pitti, and Basilica of Santa Croce. Partners in France include the Arc de Triomphe, Eiffel Tower, Louvre Museum, Palais Garnier, Château de Chenonceau, Château de Chambord, Château d'Amboise, and Puy du Fou. Partners in the United States include Disneyland California, Universal Studios Hollywood, Griffith Observatory, Natural History Museum of Los Angeles County, Museum of Modern Art New York, Statue of Liberty, Rockefeller Center, and Empire State Building. For the reputation from Alipay+ for SDGs, I have been invited to participate in sustainable finance standards across various sectors like real estate and supply chain. The Alipay+ for SDGs receives accolades from the United Nations and leaders in Singapore (Prime Minister, Mr. Lawrence Wong), Macau (Chief Executive Officer, Mr. Ho Iat-seng), Malaysia (YBrs. Tuan Samuel Lee Thai Hung, Deputy Director-General, Tourism Malaysia), and Thailand (Tanes Petsuwan, Deputy Governor, Tourism Authority of Thailand) for promoting sustainable tourism, digital transformation, and economic development.

I also extend my thanks to Professor Ambrose King and Professor Suk-Ying Wong for their encouragement and many other brilliant colleagues like Prof. Rance Lee, Dr. Kent Lee, Prof. Sara Zhong, and Prof. Nichole Cheung in the Department of Sociology at the Chinese University of Hong Kong for their generous support. As an alumnus of New Asia College, a prominent center of the New Confucian School, and a founding member of the Chinese University of Hong Kong, I have drawn immense wisdom from the works of Prof. Ch'ien Mu, Prof. Tang Junyi, and Prof. King over the past two decades. Professor King, a former president of New Asia and the Chinese University of Hong Kong, shared his diligent work ethic during a book talk, inspiring me with his dedication to academic pursuits. He was so occupied with college administrative tasks during the day that he could only dedicate his nights to writing academic papers and books, often working until 2 a.m. Professor Wong, too, serves as a role model, balancing administrative responsibilities with academic writing late into the night. Their dedication resonates with my own late-night academic endeavors.

I am grateful to the Chinese University of Hong Kong for its generous support of this project, including funding for three research projects where I served as Principal Investigator. These projects, namely, "A Transnational Private Compliance Program for SDGs: Leveraging the Digital Payments Revolution in China and Beyond" (#3134388), "A sociolegal study of Chinese environmental movements and its implication for transnational environmental governance from climate governance to ESG/SDG" (#4052319), and "Transnational Orders between China and the World" (#4937156), have been supportive in advancing our understanding of how to foster sustainable development goals through digital payments.

I want to acknowledge the invaluable contributions of the student assistants who have supported this project. Two exceptional editorial assistants, Su Ruotong (Chapters 2, 3, 5, 6, 7, 9, 12, and 14) and Fu Chengyi (Chapters 4, 8, 10, 11, 13, 15, 16, and 17), played a crucial role in editing and proofreading the early drafts of the book, ensuring the timely publication of this book. Without their valuable contributions, this book would not have been published so quickly. Furthermore, the writing of this book received support from research assistants. I am deeply thankful to the China Engagement Office of the Chinese University of Hong Kong and the University Grants Committee for providing financial support to the student assistants through the 2024 Summer Research Placement Programme. Special thanks to Professor Yu Hanzhi of Zhejiang University for recommending five outstanding student assistants. Together, we collaborated during the summer of 2024 to gather empirical evidence and evaluate the measurable social impact of this project through detailed case studies.

Their specific roles of research assistants and material editing were as follows:

- Fu Chengyi: Chapter 4 (Antom), Chapter 5 (Ant Farm), Chapter 7 (10x1000), Chapter 8 (Blue Vest), Chapter 9 (Zoloz), Chapter 12 (Malaysia).
- Meng Ketong: Chapter 8 (female entrepreneur, Girl Football), Chapter 17 (Singapore).
- Yan Jiayi: Chapter 2 (Ant Forest, GForest), Chapter 13 (Philippines).
- Liu Zhizi: Chapter 6 (MyBank), Chapter 10 (India and Pakistan).
- Su Ziqing: Chapter 4 (WorldFirst), Chapter 17 (Korea).
- Su Ruotong: Chapter 11 (Thailand), Chapter 15 (Hong Kong and Greater Bay Area), Chapter 16 (Macau), Chapter 17 (Japan, South Korea, and Singapore).
- Jiang Shuming assisted in the data collection for Chapter 14's Indonesia section.

The collaborative efforts of all those involved, including these students and other doctoral candidates under my supervision, as well as many brilliant people from Ant International, have been essential in bringing this project to fruition. I am thankful for their dedication and contributions, which have enriched the content of this book and furthered our understanding of leveraging digital payments for sustainable development.

Special thanks to Professor Yan Zhang, Professor John Braithwaite and Prof Valeria Braithwaite of the Australian National University. I extend my heartfelt appreciation to Professor Zhang for introducing me to the publication process at Springer Nature. Your guidance and insights have been invaluable in navigating the complexities of academic publishing. I would also like to express my gratitude to Professor Zhang, Professor John Braithwaite, and Professor Valeria Braithwaite for visiting Ant Group with me and engaging in discussions with us about this project. Your generous sharing of knowledge and inspiration regarding environmental restorative justice has been truly enlightening. Thank you for your time, expertise, and dedication to advancing our understanding in this important field.

I want to express my sincere appreciation to Prof. Carlos Lo from The Chinese University of Hong Kong, Prof. Ning Liu from The City University of Hong Kong, Prof. Xueyong Zhan from The Hong Kong Polytechnic University, Prof. Yongdong Shen from Zhejiang University, and Prof. Sujian Guo from San Francisco State University for hosting the conference of the Journal of Chinese Governance on sustainability governance in China. Your collaborative efforts and dedication to advancing research in this field are truly commendable. Thank you for providing a platform for scholars to exchange ideas and insights on sustainability governance in China.

I want to extend my heartfelt appreciation to Prof. Gregoire Mallard, Prof. Shalini Randeria, Prof. Nico Krisch, and Prof. Bruce Carruthers for their invaluable guidance and supervision at various stages of my Ph.D. dissertation. Your expertise and rigorousness have been essential in shaping my research journey. I want to express my gratitude to Prof. Graziella Moraes Silva and Prof. Thomas Biersteker for their unwavering support throughout my academic endeavors. Your encouragement and wisdom have been a constant source of inspiration. Thank you for being such dedicated advisors and advocates for my academic growth. I would like to express my deep gratitude to Prof. Terence C. Halliday and Prof. Susan Block-Lieb for their enlightening talk in Geneva about their remarkable book, "Global Lawmakers: International Organizations in the Crafting of World Markets." Their insights have opened a gateway for me into the fascinating world of global governance. Thank you for sharing your expertise and shedding light on how to conduct a rigorous study of international organizations in shaping world markets and global governance order.

I appreciate Prof. Sally Merry and Prof. Benedict Kingsbury for inviting me to the conference on Infrastructure as Regulation at the Institute for International Law and Justice at New York University Law School. Their insightful perspectives on Infrastructure as Regulation have inspired me to utilize the world's largest cross-border digital payments infrastructure to drive progress towards the Sustainable Development Goals.

I would like to express my gratitude to the editors of this book, Ananda Kumar Mariappan, Martina Himberger, and Jacob Dreyer at Springer Nature, for their invaluable contributions and dedication to ensuring the quality and success of this publication. Your expertise and guidance have been crucial in shaping this work into its final form. Thank you for your hard work and commitment to excellence.

Competing Interests The author has no competing interests to declare that are relevant to the content of this manuscript.

Contents

List of Figures

List of Tables

PART I

Introduction

CHAPTER 1

Exploring FinTech's Possibilities to Achieving Sustainable Development Goals Through the World's Largest FinTech: Alipay's Inclusive Partnerships in Asia and Beyond

1.1 Introduction

Mallard and Sun (2022, 2024) demonstrate that hegemony can use financialization for financial sanctions, making it an effective tool for global governance, from international peace and security to financial capitalism and international trade and economy (Halliday & Carruthers, 2007, 2009; Halliday & Shaffer, 2015). Chen and Sun (2023) further show that major powers can also utilize financialization for global carbon governance, for example, in leveraging the financialization of carbon emissions to promote low-carbon transitions, as evidenced by the carbon markets in the EU and China. This book documents a bold attempt by the world's largest fintech platform—Alipay, together with the author's team and hundreds of corporate, individual, and social partners in Asia and beyond. Through empirical evidence in China and East and Southeast Asia, it explores how financialization can be leveraged to achieve the Sustainable Development Goals (SDGs).

Evidence from 1.4 billion Alipay users participating in flagship projects in Mainland China (e.g., Chapter 3) suggests that fintech can foster the SDGs in three approaches: (1) FinTech for decarbonization, sustainability,

J. Sun, *Sustainable Financialization*, Sustainable Development Goals Series, https://doi.org/10.1007/978-981-96-3580-1_1

and the environment (Part 2: Chapters 2–3), (2) FinTech for social innovation, decent work, and inclusive development (Part 3: Chapters 4–7), and (3) FinTech for inclusive partnership and good governance (Part 4: Chapters 8–10).

Since December 2022, the comprehensive SDG initiatives under the coordination of the author's team and Alipay's global ecosystem partners in East and Southeast Asia have explored the feasibility and possibilities of fintech for SDGs in the rest of Asia. They include such latest initiatives as Alipay+ (Part 5: Chapters 11–14) and the Sustainable Travel Program (STP) (Part 6: Chapters 15–17), covering over 1.5 billion users across nine markets in Asia outside mainland China (Thailand, Malaysia, the Philippines, Indonesia, Singapore, South Korea, Japan, Hong Kong, and Macau). It shows that despite varying socioeconomic, cultural, legal, and institutional conditions in the nine markets, fintech holds promise for promoting the SDGs in the approaches mentioned above, as shown by evaluating the measurable social impacts of these initiatives on achieving the SDGs in the concluding chapter (Chapter 18).

This chapter will commence with a theoretical analysis (Section 2). This section will encompass a literature review on financialization, embeddedness, and sustainable governance and propose a theoretical framework on financial embeddedness and its leverage effects on achieving SDGs. Following this, Section 3 will outline the methods and data utilized by the author to gauge and compare the leverage effects of financialization on the SDGs within various approaches in Mainland China (Parts II–IV) and the broader Asian region (Parts V–VI). Section 4 will present empirical findings derived from these datasets, analyzing how the theoretical construct of financial embeddedness elucidates the diverse levels of leverage effects of financialization on the SDGs in different areas in Mainland China—namely, environmental, social, and governance (ESG) practices presented in Parts II, III, and IV, respectively—and across various markets in East and Southeast Asia (Parts V and VI).

Financial embeddedness refers to the degree of a field or an organization's social embeddedness (Uzzi, 1999; Granovetter, 1985, 1992) in mobilizing social resources, partners, and networks through the impacts of financialization (Carruthers & Kim, 2011; Carruthers, 2015) in the interactions of political, social and economic relations. Empirical data from Mainland China (Parts II–IV) indicate that (1) the levels of financial embeddedness in distinct sectors vary greatly. In the industry practices, the environmental initiatives (Parts II) exhibit the highest degree of financial embeddedness compared to social and governance initiatives (Parts III and IV). This distinction arises from the intersection of digital payment-supported transactions with daily lifestyle activities, enriching intensive experiences in transitioning to a low-carbon lifestyle. (2) Areas with high financial embeddedness in fintech interactions tend to yield greater leverage effects on the SDGs. Similar trends could also be observed in the fintech for SDG initiatives in the rest of Asia (Parts V–VI): (1) there are significant differences in the levels of financial embeddedness to fintech across various regions. Regions with high volumes of international travelers from fintech-embedded markets (e.g., Thailand, Malaysia, Hong Kong, Macao), showcase substantial financial embeddedness in fintech. In contrast, regions with well-established financial infrastructures (e.g., visa cards) in developed markets exhibit lower embeddedness in fintech (e.g., Japan, Singapore). (2) In regions with high financial embeddedness in fintech, initiatives launched by fintech platforms are more likely to give rise to significant leverage effects on the SDGs.

1.2 Theory of Financial Embeddedness and the Leverage Effects of Financialization on SDGs

1.2.1 Sociological and Socioeconomic Studies of Financialization

Although discussions surrounding financialization, embeddedness, climate change, and sustainable societies have been pivotal in socioeconomics and sociology in recent decades, the implications of financialization on sustainable development remain a predominantly unexplored area of inquiry. Sociologists have extensively studied the phenomenon of financialization, which refers to the increasing dominance of financial markets and institutions in the global economy and society since the 1970s. This process has not only reshaped socioeconomic practices but has also profoundly influenced social and political dynamics worldwide. At its core, financialization represents a shift in how socioeconomic activities are organized, moving away from traditional production and consumption on main street towards financial activities on Wall Street that prioritize profit maximization through financial instruments like stocks, fixed-income securities, derivatives, and futures.

Key scholars, such as Carruthers (2015) and Krippner (2005), have documented how financialization has redefined economic governance, transforming the role of financial markets from mere intermediaries to dominant players shaping economic policies and socioeconomic outcomes. This shift has been facilitated by changes in regulatory frameworks that deregulated financial markets, enabling financial institutions to engage more freely in speculative activities and capital accumulation. Consequently, financial actors gained significant influence over policymaking processes, often promoting neoliberal policies that prioritize market mechanisms over state intervention. The implications of financialization extend beyond economic restructuring; they encompass broader societal changes. For instance, Carruthers (2022) highlights how financial professionals leverage social networks and reputational capital to mitigate risks and build trust in commercial transactions, thereby embedding financial logics deeper into social relations. This integration of financial practices into everyday life not only influences economic decisions but also reshapes cultural norms and values, emphasizing profit-oriented behaviors over collective welfare. Moreover, scholars like Krippner (2005) argue that financialization has contributed to growing inequality and economic instability. By redirecting capital away from productive investments in physical capital and labor towards financial assets, financialization has exacerbated income disparities and weakened traditional welfare state protections. This trend is evident in the decline of real wages and the erosion of regulatory safeguards for workers and consumers in socioeconomic sectors heavily influenced by financial motives. Over the past few years, scholars have sought to explain the rising impacts of financialization and its neoliberal logics on world society and global capitalism. It is a vibrant and interdisciplinary field of research that has been shaped by contributions from a range of leading sociological scholars (Granovetter, 1985, 1992; Smelser & Swedberg, 2005 (1994); Carruthers & Babb, 2012 (2000); Fligstein, 2001; Krippner, 2005; Bandelj et al., 2017; Carruthers & Ariovich, 2013; Beckert, 2016; Carruthers, 2022).

Despite these profound impacts, the study of financialization (Carruthers & Kim, 2011; Carruthers, 2015) remains underexplored regarding its environmental and sustainable development implications. Less attention has been paid to how financialization intersects with global challenges such as climate change and sustainable development goals (SDGs). The ongoing prioritization of financial profit over long-term social and environmental sustainability raises critical questions about the feasibility of achieving global climate targets and fostering inclusive economic growth. In short, sociologists argue that financialization is a multifaceted phenomenon that goes beyond economic construct and restructuring to encompass profound social and political transformations. While it's economic and governance implications have been extensively studied, its impact on sustainable development and climate change remains a critical yet underexamined issue. Understanding these dynamics is essential for crafting policies that mitigate the adverse effects of financialization while promoting equitable and sustainable global development.

1.2.2 Debates About Sustainable Governance in China

The academic inquiry on sustainable governance in China, the world's largest source of carbon emissions, underscores the pressing need to address the escalating environmental crisis characterized by pollution, soil degradation, climate change, biodiversity loss, and overpopulation. This crisis has exerted unprecedented strain on natural and constructed environments, emphasizing the urgency of effective regulatory enforcement to safeguard ecosystems (Forsyth et al., 2021). In response to these challenges, the Chinese government, at central, provincial, and local levels, has undertaken significant efforts over the past decades, deploying a distinctive campaign-style regulatory enforcement approach to tackle environmental and sustainability development issues (Liu et al., 2015, 2022).

In China, the environmental governance landscape has been predominately shaped by administrative, campaign-style, and judicial enforcement mechanisms (Li & Shapiro, 2020; Liu et al., 2015; Lo & Leung, 2000; Mao et al., 2022; van Rooij & Lo, 2010). While these traditional strategies have been instrumental in addressing environmental concerns, they have also faced scrutiny for their limitations and inefficiencies. Administrative measures, a fundamental pillar of China's environmental governance strategy, involve centralized control, bureaucratic structures, and top-down directives, with government agencies responsible for monitoring and enforcing regulations, issuing permits, and ensuring compliance (Eaton & Kostka, 2018; Mao et al., 2022; van Rooij et al., 2013). Despite their effectiveness in certain areas, administrative approaches have been criticized for bureaucratic red tape, lack of transparency, and susceptibility to corruption, especially when economic growth objectives overshadow environmental considerations (Drahos, 2021; Han & Wang, 2023; Steinhardt & Wu, 2016).

Campaign-style initiatives, another key feature of China's environmental governance, are characterized by mass mobilization efforts, publicity campaigns, and targeted enforcement actions aimed at tackling specific environmental issues within set timeframes (Barbi et al., 2016; J. Li & Guo, 2023; N. Liu et al., 2018; Zhou, 2022). While effective in raising awareness and achieving short-term goals, these initiatives have been criticized for their episodic nature, tendency to prioritize short-term political objectives, and reliance on centralized decision-making, potentially overlooking long-term sustainability (Y. Li & Shapiro, 2020; L. Liu et al., 2022).

Judicial enforcement mechanisms, designed to hold polluters accountable and provide legal recourse for environmental violations, have undergone significant reforms in China to strengthen environmental protection and enhance judicial capacity (Zhang, 2016; van Rooij et al., 2016). Courts have increasingly adjudicated environmental cases, imposing fines, ordering remediation measures, and issuing injunctions against polluters. However, challenges such as limited judicial independence, uneven enforcement, and procedural barriers persist, hampering the efficacy of judicial mechanisms in deterring environmental violations and encouraging compliance (Economy, 2004). The adversarial nature of litigation may also hinder citizen participation in environmental governance (N. Liu et al., 2015; Shen & Ahlers, 2019; Stern, 2013).

While China's traditional environmental governance mechanisms have played a vital role in addressing environmental issues, there is a growing realization of the need to complement these strategies with innovative approaches that emphasize long-term incentive mechanisms to encourage voluntary environmental compliance among corporations and individuals. The emergence of fintech-enabled initiatives for environmental and sustainability governance represents a promising and yet underexplored shift towards a more holistic and sustainable paradigm in addressing the impacts of fintech-enabled technological empowerment on sustainable development and climate change in China.

1.2.3 Financial Embeddedness and the Leverage Effects of Financialization on SDGs

Financial embeddedness refers to the degree of a field or an organization's social embeddedness (Uzzi, 1999; Granovetter, 1985, 1992) in mobilizing social resources, partners, and networks through the impacts of financialization (Carruthers & Kim, 2011; Carruthers, 2015) in the interactions of political, social and economic relations. Financial embeddedness is a theoretical framework to contribute to the academic debates about financialization in sociology and the academic and public policy debates about sustainable governance in China. This framework focuses on its exploration of the environmental and sustainable development implications of financialization and its evaluation of the impacts of financialization on sustainable development and climate change. It sheds light on how financial forces intersect with political, social, and economic dynamics, shaping interactions and relationships between state, market, and society.

Mallard and Sun (2022, 2024) have demonstrated that financialization can serve as a tool for global governance, from international peace and security to financial capitalism and international trade and economy. The leveraging effects of financialization for promoting global governance in global low-carbon transitions illustrate the diverse applications of financialization in promoting sustainable development goals (Chen & Sun, 2023). This project showcases how the world's largest fintech platform, Alipay, in collaboration with the author's team and various stakeholders in China and the rest of Asia, has leveraged fintech-enabled financialization to advance the Sustainable Development Goals in Asia and beyond, based on empirical data from China and nine markets in East and Southeast Asia.

The theoretical construct of financial embeddedness underscores the importance of understanding the depth of social embeddedness in mobilizing resources and networks influenced by financialization. In Mainland China, the empirical findings highlight variations in the levels of financial embeddedness across different sectors, with environmental initiatives demonstrating higher levels compared to social and governance initiatives. This disparity is attributed to the fusion of digital payment technologies with daily activities, enriching experiences in transitioning to a low-carbon lifestyle. Moreover, the analysis extends to different regions in East and Southeast Asia, revealing disparities in the levels of financial embeddedness in fintech interactions. Regions with substantial financial embeddedness in fintech interactions, such as those with high volumes of international travelers from fintech-driven markets, exhibit significant leverage effects on the SDGs. On the contrary, regions with well-established financial infrastructures show lower levels of financial embeddedness to fintech and less pronounced impacts of sustainable financialization on the SDGs.

The study not only examines the theoretical underpinnings of financial embeddedness but also delves into the practical implications of these theoretical constructs in fostering sustainability initiatives. By analyzing the leverage effects of financialization on sustainable development in different sectors and regions, the research contributes to a nuanced understanding of how sustainable financial forces (Schoenmaker & Schramade, 2018; Popescu et al., 2021; Ziolo et al., 2019; Chiapello et al., 2023; Bernards, 2024) can drive positive outcomes across environmental, social, and governance practices. This theoretical framework offers valuable insights into the potential of financial embeddedness to catalyze sustainable governance practices and drive holistic approaches to achieve the SDGs in China and the broader Asian region (Puaschunder, 2023; Nguyen et al., 2022; Muschert et al., 2024; Alakshendra et al., 2024; Koomson & Villano, 2024; Puaschunder, 2023; Soetan & Mogaji, 2024; Mhlanga, 2023; Ojo, 2024; Braithwaite, 2024).

1.3 Methods and Data

This study draws upon a mixed-methods approach, combining qualitative analysis of fieldwork, interviews, and case studies with quantitative data from public disclosure, statistics, and ESG reports from Ant Group and its various partners in the Asian digital ecosystems. Fieldwork research was conducted in 2022–2024 under the auspices of eight research assistants from Zhejiang University and the Chinese University of Hong Kong in Mainland China and nine Asian markets outside Mainland China (Thailand, Malaysia, the Philippines, Indonesia, Singapore, South Korea, Japan, Hong Kong, and Macau). Mainland China was first explored. Evidence from 1.4 billion Alipay users participating in flagship projects in Mainland China (e.g., Chapter 3) suggests that fintech can foster the SDGs in the environmental, social, and governance (ESG) approaches (Parts II–IV). After that, nine regions in Asia were chosen for voluntary participation in the latest innovative fintech services, Alipay+ (Part 5: Chapters 11–15) and the Sustainable Travel Program (STP) (Part 6: Chapters 16–17), covering over 1.5 billion users across nine markets in Asia outside mainland China, which provide insights into the diverse implementation of sustainable governance across different contexts in and outside China.

Qualitative data was gathered through interviews with key stakeholders, including government, court, and environmental protection experts, community members, and officers from various levels of government agencies to environmental and sustainable practices. In addition, environmental law enforcement agencies, environmental lawyers and experts, as well as business leaders, have been interviewed. These interviews aimed to elucidate the corporate and individual motivations, challenges, and outcomes associated with the implementation of sustainability governance in different regions. Participant or non-participant observation of governance processes and community interactions provided valuable insights into the dynamics of local sustainability governance.

Quantitative data was obtained from official reports, public disclosure, statistics, and ESG reports from Ant Group and its various partners in the Asian digital ecosystems as well as interviews with the experts in these organizations, offering support for the sources of quantitative data. This data was analyzed to identify or verify trends, patterns, and correlations related to sustainability governance, corporate and individual behavior, community dynamics, and sustainability outcomes. By triangulating qualitative and quantitative data sources, this study aims to provide a comprehensive understanding of the role and impact of fintech-enabled financialization in sustainability and environmental practices in China and the rest of Asia.

1.4 Empirical Findings from Fintech for SDGs in Mainland China and the Rest of Asia

1.4.1 Leverage Effects of Fintech for SDGs Mainly in Mainland China

1.4.1.1 Strong Leverage Effects in Highly Fintech-Embedded Environmental Initiatives

The theoretical framework of financial embeddedness provides valuable insights into understanding the strong leverage effects observed in highly fintech-embedded environmental initiatives, particularly in the context of sustainable financialization for decarbonization, environment, and sustainability. This framework elucidates how integrating financial mechanisms within social contexts can catalyze impactful environmental actions and drive sustainable development efforts.

Case 1. Sustainable FinTech for public participation, low-carbon transition, and environmental protection: Ant Forest in China and GForest in the Philippines (Chapter 2)

In both China and the Philippines, inclusive fintech platforms such as Ant Forest and GForest have harnessed the power of e-wallets to engage over half of their populations in environmental initiatives. Through these platforms, users can translate their everyday low-carbon behaviors into tangible contributions to tree-planting projects. These platforms facilitate transparent and inclusive carbon offset projects, where individual users can directly request tree-planting activities. This innovative approach has led to one of the largest afforestation campaigns globally, showcasing the transformative potential of fintech-enabled environmental engagement at scale.

The theoretical framework of financial embeddedness explains the remarkable leverage effects observed in these environmental initiatives. These platforms have democratized environmental participation by embedding financial processes within social networks and everyday activities through e-wallets, enabling individuals to contribute to environmental protection efforts directly. The seamless integration of fintech solutions has facilitated widespread public engagement, demonstrating the power of fintech-enabled financial embeddedness in driving environmental impact on a massive scale.

Case 2. Sustainable FinTech for corporate participation and sustainable transition: InterContinental Hotels Group (IHG), Genshin Impact, McDonald's, and corporate partners in Alipay's ecosystem (Chapter 3)

Leading multinational corporations in China, including InterContinental Hotels Group (IHG), Genshin Impact, and McDonald's, have leveraged Alipay's fintech platform to advance sustainable transitions. These corporations have adopted voluntary low-carbon initiatives through strategic partnerships with Alipay, reducing costs and enhancing brand loyalty while actively contributing to climate change mitigation and pursuing Sustainable Development Goals (SDGs).

The theoretical framework of financial embeddedness underscores the role of fintech platforms in fostering corporate sustainability by integrating financial incentives and rewards into sustainability initiatives. Alipay has incentivized corporate partners like IHG and McDonald's to prioritize sustainability in their business strategies by aligning financial interests with environmental objectives. Moreover, by engaging popular entities like Genshin Impact as SDG partners, Alipay has extended the reach of sustainability efforts to diverse Z-generation audiences, encouraging young individuals to participate through gamified experiences that reward green practices.

In summary, the theoretical framework of financial embeddedness illuminates how fintech-enabled environmental initiatives can generate strong leverage effects by intertwining financial processes with social and environmental objectives. By fostering public engagement, empowering individual contributions, and incentivizing corporate sustainability, these initiatives exemplify the transformative potential of financial embeddedness in driving sustainable development outcomes and environmental protection on a broad scale.

1.4.1.2 Semi-strong Effects in Normally Embedded Social Initiatives

Financial embeddedness theory provides a robust framework for understanding the semi-strong leverage effects observed in fintech-embedded social initiatives to foster social innovation, decent work, and inclusive development. By examining examples such as those presented in WorldFirst and Antom's support for small businesses in international trade, as well as initiatives like Ant Farm and Ant Foundation's social development projects, and the efforts of MY Bank, Anext Bank, and Bettr to serve underbanked and unbanked populations, we can appreciate how financial embeddedness shapes the impact of these initiatives.

> Case 3. Sustainable FinTech for small business, developing countries, and international trade: WorldFirst, Antom (Chapter 4)

Financial services provided by fintech platforms like WorldFirst and Antom are vital in supporting small businesses from developing countries in international trade by providing accessible financial services and payment solutions. These platforms empower small businesses to engage in cross-border transactions efficiently through competitive foreign exchange rates, low-cost international money transfers, and diverse payment methods. The theory of financial embeddedness explains how these services facilitate the integration of small businesses into the global market by reducing costs, enhancing operational efficiency, and expanding their international reach. By intertwining financial processes with social and economic objectives, these platforms create semi-strong leverage effects that drive inclusive growth and sustainability in the business landscape.

> Case 4. Sustainable FinTech for social development: Ant Farm, Ant Foundation, and social partners for rural development, community facilities, and humanitarian aid (Chapter 5)

Initiatives like the Ant Foundation's projects in rural development, community facilities, and humanitarian aid underscore the transformative potential of fintech-enabled social development. By leveraging fintech platforms and expertise, these initiatives amplify the impact of philanthropic efforts through technology and inclusive financial services. By partnering with local entities and social organizations, these initiatives address critical community needs, such as access to microcredit, education, skills training, and disaster relief. The theory of financial embeddedness elucidates how these initiatives integrate financial mechanisms with social development goals, catalyzing semi-strong leverage effects that drive meaningful social impact, community engagement, and humanitarian assistance.

> Case 5. Sustainable FinTech for underbanked/unbanked social groups: MY Bank in China, Anext Bank in Singapore, and Bettr in Thailand (Chapter 6)

Fintech solutions like MY Bank, Anext Bank, and Bettr are instrumental in promoting financial inclusion among underbanked and unbanked populations. By offering tailored financial services, digital banking solutions, and mobile-based financial products, these platforms bridge the financial inclusion gap for marginalized social groups. Through collaborations with local partners and community organizations, these initiatives deliver financial literacy programs, support services, and accessible financial products to empower individuals and small businesses. The theory of financial embeddedness highlights how these initiatives embed financial services within social contexts to generate semi-strong leverage effects that promote economic empowerment, financial inclusion, and sustainable growth among underbanked populations.

Case 6. Sustainable FinTech for Decent Work and Economic Growth: 10x1000 Tech for Inclusion (Chapter 7)

"10x1000 Tech for Inclusion," an initiative by Ant Group, exemplifies how fintech can support Decent Work and Economic Growth through digital talent training and support. By partnering with international and regional organizations, this project facilitates the growth of digital technology leaders in emerging markets, promoting the development of the digital economy and fostering inclusive growth. The theory of financial embeddedness underscores how initiatives like "10x1000" embed financial resources and support within social and economic contexts, creating semi-strong leverage effects that drive workforce development, economic empowerment, and technological innovation in emerging markets.

In conclusion, the theory of financial embeddedness offers a comprehensive lens through which to understand the semi-strong leverage effects observed in fintech-embedded social initiatives. By integrating financial mechanisms with social objectives, these initiatives drive inclusive growth, social development, financial inclusion, and economic empowerment, illustrating the transformative potential of financial embeddedness in shaping impactful social change and sustainable development outcomes. Compared to the strong leverage effects typically observed in environmental initiatives (e.g., Ant Forest), semi-strong leverage effects are more prevalent in normally embedded social initiatives (e.g., Ant Village, MY bank) which are relatively distant from high-frequent daily digital payment scenarios. This difference can be attributed to the complex and multifaceted nature of social challenges, which often require a combination of financial resources, community engagement, and policy support to drive meaningful impact and more importantly, extra attention of individual and corporate participants to these initiatives in fintech applications. The diverse range of stakeholders involved in social initiatives can lead to varying levels of financial embeddedness and impact diffusion, resulting in semi-strong rather than strong leverage effects.

1.4.1.3 Weak Effects in Weakly Embedded Governance Initiatives

Financial embeddedness theory provides an effective framework for understanding the weak leverage effects observed in weakly fintech-embedded governance initiatives aimed at fostering good governance through inclusive partnership. Weak leverage effects are observed in weakly fintech-embedded governance initiatives like Ant Village, Cyber-Mulan, Wind Rider, Blue Vest, and Zoloz e-KYC applications due to the absence of direct incentives or pressures of fintech-enabled governance initiatives on public sectors to foster good governance. These initiatives, while exemplifying good practices for governance and social impact, lack the necessary mechanisms to drive strong or semi-strong financial embedment into social contexts, resulting in weak leverage effects.

> Case 7. Sustainable FinTech for inclusive partnership: Ant Village, CyberMulan for female entrepreneurs in China and Indonesia, Wind Rider for girls' football, and Blue Vest for the elderly population in China (Chapter 8)

Wind Rider supports girls' football teams in Chinese rural schools. By providing training equipment, coach training, and competition opportunities, it encourages more rural girls to participate in football and improve their physical fitness and team spirit. The first phase is to fund 100 rural girls' football teams in 5 years.

The CyberMulan Employment Training Program focuses on improving women's employability in the digital economy era, providing them with skills training and employment guidance, helping them better adapt to the work needs in the digital economy, and promoting gender equality and economic empowerment. It has supported the establishment of 17 county digital employment centers in underdeveloped counties, helping more than 15,000 rural women to obtain new professional skills training, improve digital literacy, and find jobs nearby as artificial intelligence trainers, customer service specialists, rural hotel housekeepers, social workers, etc.

The Blue Vest Action is a volunteer service to help the elderly people in need and teach them how to use smartphones and Internet services so that they can more easily enjoy the convenience brought by digitalization. It has covered more than 100 cities in 26 provinces in China till early 2023.

Good governance initiatives like Wind Rider, CyberMulan, and Blue Vest demonstrate how financial resources can support female entrepreneurs, girls' football teams, and the elderly population. However, the weak leverage effects stem from a lack of direct incentives for public sectors to actively engage in fostering good governance through fintech. While these initiatives promote empowerment and social well-being, the absence of financial pressures or incentives limits their ability to drive significant change at the societal level. Without strong financial embedment into public sector initiatives, the impact remains localized and, till now, fails to catalyze broader governance improvements.

Case 8. Sustainable FinTech for data privacy governance in the Global South: Zoloz e-KYC applications from China to over ten countries in Southeast and South Asia (Chapter 9)

Zoloz's e-KYC application has gained recognition in China and a dozen countries in Southeast and South Asia for providing efficient and secure identity authentication solutions. It utilizes advanced biometric technology and encryption algorithms to streamline the user identity authentication while safeguarding the security of personal privacy data. This not only enhances the accessibility and convenience of financial services but also addresses the challenges of identity authentication in digital governance, thereby promoting the development of privacy computing technology. As a result, it has become an essential tool for governments and financial institutions in advancing digital transformation.

Zoloz's e-KYC applications showcase effective data privacy governance solutions across Southeast and South Asia. Despite their recognition for providing secure identity authentication, these initiatives face weak leverage effects due to the absence of direct incentives for public sectors to adopt and integrate these solutions into broader governance frameworks. While Zoloz's technology enhances data security and privacy, the lack of financial embedment within public sector initiatives hinders its potential to drive systemic changes in digital governance practices in developing countries. Without strong financial pressures or incentives, the impact of these initiatives is case by case and remains limited in catalyzing widespread governance improvements in a global scale.

Case 9. Sustainable FinTech for inclusive partnership and state capability building in the Global South: Paytm in India and Easypaisa in Pakistan (Chapter 10)

Paytm in India and Easypaisa in Pakistan, both of which introduced fintech technology from Alipay, are exemplary FinTech companies for inclusive partnership and state capability building in the Global South. Paytm, with its wide range of financial services including digital payments, banking, and insurance, has partnered with Indian governments to drive financial inclusion and digital literacy. Similarly, Easypaisa, through its mobile wallet and branchless banking services, has collaborated with banks and governments to extend financial services to underserved populations in Pakistan. Both companies leverage fintech to bridge the gap between the financial sector and the unbanked, contributing to economic growth and empowering individuals and small businesses. Their partnerships with state entities enhance state capabilities in digital finance, fostering a more inclusive financial ecosystem.

Paytm in India and Easypaisa in Pakistan exemplify how fintech can drive inclusion and state capability building. However, the weak leverage effects in these initiatives arise from the limited financial pressures or incentives on public sectors to adopt fintech solutions for wider governance improvements. While Paytm and Easypaisa promote financial inclusion and empower individuals in South Asia, the lack of strong financial embedment within public sector policies and practices restricts their ability to catalyze significant changes in state capabilities and governance frameworks in developing countries. Without direct incentives or pressures, the impact of these initiatives on broader governance outcomes remains subdued. For instance, although Paytm in India and Easypaisa in Pakistan benefited from technology introduction from Ant Group, till now, they are not interested in participating in such Asia-wide SDG initiatives as Alipay+ or the Sustainable Travel Program (STP).

In summary, the theory of financial embeddedness helps explain the weak leverage effects in weakly fintech-embedded social initiatives. Despite demonstrating good governance practices and social impact, these initiatives are by large best practice examples, but they lack the necessary financial pressures or incentives to drive strong financial embedment or improve global governance within public sectors. Without direct incentives for public sector engagement, the impact of these initiatives remains limited in fostering systemic governance improvements in more countries and driving transformative change at the societal level.

1.4.2 Leverage Effects of Fintech for SDGs in East and Southeast Asia

1.4.2.1 Strong Leverage Effects for SDGs in Highly Fintech-Embedded Markets

In highly fintech-embedded markets such as those illustrated by Alipay+ initiatives in Thailand and Malaysia or the Sustainable Travel Program (STP) in Hong Kong and Macau, strong leverage effects are observed due to the comprehensive integration of fintech-enabled digital payment infrastructures within social and economic contexts, as articulated by the theoretical framework of financial embeddedness. These initiatives exhibit a high degree of financial embedment that not only drives market efficiencies but also fosters inclusive development, economic growth, and sustainability across diverse sectors.

Case 10. Sustainable FinTech for state capability and inclusive tourism: cross-border payment infrastructure for 28 million international travelers via Alipay+ ecosystem in Thailand (Chapter 11)

With a population of 70 million, Thailand attracts 28 million international tourists annually, with China as the most significant source. Despite Thailand's central bank's advancements in digital transformation, its digital payment system is open to domestic citizens only, excluding international tourists—effectively 40% of the country's population. Alipay+ offers a one-stop payment solution that integrates major digital wallets in Asia, serving as a crucial cross-border payment infrastructure that promotes inclusive tourism in Thailand.

Case 11. Sustainable FinTech for inclusive tourism and economic development: Boost, MyPB, Touch 'n Go eWallet and Alipay+ ecosystem in Malaysia (Chapter 12)

Alipay contributes to inclusive tourism and economic development in Malaysia by enhancing the country's digital payment infrastructure. Through partnerships with local players like Touch 'n Go e-wallet, Alipay provides a secure and convenient payment solution that supports government welfare programs, making digital payments more accessible to a broader population. By integrating various services such as D-store for merchants and DuitNow QR for users, Alipay helps create a unified digital ecosystem that fosters economic growth and inclusivity. In addition, Alipay's collaboration with Capital A and its focus on talent development aims to establish Malaysia as a regional digital business hub, further driving sustainable economic development and financial inclusion.

Alipay+ cross-border payment infrastructure in Thailand and Malaysia and collaborations with local players like Touch 'n Go e-wallet exemplify strong leverage effects in highly fintech-embedded markets. Through seamless payment solutions, Alipay+ promotes inclusive tourism, economic development, and state capability building. The theoretical framework of financial embeddedness elucidates how Alipay+'s extensive financial integrations in these markets facilitate enhanced digital payments, support welfare programs, and drive economic inclusivity. By embedding financial mechanisms within a diversified ecosystem of services, Alipay+ fosters robust economic growth, facilitates talent development, and enhances financial inclusion at national and regional levels.

Case 14. FinTech for sustainable travel, mega events, and best practice partners in Hong Kong and inclusive development in the Greater Bay Area (GBA) (Chapter 15)

Each week, Alipay+ facilitates convenient cross-border payments for over one million Hong Kong consumers in the GBA and another one million mainland visitors to mega events and sustainable travel in Hong Kong. The scale economy, infrastructure effect (e.g., being the leading e-wallet in Hong Kong for mainland visitors and vice versa in the GBA for Hong Kong users), and sponsorship of social and cultural mega events (e.g., the Hong Kong Book Fair) enable Alipay to run a collaborative coupon program that promotes best practices for the SDGs among leading retailers and consumer brands.

Case 15. FinTech for sustainable travel, world heritage, and economic transitions for decent work and economic growth in Macau (Chapter 16)

Since 2024, Alipay+ has become a leading vendor of digital payment solutions for public transport, catering to both mainland and international visitors to Macau. Alipay+ has facilitated Macau's rapid recovery from the pandemic by promoting sustainable travel and preserving its world heritage sites. Alipay's investment in the leading local digital firm MPay and the development of Alipay Macau's e-wallet have nurtured the local digital ecosystem. This strategic move aims to support economic transitions, foster decent work opportunities, and drive economic growth in Macau.

In Hong Kong and Macau, Alipay+ plays a pivotal role in promoting sustainable travel, economic transitions, and inclusive development through the Sustainable Travel Program (STP). By providing digital payment solutions for public transport, mega events, and local businesses, Alipay+ demonstrates strong leverage effects through its deep financial embedment within these markets. The theoretical framework of financial embeddedness highlights how Alipay+'s investments in digital ecosystems, partnerships with local firms, and support for sustainable initiatives drive sustainable economic growth, preserve cultural heritage, and create decent work opportunities. Through collaborative programs that support the Sustainable Development Goals (SDGs), Alipay+ enhances market efficiencies, promotes best practices, and fosters inclusive development across diverse sectors.

In summary, the theoretical framework of financial embeddedness sheds light on why strong leverage effects are observed in highly fintech-embedded markets, such as those driven by Alipay+'s initiatives. By intricately embedding financial infrastructures within social, economic, and governance frameworks, these initiatives enhance market efficiencies, promote inclusive development, and drive sustainable growth. The comprehensive integration of financial resources, services, and partnerships within these markets not only optimizes financial transactions but also facilitates state capability building, economic transitions, and the achievement of sustainable development objectives. Through strong financial embedment, Alipay+ exemplifies how fintech can catalyze transformative changes, create value for diverse stakeholders, and contribute to long-term sustainability and inclusive growth in dynamic market environments.

1.4.2.2 Semi-strong Leverage Effects for SDGs in Normally Fintech-Embedded Markets

In normally fintech-embedded markets such as the Philippines and Indonesia, semi-strong leverage effects for Sustainable Development Goals (SDGs) are observed due to the moderate level of integration of financial mechanisms within social and economic contexts. Despite the presence of financial embedment to some extent, the lack of scale economy, infrastructure effect, and sponsorship of social and cultural mega events makes it challenging to achieve strong leverage effects. The theoretical framework of financial embeddedness elucidates how these limitations hinder the full realization of strong leverage effects necessary for driving transformative change and sustainable development in these markets.

> Case 12. Sustainable FinTech for inclusive partnership and regional development: Akulaku PayLater, BillEase, BPI, Gcash, HelloMoney, Maya, and Alipay+ ecosystem in the Philippines (Chapter 13)

Alipay contributes to inclusive tourism and economic development in the Philippines by enhancing the country's digital payment ecosystem, open to international travelers like those from South Korea (the country's top source of travelers). Through partnerships with local players like Gcash and HelloMoney, Alipay provides convenient payment solutions that support inclusive financial services to domestic and international users (e.g., all top South Korean digital wallets are settled by Alipay+ in the Philippines). By integrating various services such as Akulaku PayLater and BillEase, Alipay helps create a digital ecosystem that fosters economic growth. Alipay's collaboration with BPI on talent development aims in further driving sustainable economic development and financial inclusion in the country.

Alipay's contributions to inclusive partnership and economic development in the Philippines showcase semi-strong leverage effects in a fintech-embedded market. Through collaborations with local players like Gcash, HelloMoney, Akulaku PayLater, and BillEase, Alipay enhances the digital payment ecosystem, catering to both domestic and international users. However, the absence of significant scale economy and infrastructure effects, such as being the dominant e-wallet provider or hosting large-scale cultural events, limits the potential for strong leverage effects. While Alipay's partnerships support financial inclusion and economic growth, the lack of extensive market dominance or high-profile sponsorships restricts the depth of financial embedment needed to drive transformative impacts on sustainable development goals.

> Case 13. Sustainable FinTech for inclusive development in the digital economy and underbanked/unbanked social groups: Doku, DANA, Akulaku PayLater, Kredivo, and Alipay+ ecosystem in Indonesia (Chapter 14)

Alipay contributes to sustainable economic growth in Indonesia. Indonesia has leapfrogged traditional banking directly entering the era of digital finance and digital economy. Alipay, through partnerships with local players like Doku, DANA, Akulaku PayLater, and Kredivo, provides secure and convenient payment solutions that support inclusive financial services to the digital economic leaders like Shopee and Lazada, helping the rapid and sustainable economic growth of the digital sector in Indonesia.

In Indonesia, Alipay's engagement in sustainable economic growth through partnerships with Doku, DANA, Akulaku PayLater, and Kredivo demonstrates semi-strong leverage effects in a normally fintech-embedded market. The country's rapid transition to a digital economy presents opportunities for financial inclusion and growth. However, the limited-scale economy and infrastructure effects, despite Alipay's collaborations with key industry players, pose challenges in achieving strong leverage effects. While Alipay's solutions support digital economic leaders like Shopee and Lazada, the lack of significant market dominance or large-scale infrastructural initiatives hinders the full realization of strong financial embedment necessary for driving sustained progress towards SDGs.

Overall, the moderate level of financial embedment in these normally fintech-embedded markets results in semi-strong leverage effects for SDGs. The absence of an extensive scale economy, dominant infrastructure effects, and high-profile event sponsorships limits the depth of financial integration needed to catalyze transformative change and sustainable development at a broader scale. While Alipay's initiatives contribute to economic growth, financial inclusion, and talent development in these markets, the lack of certain key elements impedes the attainment of strong leverage effects essential for driving lasting impact on inclusive development and regional progress. The theoretical framework of financial embeddedness provides critical insights into understanding the dynamics of financial integration and the challenges in achieving strong leverage effects for Sustainable Development Goals in these contexts.

1.4.2.3 *Weak Leverage Effects for SDGs in Weakly Fintech-Embedded Markets*

Case 16. Sustainable FinTech for sustainable tourism and Smart City in low-carbon transition: Alipay+ ecosystem in Singapore (Changi Pay, EZ-Link Wallet, ComfortDelgro, GrabPay, OCBC Digital), South Korea (Kakao Pay, NAVER Pay, Toss), and Japan (Paypay) (Chapter 17)

South Korea and Japan are popular destinations for Chinese travelers, in which sustainable tourism is promoted by Alipay+ and its local partners like Kakao Pay, NAVER Pay, Toss in South Korea and Paypay in Japan, covering various scenarios for Chinese travelers from duty-free shopping to mobile data plans to local convenient stores, ski resorts and other places of interest. In Singapore, Alipay+ becomes part of the local government's efforts towards a leading Smart City committed to low-carbon transition.

In these regions with well-established financial infrastructures (e.g., visa credit cards) like Japan and Singapore, where fintech platforms exhibit lower embeddedness, initiatives launched by companies such as Alipay+ are less likely to generate significant leverage effects on Sustainable Development Goals (SDGs). Despite Alipay+'s efforts to cater to international travelers and support sustainable tourism through partnerships with local platforms like Kakao Pay, NAVER Pay, Toss, and Paypay, several key factors contribute to weak leverage effects in these weakly fintech-embedded markets.

In Japan and South Korea, Alipay+ collaborates with local digital payment platforms to enhance the travel experience for Chinese tourists. These partnerships facilitate seamless transactions for services ranging from duty-free shopping to mobile data plans, supporting sustainable tourism initiatives. However, the weak leverage effects stem from the limited financial embedment within local economies. Despite the convenience provided to Chinese travelers, Alipay+'s initiatives lack significant scale economy and infrastructure effects that are crucial for fostering deep financial integration and impactful SDG contributions at a broader societal level.

Similarly, in Singapore, Alipay+ participates in the government's efforts towards becoming a leading Smart City committed to low-carbon transition. While these initiatives align with sustainability goals, the overall impact remains constrained due to the platform's limited financial embedment. The absence of extensive market dominance, large-scale infrastructural initiatives, and sponsorship of major social and cultural events restricts Alipay+'s ability to achieve strong leverage effects. Despite its integration into local smart city initiatives, Alipay+ lacks the embeddedness necessary to catalyze transformative changes across broader sustainable development objectives in Singapore.

In weakly fintech-embedded markets like Japan, South Korea, and Singapore, Alipay+'s initiatives primarily cater to the needs of international travelers without achieving deep integration into local economic ecosystems. The theoretical framework of financial embeddedness underscores that without substantial integration into local financial infrastructures and broader societal frameworks, fintech platforms struggle to generate strong leverage effects for SDGs. The absence of scale economy hinders the widespread adoption needed to drive significant changes in financial inclusion and economic development. Additionally, the lack of dominant infrastructure effects limits the platform's capacity to influence sustainable practices across diverse sectors. Moreover, the minimal sponsorship of social and cultural mega events further diminishes Alipay+'s potential to foster community engagement and societal impact in these regions.

In conclusion, while Alipay+'s initiatives in Japan, South Korea, and Singapore support sustainable tourism and smart city transitions, the weak financial embedment within these markets results in limited leverage effects on SDGs. Theoretical insights into financial embeddedness highlight the critical role of deep integration into local economic and social contexts for fintech platforms to drive transformative change and contribute meaningfully to sustainable development at a broader scale.

1.5 Implications of Sustainable Financialization for SDGs

This book offers a groundbreaking exploration of how FinTech, social innovation, and inclusive partnerships can drive sustainable development on a global scale. It delves into the transformative potential of Alipay, the world's largest FinTech platform, and its role in achieving the United Nations' Sustainable Development Goals (SDGs) through innovative fintech solutions. Sustainable Financialization: Leveraging FinTech, Social Innovation, and Inclusive Partnership for Sustainable Development Goals provides a comprehensive analysis of how financial technology can be harnessed to address some of the world's most pressing challenges. The book is divided into six parts, each focusing on a different aspect of sustainable development, from decarbonization and environmental protection to social innovation and inclusive governance.

Part 1 introduces the concept of sustainable financialization and sets the stage for the book's exploration of Alipay's role in achieving the SDGs.

Part 2 examines how FinTech can be used for public and corporate participation in low-carbon transitions and environmental protection, highlighting case studies such as Ant Forest in China and GForest in the Philippines.

Part 3 shifts focus to social innovation, decent work, and inclusive development, discussing how FinTech can support small businesses, developing countries, and international trade, as well as provide social development and support for underbanked and unbanked populations. It also explores initiatives like Ant Farm, Ant Foundation, and the 10x1000 Tech for Inclusion program.

Part 4 delves into inclusive partnerships and good governance, showcasing initiatives like Ant Village, CyberMulan for female entrepreneurs, and Blue Vest for the elderly, as well as addressing data privacy governance, digital payment, and state capability building in the Global South.

Part 5 introduces Alipay+, a digital infrastructure for sustainable tourism, decent work, and inclusive development, with a focus on cross-border payment systems, inclusive tourism, and economic development across various countries in Asia.

Part 6 investigates the Sustainable Travel Program, which innovates best practices for sustainable development through sponsoring mega events, protecting world heritage, and fostering economic transitions in the Greater Bay Area (GBA) as well as serving international travelers to Singapore, South Korea, and Japan.

The book culminates in a chapter that evaluates the measurable social impacts of these efforts on achieving the SDGs, providing a critical analysis of the effectiveness of FinTech in driving sustainable change. These efforts have been recognized by governments, the industry, and international organizations. For example, at the International Science and Technology Innovation Expo in 2024, Macau's Chief Executive Region, Mr. Ho Iat Seng, visited MPay, a member company of Ant International, saying: "Many thanks to MPay, and other member companies of Ant Group and Ant International for bringing new industries, new business models, and new opportunities to Macau, enabling the digital transformation of traditional industries. This will help promote Macau's six advantages and, serving as a 'golden business card' of Macau, make this international metropolis more dazzling." Similarly, at the Singapore FinTech Festival (SFF), renowned as one of the largest FinTech summits globally, Singapore's Prime Minister, Lawrence Wong, paid a visit to the booth of Ant Group. He personally engaged with the Alipay+ D-store solution, using his phone to scan and order a cup of Huggs Coffee. After seeing that Huggs Coffee, Singapore's largest local coffee chain brand with 22 chain stores across the country, has embraced Alipay+'s D-store to undergo a digital operational transformation, Mr. Wong confirmed the contribution of this project as follows: "I am thrilled to witness the local physical coffee shops in Singapore embracing digital transformation!" Indeed, prominent platforms like Chope, and the global fast food chain Burger King have already adopted Alipay+'s digital solution. Sanda Ojiambo, Assistant Secretary-General of the United Nations, and Director-General of the UN Global Compact (UNGC), said, "We sincerely hope that these cases can provide ideas and experience for global companies, especially those in developing countries." On November 29, 2024, the UNGC issued the Best Sustainable Development Cases in the 20th anniversary of the ESG principle since it has been proposed by the UNGC. Ant Group's "Green Computing and Public Participation in Low-Carbon Innovation" was featured, along with other prominent companies like China Development Bank. As examined by this book, Ant Group's sustainable development strategy is acknowledged as it "integrates commercial and social value through digital technological innovation, green inclusion initiatives, and open ecology for SMEs. Ant plans to continue investing in AI and green computing to support China's national decarbonization strategy and global SDGs."

It shows that Sustainable Financialization is essential reading for policymakers, business leaders, academics, and anyone interested in the intersection of finance, technology, and sustainability. It demonstrates that FinTech is not just a tool for financial transactions but a powerful agent for positive social change and sustainable development.

References

Alakshendra, A., Datta, A., & Reddy, B. (2024). Development, Inclusion, and Sustainability: Issues and Perspectives. Springer.

Bandelj, N., Wherry, F. F., & Zelizer, V. A. (Eds.). (2017). Money Talks: Explaining How Money Really Works. Princeton University Press.

Barbi, F., Ferreira, L. da C., & Guo, S. (2016). Climate change challenges and China's response: mitigation and governance. Journal of Chinese Governance, 1(2), 324–339.

Beckert, J. (2016). Imagined Futures: Fictional Expectations and Capitalist Dynamics. Harvard University Press.

Bernards, N. (2024). Fictions of Financialization: Rethinking Speculation, Exploitation and Twenty-First-Century Capitalism. Pluto Press.

Braithwaite, J. (2024). Simple Solutions to Complex Catastrophes. Springer Nature.

Carruthers, B. G. (2015). Financialization and the Institutional Foundations of the New Capitalism. Socio-Economic Review, 13(2), 379–398.

Carruthers, B. G. (2022). The Economy of Promises: Trust, Power, and Credit in America. Princeton University Press.

Carruthers, B. G., & Ariovich, L. (2004). The Sociology of Property Rights. Annual Review of Sociology, 30, 23–46.

Carruthers, B. G., & Babb, S. L. (2012). Economy and Society: Markets, Meanings and Social Structure. In D. Brady & J. Burton (Eds.), The Oxford Handbook of the Sociology of Finance (pp. 3–26). Oxford University Press.

Carruthers, B. G., & Kim, J. C. (2011). The Sociology of Finance. Annual Review of Sociology, 37, 239–259.

Chen, L., & Sun, J. (2023). Trans-governmental Normative Orders: Chinese Central-Local Relations in the Pilot Carbon Emissions Trading Scheme. Journal of Public Administration, 91(1), 144–160.

Chiapello, È., Engels, A., & Gresse, E. G. (Eds.). (2023). Financializations of Development: Global Games and Local Experiments. Taylor & Francis.

Drahos, P. (2021). Survival Governance: Energy and Climate in the Chinese Century. Oxford University Press.

Eaton, S., & Kostka, G. (2018). What Makes for Good and Bad Neighbours? An Emerging Research Agenda in the Study of Chinese Environmental Politics. Environmental Politics, 27(5), 782–803.

Economy, E. (2004). The River Runs Black: The Environmental Challenge to China's Future. Cornell University Press.

Fligstein, N. (2001). The Architecture of Markets: An Economic Sociology of Twenty-First-Century Capitalist Societies. Princeton University Press.

Forsyth, M., Cleland, D., Tepper, F., Hollingworth, D., Soares, M., Nairn, A., & Wilkinson, C. (2021). A Future Agenda for Environmental Restorative Justice? The International Journal of Restorative Justice, 4(1), 17–40.

Granovetter, M. (1985). Economic Action and Social Structure: The Problem of Embeddedness. American Journal of Sociology, 91(3), 481–510.

Granovetter, M. (1992). Economic Institutions as Social Constructions: A Framework for Analysis. Acta sociologica, 35(1), 3–11.

Halliday, T. C., & Carruthers, B. G. (2007). The Recursivity of Law: Global Norm Making and National Lawmaking in the Globalization of Corporate Insolvency Regimes. American Journal of Sociology, 112(4), 1135–1202.

Halliday, T., & Carruthers, B. (2009). Bankrupt: Global Lawmaking and Systemic Financial Crisis. Stanford University Press.

Halliday, T. C., & Shaffer, G. (Eds.). (2015). Transnational Legal Orders. Cambridge University Press.

Han, H., & Wang, J. (2023). Performance Management and Environmental Governance in China. Journal of Chinese Governance, 8(4), 498–532.

Koomson, K., & Villano, R. (2024). Financial Inclusion and Sustainable Rural Development. Palgrave Macmillan.

Krippner, G. R. (2005). The Financialization of the American Economy. Socio-Economic Review, 3(2), 173–s208.

Li, J., & Guo, S. (2023). Political Incentives, Bureaucratic Behaviors and Political Budget Cycles in China. Journal of Chinese Political Science, 0123456789.

Li, Y., & Shapiro, J. (2020). China Goes Green: Coercive Environmentalism for a Troubled Planet. Polity Press.

Liu, L., Tang, L., Liu, R., & Li, M. (2022). The Career Characteristics of China's Prefectural Environmental Protection Bureau Heads with Implication for the Environmental Governance. Journal of Chinese Governance, 7(2), 212–235.

Liu, N., Lo, C. W. H., Zhan, X., & Wang, W. (2015). Campaign-Style Enforcement and Regulatory Compliance. Public Administration Review, 75(1), 85–95.

Liu, N., Tang, S. Y., Lo, C. W. H., & Zhan, X. (2022). Enforcement Officials' Coping Strategies in a Changing Regulatory Environment. Public Administration, 100(2), 408–426.

Liu, N., Tang, S. Y., Zhan, X., & Lo, C. W. H. (2018). Policy Uncertainty and Corporate Performance in Government-Sponsored Voluntary Environmental Programs. Journal of Environmental Management, 219, 350–360.

Lo, C. W. H., & Leung, S. W. (2000). Environmental Agency and Public Opinion in Guangzhou: The Limits of a Popular Approach to Environmental Governance. The China Quarterly, 163, 677–704.

Mallard, G., & Sun, J. (2022). Viral Governance: How Unilateral U.S. Sanctions Changed the Rules of Financial Capitalism. American Journal of Sociology, 128(1), 144–188.

Mallard, G., & Sun, J. (2024). "International Law, Security and Sanctions: A Decolonial Perspective on the Transnational Legal Order of Sanctions." Annual Review of Law and Social Science 20.

Mao, K., Jin, S., & Hu, Y. (2022). Environmental Conservation or the Treadmill of Law: A Case Study of the Post-2014 Husbandry Waste Regulations in China. International Journal of Offender Therapy and Comparative Criminology, 66(4), 296 –326.

Mhlanga, D. (2023). FinTech and Artificial Intelligence for Sustainable Development: The Role of Smart Technologies in Achieving Development Goals. Palgrave Macmillan.

Muschert, G. W., Pereira, V., Ramiah, V., & Doker, A. C. (2024). Financial Inclusion: In Pursuit of UN Sustainable Development Goal 8.10. Springer.

Nguyen, N., Nguyen, H. V., D'Souza, C., & Strong, C. (Eds.). (2022). Environmental Sustainability in Emerging Markets: Consumer, Organisation and Policy Perspectives. Springer.

Ojo, T. (2024). Women and Finance in Africa: Inclusion and Transformation. Springer.

Popescu, I. S., Hitaj, C., & Benetto, E. (2021). Measuring the Sustainability of Investment Funds: A Critical Review of Methods and Frameworks in Sustainable Finance. Journal of Cleaner Production, 314, 128016.

Puaschunder, J. M. (2023). The Future of Resilient Finance: Finance Politics in the Age of Sustainable Development. Springer Nature.

Schoenmaker, D., & Schramade, W. (2018). Principles of Sustainable Finance. Oxford University Press.

Shen, Y., & Ahlers, A. L. (2019). Blue Sky Fabrication in China: Science-Policy Integration in Air Pollution Regulation Campaigns for Mega-Events. Environmental Science and Policy, 94, 135–142.

Soetan, T., & Mogaji, E. (2024). Financial Services in Nigeria: The Path Towards Financial Inclusion, Economic Development and Sustainable Growth. Palgrave Macmillan.

Steinhardt, H. C., & Wu, F. (2016). In the Name of the Public: Environmental Protest and the Changing Landscape of Popular Contention in China. China Journal, 75(75), 61–82.

Stern, R. E. (2013). Environmental Litigation in China: A Study in Political Ambivalence. Cambridge University Press.
Smelser, N. J., & Swedberg, R. (2005). Introducing Economic Sociology. The Handbook of Economic Sociology, 2, 3–25. Princeton University Press.
Uzzi, B. (1999). Embeddedness in the Making of Financial Capital: How Social Relations and Networks Benefit Firms Seeking Financing. American sociological review, 64(4), 481–505.
van Rooij, B., Fryxell, G. E., Lo, C. W. H., & Wang, W. (2013). From Support to Pressure: The Dynamics of Social and Governmental Influences on Environmental Law Enforcement in Guangzhou City, China. Regulation and Governance, 7(3), 321–347.
van Rooij, B., & Lo, C. W. H. (2010). Fragile Convergence: Understanding Variation in the Enforcement of China's Industrial Pollution Law. Law and Policy, 32(1), 14–37.
van Rooij, B., Stern, R. E., & Fürst, K. (2016). The Authoritarian Logic of Regulatory Pluralism: Understanding China's New Environmental Actors. Regulation and Governance, 10(1), 3–13.
Zhang, X. (2016). Judicial Enforcement Deputies: Causes and Effects of Chinese Judges Enforcing Environmental Administrative Decisions. Regulation and Governance, 10(1), 29–43.
Ziolo, M., Filipiak, B. Z., Bąk, I., & Cheba, K. (2019). How to Design More Sustainable Financial Systems: The Roles of Environmental, Social, and Governance Factors in the Decision-Making Process. Sustainability, 11(20), 5604.
Zhou, X. (2022). The Logic of Governance in China: An Organizational Approach. Cambridge University Press.

PART II

Sustainable Financialization for Decarbonization, Environment and Sustainability

CHAPTER 2

Sustainable FinTech for Public Participation, Low-Carbon Transition, and Environmental Protection: Ant Forest in China and GForest in the Philippines

2.1 Sustainable Green Practices Confronting Challenges Directly

Global environmental issues have become one of the major challenges faced by the world. Since the advent of human industrial civilization, the productive forces created in just a few centuries have surpassed the sum of those created in all previous eras. Such rapid advancements in productivity and production efficiency have led to swift development in human society. However, this development has been predicated on the exploitation of the environment and the unrestrained extraction of natural resources, inflicting irreparable damage on the Earth's environment over time. This is evident in various domains, including climate, oceans, and biodiversity and is closely related not only to the current survival and development of humanity but also to its future concerns.

Since the first United Nations Conference on Environment and Development in Rio de Janeiro in 1992, the international community has begun to link economic development with environmental protection, proposing the concept of sustainable development. Through continuous international conferences and long-term negotiations and coordination, a series of consensuses on international cooperation have been reached, calling for various actors to participate in global environmental protection.

J. Sun, *Sustainable Financialization*, Sustainable Development Goals Series, https://doi.org/10.1007/978-981-96-3580-1_2

The Paris Agreement, reached at the 2015 Paris Climate Conference, is one of the most representative outcomes of climate governance in recent years. It formally established a top-down governance approach in law and set a target of controlling global temperature change within 1.5 °C.

Governments, businesses, and the public have recognized the importance of environmental protection and sustainable development. Collective efforts are being made to curb greenhouse gas emissions and safeguard our natural surroundings. Businesses, integral to society, are stepping up to their social responsibilities in environmental protection. Ant Group, a prominent fintech leader, originated from Alipay, which was founded in 2004 and has since remained dedicated to its social responsibilities. Recognizing the power of small actions in tackling environmental challenges, Ant Group has been advocating for green and low-carbon practices from the start. Aligned with national and global initiatives, the company is leveraging technological innovation and transforming business models to steer millions towards a sustainable and eco-friendly future.

In this chapter, we'll delve into Ant Group's commitment to green and low-carbon initiatives, which are not only shaping sustainable fintech practices in China but also making a global impact. These initiatives represent Ant Group's significant strides in sustainable development and serve as a model for international efforts. In the meantime, the successful strategies employed by Ant Group offer invaluable lessons and insights that can guide other businesses on their journey towards sustainability.

2.2 Green Economy Driven by Enterprises: Ant Forest's Contribution to China's Carbon Neutrality

Ant Group's green and low-carbon efforts come mainly from two important groups: the public and businesses. They work together to reach carbon neutrality. Among the public initiatives, Ant Forest stands out as the most influential and well-known. Since it started, Ant Forest has caught the attention of the government, media, and non-governmental organizations both in China and around the globe. We will also talk about the Ant Forest project in the next part. It will show how Ant Group is making China greener and more eco-friendly by looking at the project's main points, how it works, and what it has achieved.

2.2.1 Ant Forest: Goals and Vision

Ant Forest is an ecological conservation initiative launched by Ant Financial on Alipay in August 2016, aimed at stimulating public environmental awareness and action through digital technology and innovative interactive models. The project centers on reducing individual carbon footprints by allowing users to record their daily low-carbon behaviors on Alipay, generating virtual energy that can be used to plant real trees. Users' low-carbon activities include walking, cycling, and making online payments, all of which contribute to reduced carbon emissions; the system generates corresponding green energy based on these reductions. A relevant official from Ant Group has emphasized that Ant Forest is a purely philanthropic project that does not yield any financial returns for the company, with the annual investment of hundreds of millions by Ant Group being classified as pure charitable donations. The original intention behind the establishment of Ant Forest was to promote widespread participation in low-carbon environmental actions across society through an "Internet + Environmental Protection" model, effectively reducing carbon emissions in daily life, improving the ecological environment, and enhancing public environmental awareness. This initiative enables individuals to contribute to the sustainable development of the planet through their small efforts, thereby inspiring more people to engage in green actions.

2.2.2 Integrating into Daily Life: User Participation Mechanism and Innovative Expansion of Ant Forest

People join Ant Forest by doing low-carbon activities in their everyday lives. These activities include taking the subway, paying bills online, and buying tickets on the internet. By doing these things, they collect green energy points. When they get enough points, they can grow a virtual tree on Alipay. As the virtual tree grows up, Alipay works with charity groups to plant a real tree somewhere on Earth. Users can watch their trees grow using satellite images. Besides planting trees, users can also protect them. The newly launched "Protection Area" program allows users to "adopt" designated natural reserves through the application platform, thereby achieving the goals of environmental protection and enhancing public environmental awareness.

Ant Forest User Terminal Interface (*Source* China Green Carbon Foundation)

Ant Forest also possesses certain social gaming characteristics. The green energy of users is designed to disappear after 72 hours of being generated. Users are prompted to click to "collect" their energy in a timely manner, and they can also help friends harvest energy that is about to vanish or "water" it. For users, the concept of a low-carbon lifestyle has transformed from an abstract idea into tangible forests, and Alipay has gained increased user activity and stickiness along with the popularity of Ant Forest.

Since its launch, Ant Forest has been extremely popular, with the number of users exceeding 200 million by January 2017, and by April of this year, the user count had already surpassed 500 million. An increasing number of lifestyle services are also achieving data integration with Ant Forest, such as mobile phone recycling, green office practices (paperless offices or online meetings), green packaging (shopping at merchants using environmentally friendly courier bags or tape-less cardboard boxes), electronic reading, and so forth.

To further incentivize the public to adopt green and low-carbon actions, the interactive formats of Ant Forest are continuously being enriched. On the "National Low-Carbon Day" in 2021, the "Green

Energy Action" was officially launched, partnering with industry collaborators to provide low-carbon products and services. Together, they have built a rich array of green consumption and lifestyle scenarios, including green travel, green accommodation, green shopping, green recycling, green government affairs, ecological protection, and restoration. This has formed an initially scaled green open cooperation community, using Ant Forest's "green energy" as a point-based reward to encourage consumers to choose energy-saving, low-carbon products and services, thereby promoting a society-wide adoption of a moderate, green, and low-carbon lifestyle and environmental protection trend.

In addition to consumer behavior on the C-end, for businesses on the B-end, this is also one of the important forces in promoting corporate social responsibility and green-digital economy. "The 'driving force' of carbon emissions comes from the demand side of consumption. If consumers prefer products with low emissions, it will force companies to adopt low-emission production methods," said Yang Pingjian, director of the Environmental Sociology Division of the Chinese Academy of Environmental Sciences. To achieve the goal of carbon neutrality, efforts must be made in both "modes of production" and "lifestyles." He believes that Ant Forest has created a popular way to promote green consumer choices, and its significance lies in awakening consumers' green consumption awareness, innovating public participation methods, and demonstrating the effect of other companies taking on environmental responsibilities better.

On August 25, 2021, also "National Low-Carbon Day," Ant Forest officially launched the "Green Energy Action." Not only Hai Di Lao but also over 100 companies including Haier, KFC, FAW-Volkswagen, Muji, and Ele.me participated as the first batch. Together, they used Ant Forest's "green energy" as a point-based reward to encourage consumers to choose more energy-saving and low-carbon products and services, promoting a society-wide adoption of a moderate, green, and low-carbon lifestyle and environmental protection trend. This also means that from this year's "National Low-Carbon Day," the public has more opportunities to earn Ant Forest's "green energy" through green and low-carbon consumer choices. For example, by choosing a Class 1 energy-efficient air conditioner through the carbon emission reduction methodology provided by the Beijing Green Exchange, consumers can earn 18.4 kg of "green energy." Consumers can use this energy to apply for "tree planting" or to claim protected areas in Ant Forest. Companies will then donate funds to public welfare organizations to plant suitable local tree species in areas that urgently need ecological restoration or to protect endangered wild animals and plants.[1]

Wang Xiaoying, the person in charge of the "Green Energy Action" project, once said that the positive response and participation of over 100 companies means that more socially responsible enterprises have joined Ant Forest's "public welfare incentive mechanism" that aims to link people and nature: that is, using corporate donations to support "visible green" and to encourage the public to choose a green and low-carbon lifestyle, to create "invisible green" with practical actions, and to reshape low-carbon environmental protection and green consumption behaviors in the context of the digital transformation of the economy and society.

In June 2022, while Alipay continued to adhere to the "green energy" distribution rules based on carbon emission reduction assessments, it also launched a green action mini-program based on the "Citizens' Environmental Behavior Norms." It called on the entire population to participate in green actions such as saving water, saving electricity, saving food, green travel, reducing paper and plastic, and recycling. After meeting the corresponding standards, they can obtain an "Action Taker for a Beautiful China" electronic certificate, which greatly stimulated the public's enthusiasm for participation. At the same time, it has also made continuous efforts on the path of quantifying individual carbon emission reduction standards. In November 2022, the group standard "Personal Carbon Emission Reduction Incentive Management Standards Based on Internet Platforms," led by Ant Group, was officially released. It proposed for the first time a basic framework for quantifying and incentivizing individual carbon emissions through the internet, providing support for various entities to better encourage the public to participate in green and low-carbon behaviors through internet platforms.

The fields of environmental protection involved in Ant Forest are also continuously expanding. From vegetation restoration to biodiversity conservation, in April 2020, Ant Forest officially launched the "Everyone's One Square Meter, Jointly Protecting Biodiversity" activity. The activity was a collaboration between Ant Financial, the China Environmental Protection Foundation, and the Shanshui Nature Conservation

[1] https://www.eco.gov.cn/news_info/48250.html.

Center and was guided by the Executive Office of the 15th Conference of the Parties (COP15) to the Convention on Biological Diversity. Users can "redeem" 1 square meter of protected area by using 350 g of green energy, establishing a close connection with the "largest area, most species, highest altitude" Jiatang protected area in Ant Forest, supporting biodiversity conservation in the Sanjiangyuan area. At the same time, it developed an interactive model of "online patrol" of protected areas, allowing users to experience the daily patrol work of each protected area according to the route based on their daily number of steps. It integrated real and vivid offline patrol scenes such as removing snares and rescuing animals, to understand the patrol work of protected areas and pay attention to the dynamics of public welfare protected areas. In collaboration with China National Geography, the "Magical Species" project was also launched, with a total of 485 rare Chinese wild animals going online and publishing the popular science book "Magical Species: A Century of Wildlife Conservation in China" in cooperation with the China Environmental Protection Foundation and China National Geography, telling the story of biodiversity conservation in China. In addition, Ant Forest has expanded from terrestrial ecological protection to marine ecological protection. On World Oceans Day in 2022, "Ant Forest | Magical Ocean" was officially launched, allowing users to participate in the cultivation and planting of marine plants and the cleanup of marine debris through online virtual methods, promoting the sustainable use and protection of marine ecosystems.

2.2.3 Dual Incentives: Achievements and Impact of Ant Forest

As of 2023, Ant Forest has spanned seven years, during which countless efforts have accumulated like sand into a tower, transforming this project from a small environmental concept into a green action participated in by over 690 million people. It has not only achieved significant results in low-carbon emission reduction and ecological protection but also has subtly guided domestic citizens to establish environmental awareness and the pursuit and persistence of a green lifestyle, widely praised both domestically and internationally.

In its seven years since inception, Ant Forest has partnered with more than 20 public welfare organizations, planting 475 million trees and participating in the construction of 31 nature reserves, successively supporting the ecological construction of 22 provinces such as Inner

Mongolia and Gansu. Various ecological public welfare projects of Ant Forest have created over 3.7 million cumulative labor income opportunities in planting, maintenance, and patrol. Starting from 2022, Ant Forest further extended its public welfare exploration of ecological protection and restoration from land to the ocean, restoring seagrass beds and mangroves in the coastal areas of Shandong and Fujian. To date, it has completed the cultivation and planting of 66.6 million eelgrasses, 2.72 million autumn eggplants, and the restoration of 2,000 acres of mangroves. Ant Forest, in collaboration with public welfare organizations, has supported ecological construction in 22 provinces including Inner Mongolia and Gansu, with a total agreed donation of more than 3.4 billion yuan, and an actual expenditure of 2.053 billion yuan. These funds are directly used for local ecological restoration and protection projects, driving local economic development.[2]

On August 28, 2023, Ant Group invited over 200 ecological partners, including frontline tree planters, mobile phone tree-planting netizens, experts in related fields, and representatives of public welfare organizations, to participate in the "Ant Forest Green Ecological Partners Conference." They announced the overall situation of the "Ant Forest" operation over the past seven years: since its launch in August 2016, Ant Group has cumulatively agreed to donate 3.465 billion yuan through "Ant Forest." As the ecological projects are implemented across various regions, 2.053 billion yuan has been actually expended, with the remaining funds to be paid by Ant Group and the Ant Forest Foundation according to the progress of planting and maintenance and the results of acceptance inspections.

Zhang Zhanhai, Special Envoy of the Chinese Government to the United Nations Ocean Conference and former Chief Engineer of the Ministry of Natural Resources, stated in his speech at the event: By supporting ecological construction through corporate philanthropic funding, Ant Forest has attracted hundreds of millions of people to pay attention to nature and practice green and low-carbon living, reflecting the company's high degree of social responsibility.

[2] Ant Group, 2023, https://www.antgroup.com/news-media/press-releases/1693215256000.

Ant Forest Partner Conference 2023 (*Source* Ant Group)

2.2.3.1 Green Practices in Desertification Areas in China

Minqin County, under the jurisdiction of Wuwei City in Gansu Province, is surrounded by deserts on three sides, with strong winds and barren trees, to the point where speaking against the wind means ingesting sand. Only the sea buckthorn tree, which is particularly adaptable to saline-alkali land, can survive. Since 2016, every time someone stamps a sapling in Alipay's Ant Forest, a sea buckthorn tree is planted in the sea buckthorn forest here. Over the past three years, about 500 million people have "planted" 122 million real trees through this mobile application, with an average of about 110,000 trees planted daily in desert areas like Minqin County. For most tree planters in Minqin County, tree planting is not so much a belief as it is an opportunity and hope to solve the family's livelihood. There is a resettlement village near the county town; it was relocated from the mountains earlier this year, and children's schooling is no longer a problem, but it is still difficult for the family's labor force to find work. However, now all adults in the village have a new goal—planting trees in Ant Forest. From the actual effect, Ant Forest has to some extent driven employment and economic development in China's impoverished areas.

In Alxa, Inner Mongolia, a real sea buckthorn tree was also planted in this severely desertified area at almost the same time. Nie Yusheng, a 57-year-old herder from Alxa, Inner Mongolia, is now a forest ranger for Ant Forest, looking after more than 2,000 hectares of forest. He once said, "I never thought that an ordinary herder like me could participate and help change the environment. In just three years, the land where I live has become greener, and I have made my contribution here."

Ant Forest's vision is providing an experience for green governance by bringing every individual closer to environmental protection methods. In March 2019, a couple from Zhengzhou, Henan, went to the Alxa Desert in Minqin, Gansu, to take wedding photos. The bride said that after collecting energy for half a year to plant a tree, she wanted to go and see it, among so many trees, there is one tree that she planted. Similarly, You Shengde, who runs a travel company in Hangzhou, found that many people who play Ant Forest will have the idea of going to "see the site." He seized this opportunity, negotiated with Ant Forest, and organized a group of tourists to visit Ant Forest in Alxa, Inner Mongolia. In this way, the tourism industry, which relies on green environmental protection actions, can also be gradually integrated into local economic development, which is extraordinary for those underdeveloped areas with poor natural conditions.

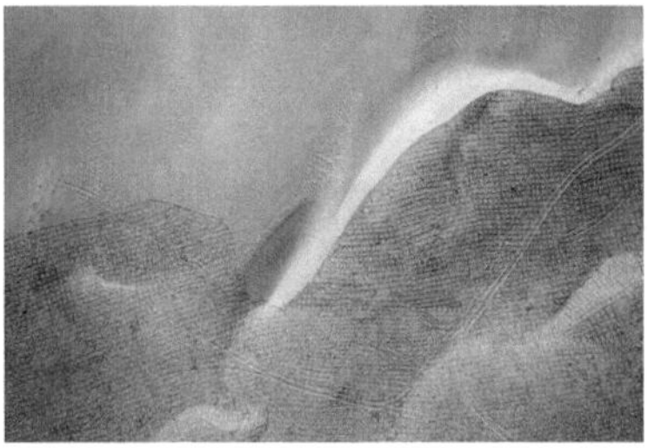

Saving degraded land in Tennger Desert, Gansu Province by Alipay Ant Forest (*Source* UNFCCC)

2.2.3.2 *Multi-dimensional Evaluations of Ant Forest*

As the largest public welfare organization collaborating with Ant Forest on afforestation efforts, Chen Shuxian, the chairman of the China Green Foundation, asserts that Ant Forest has achieved innovations in three key areas over seven years of philanthropic practice: project management models, ecological outcome consolidation, and public funding oversight. This initiative has become one of the significant platforms for social forces to participate in land greening and ecological construction in the new era. It not only broadens employment and income opportunities for farmers and herders but also serves as an exemplary model for ecological prosperity.

Lee SungAh, Deputy Secretary-General of the International Union for Conservation of Nature (IUCN), expressed high appreciation for the achievements of Ant Forest. "IUCN will continue to collaborate with Ant Group and partners in China to create more innovative solutions," said Lee SungAh. IUCN has partnered with Ant Forest to provide scientific and technical support in assessing the ecological outcomes of public welfare projects.

Ant Forest's results are seen in carbon reduction and ecological protection. It also raises public environmental awareness. Ant Forest also plays a crucial role in enhancing public environmental awareness. By turning eco-friendly actions into green energy points, Ant Forest offers a new and fun way to interact. This way of interacting lets users help the environment through their everyday choices, which motivates them to engage more in low-carbon activities. This innovative method combines technology with environmental protection, making it easier for people to join in and feel a sense of accomplishment. It turns environmental protection into something real and possible, not just a distant idea.

Moreover, Ant Forest fulfills its social responsibility by translating virtual trees into real-world outcomes, allowing users to witness the lush results both online and offline. This interaction and feedback mechanism not only strengthens users' environmental awareness but also cultivates low-carbon behaviors as lasting habits. While enjoying the gamified experience, users continuously enhance their environmental consciousness and sense of responsibility. Under the dual influence of Ant Forest's guidance in green energy and significant ecological protection outcomes, public enthusiasm and awareness for environmental action are effectively stimulated. This innovative method, which merges environmental protection with digital technology, transforms environmental protection into a daily practice achievable by everyone through their efforts. More individuals recognize the importance of a low-carbon lifestyle through Ant Forest and consciously practice low-carbon behaviors in their daily lives.

Peng Yijie, President of Ant Group's Social Welfare and Green Development Division, summarizes the Ant Forest model as a "dual-end incentive behavior," wherein the visible "green" in the distance inspires the invisible "green" nearby. Following actions that generate green emissions reductions, Ant Forest provides incentives for these behaviors. Specifically, the energy calculation system of Ant Forest encompasses five sectors: green travel, reduced travel, paper and plastic reduction, energy efficiency, and recycling. Within these sectors, there are over 40 low-carbon scenarios, each associated with a corresponding "green energy" value. The logic of energy calculation follows a fixed methodology; for example, when using an eco-friendly bag, the calculation considers the emissions of plastic bags over their entire lifecycle and the proportion of individuals using plastic bags, ultimately yielding an energy value.[3]

Ant Forest is well-known and praised for its creative ways to be green and low-carbon, and for the big difference it makes for the environment. In China, Ant Forest has won many important awards, like the "National Poverty Alleviation Innovation Award" and the "National Environmental Protection Science and Technology Award", showing a recognition of Ant Forest's great job in saving the environment and helping people out of poverty. It also shows how important Ant Forest as a program of the private sector is in making the economy greener and more sustainable. Moreover, Ant Forest has been named an "Outstanding Case of China's Internet Plus Public Welfare", which makes it a leader in China's efforts to do good to the environment and promote a green future.

Ant Forest is also famous around the world for winning many important awards. One of these is the UN Champions of the Earth award from the United Nations Environment Program, which is the top global prize for protecting the environment.[4] This award indicates that Ant Forest has done an amazing job in making new ways to protect the environment and getting people involved. Also, the World Economic Forum has called Ant Forest a "Global Lighthouse Project," showing that it has an important impact worldwide and is an exemplary example for others to follow.

Deng Guosheng, the Deputy Dean of the Academy of Philanthropy at Tsinghua University, evaluates Ant Forest's approach and says that companies can increase customer favor and even gain extra business value by being socially responsible. This is a sustainable strategy that wins for everyone and can last a long time, achieving both charity and environmental protection goals, and also giving a boost to corporate giving.

[3] Jiemian, 2021, https://www.jiemian.com/article/6553426.html.

[4] UNEP, 2019, https://www.unep.org/news-and-stories/press-release/chinese-initiative-ant-forest-wins-un-champions-earth-award.

However, some experts have expressed concerns about the sustainability of artificial afforestation in water-scarce deserts. There are also opinions that restoring degraded forests is a more sustainable strategy.

Deng Guosheng stated that the "technology + environmental protection" model has certain limitations. Firstly, it concerns the type of audience, with young people clearly dominating in Ant Forest. Data on Ant Forest users disclosed in 2017 showed that users under 28 years old accounted for 60% of the total number of Ant Forest users, while those over 50 years old accounted for less than 20%. He mentioned, "This model involves more young people, who are also more likely to master the rules of the game. However, older individuals or people in some rural areas are not as involved." Deng Guosheng also pointed out that operating such technological products requires the continuous introduction of new features and sections to maintain user interest.[5]

In addition, some analysts believe that the "carbon account" recording model of Ant Forest, which only accounts for users' low-carbon behaviors and ignores their high-carbon actions, is unreasonable. Users who are particularly high-carbon during normal times can deliberately obtain carbon points rewards by engaging in some low-carbon actions, while serious carbon account management requires accounting for and controlling the total personal carbon emissions.

Wang Ling, a staff member of Ant Forest, once introduced in a media interview that Ant Forest is constantly being updated, and new ways of playing are continuously being launched. The "Public Welfare Forest" feature launched on World Environment Day in 2019 uses the influence and appeal of celebrities, companies, universities, or social organizations to encourage users to allocate some of their energy to water their public welfare forests, and the trees planted are not attributed to individual users. As of August 2019, the number of users participating in Ant Forest's Public Welfare Forest has exceeded 20 million, and the cumulative number of watering has exceeded 93 million times.

Ant Forest's actions show that Ant Group is a leader in going green and reducing carbon. They also show other companies around the world how to think about growing in a way that's good for both business and planet. Ant Forest inspires more businesses to join in and protect the earth using new ideas and technology. Together, they can face the problems of

[5] Thepaper, 2019, https://www.thepaper.cn/newsDetail_forward_5030741.

our changing climate and work towards a future that's both green and low in carbon.

2.3 The Internationalization of Ant Group's Green and Low-Carbon Initiatives

Facing the challenge of making green and low-carbon efforts global, Ant Group takes a step-by-step approach. When reaching out to people in different countries, they respect the local ways and values, and they support green habits that fit each place. A great example of this is when they started GForest in the Philippines. At the same time, for Chinese people traveling overseas, Ant Group works with businesses in other countries to give green travel rewards through Alipay's "Hui Chu Jing" (惠出境) program, which is also linked to Ant Forest. These efforts indicate how they bring their sustainable actions from China to the world.

2.3.1 Greenwashing and Challenges in Corporate Social Responsibility

"Greenwashing" is a portmanteau of "green" and "whitewashing," referring to the practice of covering up one's non-green practices by boasting environmentally friendly actions. The term was first coined in the mid-1980s by an American environmentalist, Jayson Westerveld, who noticed that a hotel he stayed at would remind guests in their rooms to reuse towels to "save our planet", claiming that millions of gallons of water would be wasted daily by laundering towels used only once. However, he later discovered that the hotel's practice was merely a cost-saving measure to reduce the number of towels washed. Thus, he termed this hypocritical behavior "greenwashing." In 2010, "greenwashing" was officially included in the Oxford English Dictionary. The non-profit organization "Corporate Watch" defines it as: "Corporations disguise themselves as 'friends of the environment', attempting to conceal their destruction of society and the environment, in order to preserve and expand their market or influence."

As public environmental awareness continues to grow, some companies, in an effort to beautify their image or for market expansion, frequently associate themselves with green and low-carbon initiatives, but their ultimate goal is not environmental protection but economic benefit. Scientific American magazine has pointed out that the most common pattern of greenwashing is companies actively promoting environmental plans or products while their core business continues to damage the ecosystem. There is no shortage of corporate cases, and public and media attention to greenwashing is also increasing: the Massive Arithmetic Index shows that compared to 2021, the comprehensive index for "greenwashing" in 2022 increased by 100% year-on-year, and the search index increased by 54.55%. This poses a trust challenge for Ant Group's promotion of green and low-carbon actions on the international stage.

On the path of greenwashing, there have been numerous companies that have flown the flag of environmental protection only to be mercilessly exposed by reality. For example, in the 1980s, the oil giant Chevron was once the subject of a case study in business schools for its advertisements: in a fairy tale-like snowy forest, a hibernating bear snores away, and Chevron seizes the opportunity to extract oil, quietly leaving before they wake up in spring, allowing the bear mother and her cub to play worry-free on the pristine grassland. In 1989, however, the once-cleansed image of the industry was shattered by the Exxon Valdez oil spill in Alaska, with countless wild animals losing their lives due to the oil spill, and the nature of pursuing profits under the green veneer was also brutally exposed by the harsh reality.

In addition to oil companies, there are also cases of greenwashing in the forestry industry. The American forestry company Weyerhaeuser once released an advertisement in which scientists rowed a boat into an underground river surrounded by schools of fish. "Weyerhaeuser works with scientists and environmental organizations to protect fish," the voiceover said, "We take this very seriously." However, according to The Guardian, the company's excessive logging has led to a sharp decline in salmon numbers. For the forestry giant Weyerhaeuser, the fate of the salmon is insignificant. "Weyerhaeuser has no plans to stop logging forests," Steven Whitney of the American Wildlife Association told The Guardian. "Nor does it promise to leave forest areas near salmon spawning sites as a buffer."

There are many such corporate cases, and the practice of greenwashing not only exacerbates environmental degradation but also diverts resources from genuine environmental initiatives, thereby hindering effective sustainable development measures and subtly eroding public concern and emphasis on environmental causes. The established precedents serve as an important lesson and a significant challenge for Ant Group in promoting green and low-carbon actions.

2.3.1.1 GForest: "Philippines' Version of Ant Forest"

To reach its international users, Ant Group uses localized promotion strategies tailored to each region. It introduces the green and low-carbon model of the Ant Forest program to other countries. Ant Group also starts projects that match the local customs and what people need. An exemplary case is when they brought GForest to the Philippines.

The Philippine archipelago is one of the most biodiverse regions on the planet. However, due to improper human activities such as deforestation, land development, and agricultural practices, the ecosystem of the Philippine islands has been severely impacted, with an average of 52,000 trees disappearing daily and the primary forest habitats of some endemic species reduced to only 4%, posing a significant threat to the survival and reproduction of wildlife.[6] Faced with this dire situation, GForest has teamed up with several organizations, including Globe, the Ramon Aboitiz Foundation, the Biodiversity Finance Initiative, the World Wildlife Fund Philippines, Century Pacific Food, Inc., Friends of Hope, Inc., and Ayala Land, Inc. to leverage the evolving technological landscape to take measures to aid the environment.

On June 25, 2019, GCash, a mobile payment platform with 15 million users and known as the "Philippine version of Alipay," also launched "GCash Forest." It introduced a model similar to Alipay's Ant Forest, capable of tracking users' green lifestyle choices and encouraging citizens to participate in environmental protection and afforestation. This marked the first time that a Chinese enterprise's digital technology public welfare model was extended overseas.

Specifically, GForest allows users to accumulate "green energy points" through daily digital transactions, such as bill payments and transfers, which can be used for tree planting. Every transaction or eco-friendly activity completed by GForest users through the Philippine digital payment application that is a Globe transaction can earn green energy in the GForest section. Once enough energy is accumulated, users can plant virtual trees using their green energy. GCash then plants a real tree in the water source areas supplying the Manila metropolitan region in the Philippines. As of July 2024, GForest has 18 million registered users who have planted enough trees to cover approximately 11,600 hectares,

[6] https://thebeat.asia/manila/terra/nature/explainer-how-does-gcashs-gforest-work.

an area nearly equivalent to the size of San Francisco.[7] Additionally, users can adopt a local tree individually or with other GCash users through the app, contributing to the reforestation efforts in the Philippines.

GForest User Terminal Interface (*Source* Gcash)

The afforestation initiatives of GForest are predominantly focused on critical ecological zones in the Philippines, such as the Ipo Watershed and the Alaminos Carbon Forest. The tree-planting activities in these regions not only contribute to the restoration of the ecological environment but also enhance biodiversity and safeguard the local natural resources. Presently, GForest collaborates with esteemed partners including WWF Philippines, Ramon Aboitiz Foundation Inc (RAFI),

[7] Gcash, 2024, https://new.gcash.com/services/gforest.

Ayala Land Inc (ALI), Friends of Hope Inc (HOPE), Century Pacific Food Inc (Century), Philippine Coffee Board Inc., WWF, and Culion Foundation. These partners work directly with local farming communities to plant seedlings and monitor their growth.

Cooperation of green project between RAFI and GCash in Cebu (*Source* Inquirer)

By December 19, 2021, the cumulative virtual tree planting by GForest's user base reached 1,060,000, with 1,000,000 of these transitioning into physical trees established across diverse geographical sites. This achievement is monumental for the contributors who have been actively engaged in the program, demonstrating a commitment to environmental conservation. Indeed, the aggregate carbon emissions averted by GForest's user community are so substantial that the green energy accumulated by GCash could theoretically power more than 25 circumnavigations of the globe by aircraft.

GForest is growing and doing more than just planting trees. It is now helping local farmers with projects that mix farming and trees, which creates ways for them to make a living that is sustainable for the environment. Through cooperating with multiple partners, GForest empowers local farmers in their communities. More than 2,000 farmers who grow coconuts have been given seeds, to get more fruit and increase their income. Moreover, GForest also has offered 150 local farmers tree-planting training, thereby enhancing their planting skills and livelihood capabilities.

As the project continues to evolve, GForest has expanded its scope beyond reforestation to include support for agroforestry projects that benefit smallholder farmers, thereby aiding in the creation of sustainable livelihoods for local small-scale agriculturalists. The GForest initiative, through collaboration with multiple partners, supports, and empowers local farmers. For instance, the project has provided seeds to over 2,000 coconut farmers, assisting them in increasing their yields and income. Furthermore, GForest has also offered tree-planting training to 150 farmers, thereby enhancing their skills and livelihood capabilities.

Practice and Influence of GForest (*Source* Gcash)

2.3.1.2 Targeting Domestic Users: Sustainable Tourism Based on Ant Forest

For Chinese users abroad, Ant Group has launched the "Hui Chu Jing" program, which integrates existing domestic green and low-carbon projects and collaborates with overseas merchants to provide green and low-carbon travel assurance for Chinese tourists, aiding in sustainable travel.

Taking the "Stroll Globally, Enjoy the World" activity launched by Alipay during the May Day holiday this year as an example. To meet the demand for sustainable outbound travel among Chinese nationals, starting from April 22, Alipay joined forces with millions of overseas merchants to offer a three-week "Stroll Globally, Enjoy the World" May Day discount for Chinese outbound tourists. The activity covers the most popular destinations for Chinese outbound travel, including Hong Kong and Macau, Japan and Korea, Southeast Asia, Europe, Australia and New Zealand, and North America. The initial launch includes a 0.51-yuan global travel card with a variety of services and a high green content. Domestic users can purchase a global travel card worth over 200 yuan for only 0.51 yuan in the "Hui Chu Jing" section of Alipay. The card includes a collection of essential rights for outbound travel, such as discounts, duty-free shopping, flights, taxi rides, and data usage, facilitating sustainable travel abroad. The specific forms include collaborating with overseas merchants to access Alipay+ payment services. Data shows that Alipay's offline coverage abroad has exceeded 70 countries and regions, with over 5 million various overseas merchants offering dining, entertainment, transportation, and more, allowing domestic users to travel seamlessly with a single wallet and enjoy a convenient mobile payment experience. At the same time, Alipay also partners with local merchants to offer discounts, providing cost savings for users while encouraging the use of mobile payments overseas and achieving the goal of low-carbon tourism. In addition to shopping, Alipay's transit code can also be used to ride local public transportation abroad, achieving "one code to go anywhere." In Hong Kong, the "one code to go anywhere" experience continues to be upgraded, with public transportation not only allowing for scanning to ride but also earning Ant Forest green energy. Scanning to ride a bus yields 80g of energy, and taking the subway yields 52g of energy, with a maximum of 5 accumulations per day for each mode of transportation. This is the first-time Alipay users can accumulate Ant Forest energy when traveling abroad, and in the future, this green energy is expected to accompany Chinese tourists to more places around the world.

With diverse "Hui Chu Jing" activities, Ant Group makes it easier for Chinese travelers to pay with their phones when they are overseas. Besides, they have teamed up with businesses in other countries to give deals on green transportation and tourism. In this way, Ant Group has not just helped people save money and time but also been trying to push for a more sustainable globe.

2.4 Conclusion

This chapter showcases Ant Group's efforts and contributions to sustainable development through an introduction of its domestic and international green and low-carbon practices. Domestically, Ant Forest has stimulated public environmental awareness through innovative interactive models, achieving remarkable results in low-carbon emissions reduction and ecological protection. In the process of internationalization, facing international skepticism about the potential "greenwashing" of large corporations, Ant Group has carved a unique international path through targeted actions and strategic innovation, achieving significant outcomes in carbon emission reduction and ecological conservation.

Ant Group has been at the forefront of ecological conservation, implementing targeted initiatives that have yielded substantial environmental benefits. One of its flagship programs, Ant Forest, has catalyzed a green movement among the public, encouraging the adoption of low-carbon lifestyles. According to Ant Group's 2023 Sustainable Development Report, the platform has mobilized over 690 million individuals to participate in carbon reduction efforts, accumulating more than 31 million tons of "green energy" through everyday eco-friendly actions.

Ant Forest planted 475 million trees in 11 different places in China, adding up to an area as big as 5.08 million mu. A lot of these trees are helping the "Three-North Shelter Forestation Project" (三北防护林), which indicates great significance for environment protection and renewable resources. Besides, Ant Group has set up 31 areas to protect nature in 13 provinces and cities, keeping over 4,800 square kilometers of land safe. In Weihai, Shandong, 66.6 million seagrasses have been planted. In Ningde, Fujian, 2.72 million Kandelia candel trees have been added to help bring back mangrove areas. These efforts are making a big difference in restoring seagrass beds and mangroves, which are crucial for marine life.

Ant Forest has joined forces with organizations and platforms to promote green and low-carbon cooperation in large-scale events such as the Asian Games, with over 100 million people participating in the Asian Games' carbon neutrality, earning a Guinness World Record; by the end of 2023, Ant Group had launched 60 marine areas and 605 marine life knowledge popularization, with nearly 130 million people participating, while the Ant Forest ancient tree protection project supported the protection of ancient and famous trees in six cities including Guangzhou, Hangzhou, Nanjing, Fuzhou, Jinan, and Changsha.

Furthermore, in April 2023, Ant Group established the Ant Forest Green Development Foundation, which aims to collaborate with government departments and public welfare organizations to conduct a diverse range of ecological protection, restoration, customer service, and environmental research initiatives.

Regarding Ant Forest, in terms of carbon emission reduction, according to Ant Group's 2023 Sustainable Development Report, over 690 million people have participated in Ant Forest, adopting green and low-carbon lifestyles in their daily lives, accumulating over 31 million tons of "green energy" in carbon emission reductions. Ant Forest has provided the public with new scenarios for green living and green consumption, with 638 "Green Energy Action" partner openings and more than 60 types of cooperative low-carbon scenarios;

As for GForest, according to the official website data of GCash, GForest has planted over 2.7 million trees. By December 2022, GForest had accumulated 12 million registered users, becoming one of the largest science-based environmental protection movements in the Philippines; the trees planted by users are enough to cover about 5,600 hectares of land in the Philippines; GForest users have reduced carbon emissions by over 30.5 billion grams, and the saved carbon emissions are enough to fuel a voyage around the Earth 25 times. GForest, modeled after Ant Forest, has made a tremendous contribution to the environmental protection actions in the Philippines, allowing the Philippine public to collectively participate in green actions and greatly enhancing their awareness of sustainable development.

It is obvious that Ant Group's eco-friendly moves have made a real difference for our planet. Their projects have gotten people excited about taking care of the environment and have cut down on carbon pollution and protected nature right here at home with their creative ideas. On the global scale, Ant Group's efforts have put it at the front of the green revolution and shown other companies around the globe how to take sustainable development seriously. Their experiment is now inspiring more businesses to jump in and use creative ideas and technologies to fight climate change together and work towards a sustainable future.

CHAPTER 3

Sustainable FinTech for Corporate Participation and Sustainable Transition: InterContinental Hotels Group (IHG), Genshin Impact, McDonald, and Corporate Partners in Alipay's Ecosystem

3.1 Corporate Participation and Sustainable Transition

Sustainable development has gained international attention in the modern era, and many businesses are promoting constructive social and environmental change through creative partnerships. As one of the world's top technological companies, Ant Group has started a number of significant sustainable development initiatives with major players in the market like McDonald's, Genshin Impact, InterContinental Hotels Group (IHG), and business partners in Alipay's ecosystem.

The Ant Forest project, a public welfare innovation on the Alipay platform, encourages users to accumulate "green energy" through daily green behaviors to plant virtual trees. The growth of each virtual tree corresponds to the planting of a real tree, thereby motivating users' environmental actions. Since its launch in 2016, Ant Forest has become an influential environmental protection project in China and globally. It encourages users to adopt low-carbon behaviors in daily life, such as walking, cycling instead of driving, and paying utility bills online, to accumulate "green energy." Once users have collected a certain amount of

J. Sun, *Sustainable Financialization*, Sustainable Development Goals Series, https://doi.org/10.1007/978-981-96-3580-1_3

energy, they can plant a sapling in the virtual space. Ant Forest, in collaboration with partners, will then plant a real sapling in decertified areas or other places in need of greening.

Engaging corporate participation is an essential aspect of Ant Forest's mobilization of the entire society to participate in low-carbon emission reduction and environmental protection practices. The success of Ant Forest lies not only in its innovative concept and technological implementation but also in its successful integration of corporate social responsibility (CSR) with public welfare activities, attracting numerous enterprises to participate. These enterprises not only provide financial support but also actively participate in specific tree-planting activities, creating a positive social demonstration effect. This project not only promotes afforestation but also provides a quantification standard for individual carbon emission reduction through scientific methodologies.

Coca-Cola China is a shining example of how businesses can team up with initiatives like Ant Forest to boost sustainable practices. The goal is to discover a sweet spot where both the environment and the bottom line benefit from repurposing empty drink bottles. They started an innovative new strategy in June 2022: labeling their bottles with recycled materials. Going green was not the only reason for this action. Some of their most well-known brands, including Schweppes, Minute Maid, Sprite, and Coca-Cola, were affected. The labels are crafted with super slick recycling tech that makes them look as good as new but with a 30% reduction in virgin plastic use. This is a big step towards a circular economy, where resources keep flowing and waste is a thing of the past.

Coca-Cola China's initiative with recycled material bottle labels is a concrete step in their "World Without Waste" sustainable packaging strategy. This move not only cuts down on the need for new plastic but also slashes carbon emissions in the production process. According to the Beijing Green Exchange, each package with a recycled label cuts carbon emissions by 2 g. While that might not sound like much, when you think about the national or global scale, the total impact is huge.

Teaming up with Ant Forest, Coca-Cola China has been spreading the word on environmental protection. When you buy a product with recycled labels from Coca-Cola's Tmall store, you earn Ant Forest green energy. This incentive gets people more interested in recycling and encourages them to take real environmental action. Every purchase becomes a little environmental gesture, letting consumers help the planet one drink at a time.

As early as 2021, Coca-Cola China added a "Recycle Me" sustainable packaging logo to its product packaging, reminding consumers of the recycling and utilization of beverage bottles in a creative way. This logo not only conveys the concept of beverage bottle recycling but also enhances the interaction between consumers and products, allowing environmental protection messages to be conveyed to the public in a more intimate and life-close manner.

By joining forces with Ant Forest, Coca-Cola China is showing its commitment to sustainability and taking concrete steps towards its goal of making all its packaging recyclable by 2025. They are setting a precedent for other consumer goods companies even other industries to follow suit.

In its recent sustainability reports, Ant Group has highlighted its "Dual Value Creation" strategy, which tackles social issues with tech innovation and has made strides in areas like research investment, supporting small businesses, and digital services. Ant Group's initiatives are not just driving their own sustainable growth but also aligning with global sustainability objectives. They are proving that businesses can tackle worldwide challenges through industry partnerships and work towards a balanced development that benefits society, the economy, and the environment.

3.2 Ant Forest and InterContinental Hotels Group (IHG): Exploring Sustainable Tourism Models in the Hospitality Industry

3.2.1 Green Energy Action: Reducing Carbon Footprint

As the world gets more serious about tackling climate change, businesses are looking for ways to shrink their carbon footprints. In this push for sustainability, Ant Forest, a brainchild of Ant Group through its Alipay platform, has emerged as a key player in encouraging eco-friendly living. Since its debut in 2016, Ant Forest has not only raised environmental awareness among hundreds of millions but also drawn in many companies eager to do their part. By the end of 2022, the global hospitality giant IHG joined forces with Alipay's Ant Forest for the "Green Energy Action", marking the first time the initiative has extended into the hotel sector. Together, they are championing the "Green Stay" concept, taking the low-carbon movement deeper into the hospitality industry and opening a new chapter in corporate partnerships.

IHG Joins Ant Forest's Green Energy Action (*Source* Internet)

The "Green Energy Action" is an initiative launched by Ant Forest in 2021, aiming to unite enterprises with social responsibility to encourage consumers to choose low-carbon, environmentally friendly products and services in various consumption scenarios using "green energy" as a point-based reward. This action covers not only green travel and green packaging but also extended to the hotel industry for the first time in 2022, with the cooperation with InterContinental Hotels Group (IHG) being a typical example.

An industry leader in hotel management, IHG stays mindful of its responsibility to promote environmentally friendly travel. They were not simply responding to a call to action when they joined Ant Group's "Green Energy Action." IHG also made a strong statement about its environmental commitment. IHG's "Green Stay" program is all about encouraging guests in the Greater China area to cut back on using those little single-use items that hotels typically offer, like toothpaste, toothbrushes, combs, and soap. By doing so, participants can earn 92 g of "green energy" on Ant Forest. This virtual energy doesn't just let users grow digital trees; it also has the potential to fuel real-world greening initiatives. It is a small step that makes a big difference.

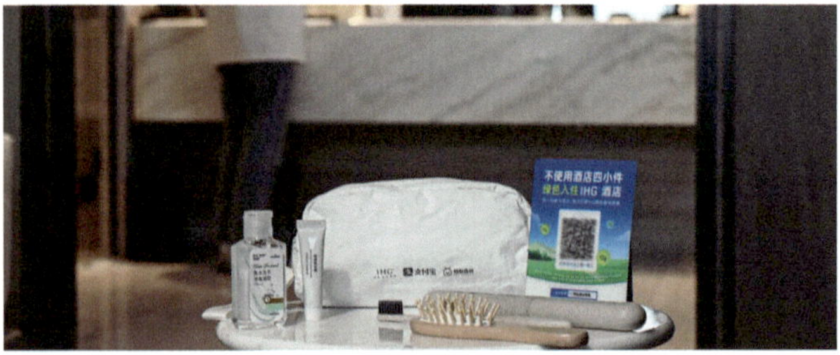

IHG's "Green Stay" (*Source* Ant Group)

The initiative of "green stay" has been actively responded to by enterprises and consumers and has achieved initial results. Statistics show that if all members of IHG in the Greater China region can complete at least one "green stay" every year, the accumulated Ant Forest "green energy" will reach an astonishing 1380 tons.[1] This figure not only reflects the tremendous energy gathered from individual small actions but also highlights the importance of enterprises guiding consumers to participate in environmental protection.

At the beginning of 2022, the Chinese government issued the "Plan to Promote Green Consumption," which clearly states that green consumption is a profound change that requires the integration of green concepts in all consumption links. The cooperation between Ant Group and InterContinental Hotels Group is a specific implementation of this policy direction, in which the awareness of green consumption among consumers and enterprises has been enhanced, jointly promoting the entire society to develop in a more economical and low-carbon direction.

Faced with the severe challenges of global climate change, the business community has an unshrinkable responsibility. Zheng Jiang, General Manager of Ant Group's Sustainable Green Public Welfare Department, emphasized that enterprises should not only strive to reduce their own carbon emissions but also become pioneers in advocating green consumption. By influencing more consumers through their actions, they can drive the low-carbon transformation of the entire production side. In 2022, the "Green Energy Action" penetrated the supply side, attracting 636 enterprise brands, including Muji, Nike, Starbucks, Nestle, Wanda, FAW-Volkswagen, Unilever, Haier, and SF, to join. The low-carbon scenarios covered by the cooperation reached more than 60.[2]

These enterprises cover all aspects of people's daily lives, from clothing, food, housing, and transportation to entertainment and leisure, jointly forming a powerful green consumption enterprise partner low-carbon

[1] https://company.stcn.com/gsdt/202209/t20220928_4885211.html.

[2] Ant Group 2022 Sustainability Report, 2022, https://mdn.alipayobjects.com/huamei_f5zjmk/afts/file/A*ixdpSL7H6BUAAAAAAAAAAAAADgx7AQ/%E8%9A%82%E8%9A%81%E9%9B%86%E5%9B%A22022%E5%8F%AF%E6%8C%81%E7%BB%AD%E5%8F%91%E5%B1%95%E6%8A%A5%E5%91%8A.pdf.

action alliance. They encourage users to practice more green and low-carbon behaviors in daily life and reduce their own carbon footprint. The cooperation between InterContinental Hotels Group and Ant Group is not only a strong practice of the green consumption concept but also an exploration of the future path of sustainable development.

3.2.2 *Cooperation with Alipay: Digital Payment Empowers Sustainable Travel*

Mobile payments have changed the way people pay in the digital era, but they are also a force for social advancement. Through the collaboration with IHG, Alipay has not only transformed the hotel industry but also boosted sustainable tourism. This section will investigate in detail how Alipay has used digital payment technology to help IHG boost its business competitiveness and seize this opportunity to support the long-term transformation of the hotel industry as a whole.

The background of the global partnership cooperation between Alipay and IHG can be traced back to 2016.[3] On the basis of the original cooperation, InterContinental Hotels Group and Ant Group announced a strategic cooperation in Shanghai in 2020, making IHG the first international hotel group to open an official booking channel payment through Zhima Credit. This means that guests with a Zhima Credit score of 600 or above can enjoy the convenient service of "0-yuan booking, 0-yuan deposit, 0-s check-out" during the hotel reservation process in the Greater China region of IHG, highlighting IHG's determination in the digital layout of the Chinese market.

This digital collaboration has taken the user experience to new heights. With mobile internet becoming the norm, the way people pay has changed a lot. Alipay, being one of China's biggest mobile payment platforms, caters to users of all ages. IHG's partnership with Alipay is all about keeping up with this shift, making stays easier and more convenient for guests. With "Zhima Credit Stay", guests can skip upfront payments and deposits, and the bill is settled automatically at checkout, which streamlines the process and saves both time and money. For business travelers, this means no more tying up capital in deposits and room charges before check-in, a real boon for those always on the move.

[3] IHG, 2016, https://www.ihgplc.com/en/news-and-media/news-releases/2016/ihg-announces-global-partnership-with-alipay.

For the hotel front desk, work efficiency has soared, reducing guest wait times and boosting overall satisfaction. By partnering with Alipay, IHG strengthens its brand and gains a competitive advantage in the hotel sector. This collaboration attracts new clients in addition to providing guests with greater convenience. As Emir, COO of IHG Greater China, puts it, this model enhances the guest experience and shows IHG's "genuine hospitality."

IHG has also seen a boost in customer loyalty. Beyond a smooth payment experience, IHG has rolled out promotions on the Alipay mini-program, like discounts of up to 299-yuan post-booking. These moves attract new customers and strengthen ties with the existing ones. Digitally, IHG can get a better grasp of what customers want and offer more personalized services, building stronger relationships.

Since fully jumping on board with Alipay's green initiative in 2022, InterContinental Hotels Group has seen steady growth in the Greater China market. Official numbers show that in 2023, their RevPAR in this market surpassed the same period in 2019, with a third quarter year-on-year jump of 43.2%. With 21 new hotels opened in the region in the third quarter, up 5.5% year-on-year, Greater China has become one of InterContinental's fastest-growing markets globally. This mirrors the tourism market's quick rebound and highlights IHG's robust competitiveness in China.

"InterContinental Group's strategy in the Chinese market is first reflected in the digital experience." said Lu Haiqing, Chief Corporate Affairs and Strategic Relations Officer of IHG Greater China.[4] Digital development is regarded as one of the key factors to promote economic growth, and digital platforms such as Alipay are widely popular, with the Chinese market at the forefront of global digital development. The hotel industry is also more actively embracing digital transformation and upgrading. In 2023, IHG launched the IHG Douyin Travel Industry Solution, becoming the first international hotel group in the Greater China region to directly connect with Douyin's coupon products. As of August 31, 2023, the contribution rate of digital channels has approached 20%. The second is to promote the deep integration of culture and tourism. "Promoting tourism with culture and showcasing culture with tourism" has become an inevitable choice for developing a modern

[4] https://m.mp.oeeee.com/a/BAAFRD000020231108872657.html.

tourism industry and promoting cultural dissemination. IHG created the Hualuxe brand in 2012 for guests who like Chinese culture, meeting the needs of Chinese consumers for high-end hotels and the inheritance of Chinese culture, while also reflecting InterContinental's "in China, for China" strategy and its long-term commitment to the Chinese market. In the Greater China region, which includes well-known cities and resort locations, Hualuxe Hotels & Resorts now operates 20 hotels and has 24 hotels under construction as of November 8, 2023. Furthermore, the InterContinental Hotels Group is dedicated to supporting sustainable development by modernizing its brand and transforming its inventory.

For instance, the launch of the Vignette Collection brand aims to provide guests with a unique personalized luxury stay experience and has also become a new anchor for the group to develop the high-end hotel stock market. In this way, InterContinental not only improves the quality of hotels but also promotes the effective use of resources.

IHG and Ant Group have teamed up to answer Alipay's push for eco-friendly travel with the "Green Stay" initiative. By cutting back on single-use items like toothpaste and toothbrushes, guests can earn "green energy" from Ant Forest. This energy can be used to create real green projects in addition to growing virtual trees. It's a win–win situation that increases visitors' awareness of environmental issues and demonstrates green practices for the hotel sector.

The alliance between InterContinental Hotels Group and Alipay isn't just about changing how we pay; it's about charting a new course for the future of hospitality. With digital tools, InterContinental Hotels Group has sharpened its competitive edge and set a new standard for sustainable practices. They're not just talking the talk; they're leading the charge in digital innovation, elevating guest experiences, and steering the hotel industry towards a smarter, greener future.

Behind the digital transformation of InterContinental Hotels Group, Alipay's digital ecosystem partner - Green Cloud, has grown into China's leading digital ecosystem service provider in the large accommodation industry. The Green Cloud team has more than 30 years of hotel informatization experience and an advanced technical background. Based on the four core businesses of Green Cloud PMS, Oracle Hospitality, data platform, and e-commerce platform, it has formed a digital solution for four major sectors: hotel accommodation, hotel catering, hotel groups, and cultural tourism complexes, including accommodation, catering, marketing, finance, operations, group management, and cultural tourism services, covering the entire scene and the entire chain of digital solutions. Green Cloud, together with Alipay and upstream and downstream partners in the hotel industry, interconnects and empowers the digital transformation of the industry. To date, Green Cloud has served more than half of the top 50 hotel groups in China, with nearly 30,000 signed hotels.[5] Its embedded digital ecosystem in Alipay has brought significant value to the Chinese hotel industry: first, the efficiency of the hotel front desk has been improved, achieving one-click check-in completion in 30 s, which is 800% more efficient than the average manual processing time of 3 min. Second, it increases customer stickiness and enhances hotel revenue. After customers check-in, the hotel's mini-program in the Alipay APP automatically links to the customers, and the opening rate of the mini-program is higher. Third, it quickly develops members. Through the smart stay facial recognition to join the membership, the hotel group's membership level is matched with the Alipay membership level, which not only helps the hotel acquire more members but also enhances the hotel group's benefits for Alipay's level members.

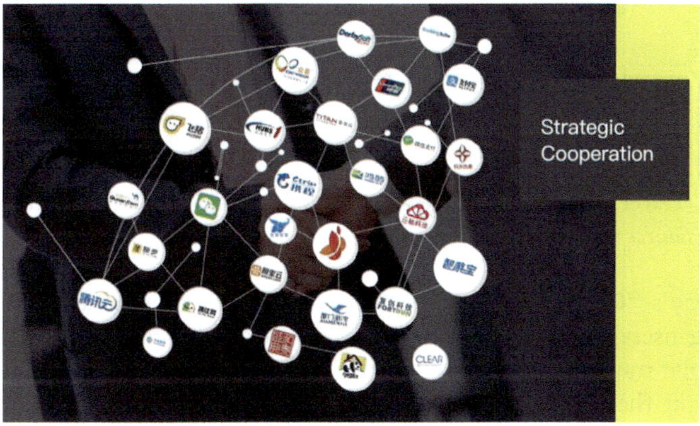

Green Cloud Partnership (*Source* Green Cloud)

[5] Green Cloud, https://www.ihotel.cn/category/5351.

3.3 Genshin Impact: Cross-Industry Collaboration with the Gaming Industry

Genshin Impact has gained a huge fan base and has been a global success since its September 2020 premiere. With more than 100 million registered users in China alone by 2023, it had established itself as the most popular game in the nation. Internationally, it is just as popular, with over 56 million players outside China, making it the most successful Chinese game to grace the global stage.

Gamers all over the world adore Genshin Impact, which has received great reviews for its breathtaking graphics, compelling storylines, and open-world adventure. By collaborating with Alipay for the "Green Travel Week" (绿色出行周) event in 2022, the game took a bold move. They established a new benchmark for inter-industry cooperation by deftly fusing the excitement of gaming with practical environmental initiatives.

"Green Travel Week" (绿色出行周) Activity Launched by Genshin Impact and Alipay (*Source* Genshin Impact)

Alipay's "Green Travel Week" happens at a set time each year, challenging users to go green on the go by biking shared cycles or hopping on public transit. To join in, players just need to look up "Green Travel Week" in the Alipay app, where they can find the event page and take on eco-friendly tasks to rack up "green energy." Consider using Alipay's QR code to pay for bus or metro rides or riding a shared bike—each task earns you points.

The green energy you gather can then be traded in for Genshin Impact goodies on Alipay. That might mean snazzy character skins, game currency, or other virtual treats. For instance, you might swap your energy for exclusive character outfits or save up to get a bundle of in-game cash to try for new characters or weapons.

This team-up between Genshin Impact and Alipay blurs the lines between gaming and real-life action, creating a fresh kind of interactive adventure. Players step out of the digital realm and into the great outdoors, embracing green travel and merging their online and offline lives. It is about developing a green mindset, not simply about the loot. Players are encouraged to live more sustainably by the event, which benefits the environment. Furthermore, it creates a close-knit community where users may share experiences and encourage one another to adopt more environmentally friendly practices. In addition to strengthening the ties between players, this sense of community increases the event's visibility and influence.

In 2023, the cooperation between the two parties continues. The green travel activity jointly released by Genshin Impact and Alipay has four phases, each with different themes and rewards.

Green travel activity jointly released by Genshin Impact and Alipay in 2023 (*Source* Genshin Impact)

In this activity, travelers can search for "Genshin Impact" through the Alipay APP home page to enter the activity page and click on "accumulate mileage" to complete designated tasks and obtain green mileage. In addition, each time travelers use Alipay for travel, they will generate additional green mileage, which travelers need to collect at the mileage mall.

The green mileage obtained by travelers can be used to exchange for Genshin Impact-themed prizes, including but not limited to character skins, game currency (limited to 8 million copies), and other game props. In addition, they can also exchange for peripheral lottery opportunities to win game co-branded peripherals, series badges, travel coupons, and other physical rewards.

In January 2024, Genshin Impact and Alipay once again launched a new joint activity. During the event, players can obtain a Genshin Impact-themed Alipay full-end skin by completing tasks, which is a permanent skin, along with a rich array of in-game props. From January 20th to 22nd, travelers can use 20 miles to sign up for the draw for a free benefit on the Genshin Impact-themed flight on February 3rd. The lucky winners can have their airfare waived, only needing to pay the fuel and construction fees to take the themed flight on the event day. In addition, players can also search for designated keywords through Alipay's flight channel to draw game currency and physical peripheral prizes, including co-branded gift boxes, commemorative stamps, etc., greatly enriching the player's participation experience.

Green travel activity jointly released by Genshin Impact and Alipay in 2024 (*Source* Genshin Impact)

In summary, the Genshin Impact and Alipay green travel activity is a successful cross-industry cooperation case. This innovative cross-industry cooperation between Alipay and the gaming industry provides more enterprises and a large number of game developers with new and valuable sustainable development ideas. Genshin Impact, as a game with a huge fan base, not only benefits a wide range of players through in-game rewards but also attracts more public attention to real-life environmental protection issues, which in fact helps to reduce carbon emissions and environmental actions. By combining the fun of the game with environmental protection actions, Genshin Impact and Alipay have created a positive interactive platform together, encouraging more people to participate in green actions, thus creating a positive social effect. This cooperation model demonstrates the possibility of interaction between games and the real world and provides new ideas for future green travel advocacy activities.

3.4 Sustainable Practice in the Catering Industry: McDonald's China

3.4.1 Multi-Dimensional Practice of ESG Strategy

In China, McDonald's serves 500 million hamburgers, uses 400 million paper takeout bags, and has 1.3 billion patrons annually. McDonald's now has 5,700 stores in China and aims to have 10,000 by 2028. It currently has 260 million users on its internet platform and employs 200,000 people. This well-known fast food chain McDonald's has demonstrated a pro-ESG stance. McDonald's multifaceted ESG approach and particular practices in China and around the world will be covered in detail in this section.

In China, McDonald's has proven its leadership in sustainable construction. From 2018 to 2022, 1,800 (about 95%) of the new restaurants opened by McDonald's in China met the "Leadership in Energy and Environmental Design" (LEED) standards issued by the U.S. Green Building Council (USGBC).[6] Currently, McDonald's has 2,500 green restaurants in China, with an annual reduction of 80,000 tons, and an average of one new green restaurant opening every 9–10 h.[7] McDonald's China is the largest chain enterprise in the Asia–Pacific region that meets

[6] McDonald, https://www.mcdonalds.com.cn/news/mcdonalds-china-announces-scale-for-good-green-package-and-green-restaurant.

[7] Foodinc, 2023, https://www.foodinc.com.cn/2023/03/24/20230324001/.

LEED standards, marking McDonald's as a leader in green building practices in the Asia–Pacific region. The zero-carbon restaurant in Shougang Park, Beijing, has even obtained LEED zero-carbon and zero-energy certification, making it the first green restaurant in Asia to obtain dual net-zero certification, highlighting McDonald's unremitting efforts and innovative spirit in sustainable development. Mahesh Ramanujam, President and CEO of USGBC and Green Business Certification Inc., once said, "McDonald's China has created a model for green restaurants, not only in restaurant design but also in daily operations—energy-saving and emission reduction and creating a healthier and more comfortable indoor environment for customers and employees."

Zero-carbon restaurant in Shougang Park, Beijing (*Source* McDonald)

Green Recycling is also a very important link. McDonald's trays have turned green, made from waste plastic. In terms of creativity, it's not just about the trays; the "Reshape Goods" project recycles waste plastic, designs, and creates a series of green experience devices, such as environmentally friendly charging bicycles that generate electricity by "pedaling." At the China headquarters, the dining area tables and chairs are all made from the results of "Reshape Goods."

On top of that, McDonald's China is also a leading enterprise in promoting green packaging and material recycling. In 2018, McDonald's China announced three green packaging action plans, aiming to use packaging more responsibly while ensuring customers' dining experience, continuously optimizing and reducing packaging, especially plastic packaging materials.

Action 1: Actively reduce the use of packaging materials. In 2015, McDonald's China successively canceled the packaging boxes for Spicy

Chicken Sandwich, Filet-o-Fish, and Grilled Chicken Filet Burger, changing to single-layer packaging paper, reducing the paper usage for related products' packaging by nearly 80%. For packaging such as French fries boxes, pie boxes, hamburger packaging paper, napkins, and paper boxes, McDonald's optimized the cutting process and size design, reducing the required paper amount by about 20%.

Action 2: Reduce plastic. Starting from November 1, 2018, McDonald's China began piloting "no straw lid" and not actively providing straws in 10 restaurants in Beijing to encourage consumers to reduce straw usage. As early as 2007, McDonald's China took the lead in providing paper takeout bags, reducing more than 2 billion plastic bags over the past decade. In addition, all restaurants have also replaced the plastic cups for McFlurry and adjusted the size of knives and forks to reduce plastic usage.

Action 3: 100% use of internationally forest-certified virgin paper. Before 2020, 100% of the paper food packaging used by McDonald's China came from virgin paper certified by the Forest Stewardship Council or the Program for the Endorsement of Forest Certification to support sustainable forest management.

McDonald's has also taken active measures in data security protection, establishing a dedicated data security governance office to prevent risks and enhance business capabilities, reflecting McDonald's emphasis on customer privacy protection.

McDonald's is no slouch when it comes to giving back to the community and fostering the next generation's career growth. They are all in on global community service, with a special focus on boosting young people's career prospects. By hooking up with schools and colleges to offer learning and internship opportunities, McDonald's is equipping the youth with the skills they need to thrive in a world that is always changing.

And that is not all. McDonald's community service efforts are further enhanced via the "Ronald McDonald House" project. Offering lodging and other forms of assistance to families whose children are battling health issues is the main goal of this operation. McDonald's is also able to support families during times of need.

3.4.2 *McDonald's Sustainable Food Supply Chain Standards*

McDonald's is all about keeping food safety top of mind and going green in its supply chain. They are big on setting the bar high for sustainable food practices. In China, McDonald's has a special Food Safety and Quality Committee that keeps a close eye on the whole process, from farm to table. They use digital tricks to keep tabs on every step of food production, making sure things run smoothly and safely. Also, they make sure their crew is well-trained, so everyone is on the same page about food safety.

In line with the healthy eating trend and going green, McDonald's set a goal in 2022 for at least half of their Happy Meal meals to meet their global nutrition standards, which means keeping it under 600 cal. This demonstrates their concern for the health of families, particularly the young ones. McDonald's is constantly creating new, healthier menu items for its patrons to satiate their desire for both tasty and environmentally friendly options. In order to reduce waste, they also make sure to use ingredients meticulously which is a good thing for the environment.

In terms of specific product supply standards, McDonald's star products, the Filet-O-Fish and Double Filet-O-Fish, bear the Marine Stewardship Council (MSC) sustainable fishery logo, ensuring that the fish used in McDonald's fish burgers is 100% sourced from fisheries certified by the MSC for sustainability. In addition, including coffee beans, packaging paper, and even chicken feed soybeans have also obtained corresponding green sustainable procurement certifications.

In 2023, McDonald's, together with its nine major suppliers, launched the McDonald's China Regenerative Agriculture Program. McDonald's China purchases nearly 300 kinds of food ingredients every year, more than two-thirds of which come from these nine suppliers. In the first phase of the program from 2023 to 2028, McDonald's will mainly focus on four actions: (1) promoting and popularizing the concept of McDonald's China's regenerative agriculture; (2) promoting the upgrade of pilot projects for McDonald's China's regenerative agriculture; (3) promoting the continuous implementation of McDonald's China's regenerative agriculture in line with Chinese agricultural realities; (4) establishing a cooperative ecosystem for McDonald's China's regenerative agriculture.[8]

[8] Foodinc, 2023, https://www.foodinc.com.cn/2023/03/24/20230324001/.

Taking potatoes as an example, McDonald's is all about growing them the right way, with tight rules on planting that cut down on chemicals. This action helps keep the soil healthy, maybe even better than before. Obviously, it is a big part of their plan for greener agriculture, which is all about looking after the earth. To make this happen, McDonald's works hand in hand with farmers, showing them how to switch up crops with the seasons to keep the soil in tip-top shape.

McDonald's China Regenerative Agriculture Program (*Source* Foodinc)

Now flip to the chicken side of things. McDonald's in China is leading the charge for a cage-free chicken policy, setting a new standard for sustainable food chains. They have put a "Cage-Free Chicken" logo on their chicken boxes and placemats, making a big statement about their commitment to a kinder way of raising birds. When they say "Cage-Free Chicken", they are not talking about a special breed, but talking about a way of farming that puts animal welfare front and center.

"Cage-Free Chicken" logo in McDonald's ad (*Source* McDonald)

Traditional broiler farming pursues high-density farming to improve space output, which usually means keeping chickens in cages to restrict their range of movement, thereby reducing exercise consumption and accelerating growth. However, this method is not only criticized by animal welfare organizations but may also not be as tasty as free-range chickens from a meat quality perspective. McDonald's "Cage-Free Good Chicken" adopts a more relaxed breeding environment. For the breeding method of White Feather Chickens, McDonald's has reached an agreement with breeders to ensure that the chickens are in a good environment of flat-layer broiler houses and are free to move, and the use of hormones is strictly prohibited. The chickens are placed in spacious flat-layer broiler houses, with 5 to 10 cm of rice husk bedding on the ground to provide a comfortable living condition for the chickens and allow them to naturally peck and scratch. According to statistics, these chickens can walk thousands of steps a day, which is a relatively large amount of activity compared to traditional caged chickens.[9]

In addition to cage-free breeding, McDonald's also adheres to high standards of hygiene management. Breeders must prepare the bedding before a batch of chicks start and must carry out cleaning and disinfection after the chicks are sold, which takes more than 6 weeks. They live

[9] Mcdonald, 2024, https://www.mcdonalds.com.cn/news/20240619-Good-Chicken-Cage-Free#:~:text=%E9%99%A4%E4%BA%86%E2%80%9C%E9%9D%9E%E7%AC%BC%E5%85%BB%E6%AE%96%E2%80%9D%E7%9A%84%E7%89%B9.

a two-point一线 life between the living area and the production area, and they must also disinfect twice a day when entering the production area from the living area. The design and management of the chicken farm strictly follow hygiene standards, striving to reduce the occurrence of diseases from the site selection to daily operations, thereby reducing the use of antibiotics. The site selection for the chicken farm needs to be far away from the hustle and bustle of the city, and the design distinguishes between the outer area, living area, and work area, with different blocks separated and restricted access to outsiders. The chicken farm monitors temperature, humidity, lighting, ventilation, etc., according to the set standards, coupled with a carefully formulated feed like infant formula, is also divided into stages according to the age of the chicken, and a scientific and reasonable vaccine to minimize or even avoid chicken diseases. During the slaughter and processing, McDonald's has also formulated detailed animal welfare guidelines to ensure that the chickens are treated humanely in the last stage of their lives.

By using these stringent breeding and management practices, McDonald's not only raises the caliber of its chicken but also satisfies its CSR obligations and advances the idea of sustainable development. This strategy encourages more businesses to join the development of a sustainable food supply chain and serves as an example for the sector.

3.4.3 McDonald's and Alipay's Digital Finance Cooperation: High-End Product Marketing Case

In the Chinese market, McDonald's has established a close partnership with Alipay and introduced the Angus MAX Beef Burger, a flagship product line. It has successfully raised the product's market recognition and improved sales performance through a number of creative marketing techniques.

Cooperation among McDonald’s, Alipay, and Zhima Credit (*Source* Alipay)

McDonald’s recognized the unique attributes of Alipay’s mini-program and its user group, deciding not to simply copy other platforms, but to closely integrate with the characteristics of Alipay and customize the development of the mini-program. Through close cooperation with the Alipay team, McDonald’s innovatively designed a promotion plan combined with Zhima Credit, making the promotion of new products more targeted and attractive.

Zhima Credit is a personal credit score launched by Ant Financial, a subsidiary of Alibaba Group, which is a third-party credit assessment and credit management agency. Considering the relatively high pricing of the

Angus MAX Beef Burger (about more than 30 yuan), McDonald's innovatively proposed the concept of "0 yuan to enjoy first." Customers can taste this burger for free in advance through their Zhima Credit points, but they need to complete the task of repurchasing within a certain period of time, otherwise, they need to make up for the first consumption cost. This approach not only attracts the interest of a large number of customers but also effectively prevents pure activity in exploiting discounts.

The activity is divided into three steps. First, customers enter the activity page through offline store QR code scanning or Alipay search. Second, users will open the task to view the activity rules: customers with a Zhima Credit score of 600 or above can participate, obtaining two coupons, one for the first stage of 0 yuan to purchase the Angus MAX Beef Burger, and the other for the second stage discount coupons. Finally, customers use the coupons to purchase meals in the store, the activity page displays the task progress and reminds customers to complete the task through various channels.

By setting a threshold of 600 points for Zhima Credit, McDonald's ensures that participants in the activity have a higher level of credit, thereby increasing the fulfillment rate of secondary consumption.

To ensure the success of the activity, McDonald's adopted a comprehensive publicity strategy, including online live broadcasting, social media promotion, and offline store promotion. McDonald's warmed up the activity in the Tmall live broadcast room, inviting the chief marketing officers of the two companies and well-known athlete Su Bingtian to participate in the live broadcast, attracting widespread attention. At the same time, McDonald's also created a topic on the Chinese social media platform Weibo, #How Happy Can Saving Money Be# (#省钱能有多快乐#), which successfully ranked sixth on the hot search list. Many loyal fans spontaneously shared activity information on social media platforms, made tutorial videos, and further expanded the influence of the activity. At the same time, more than 3,000 offline stores nationwide unified the layout of promotional materials to coordinate with the online activity and formed a strong momentum. After the activity went live, more than 6,000 people participated within just one minute. By the end of the activity in November, more than 500,000 people participated, bringing more than 300,000 new users to McDonald's. Among them, more than 30% of the customers participated in the activity through Zhima GO, and the redemption rate was five times that of usual activities, and the 7-day repurchase rate was more than three times that of usual activities, with the customer unit price increasing by 32%.[10]

McDonald's offline store promotional activities (*Source* Alipay)

The triumph of this partnership is a testament to two things: McDonald's teamwork with Alipay is on point, and the Zhima Credit system is a solid way to find the cream of the crop when it comes to customers. With some out-of-the-box marketing and a full-court press on promotions, McDonald's has turned the Angus MAX Beef Burger into the go-to high-end burger in China, giving the brand a serious boost.

Besides, McDonald's is now pushing for greener and top-shelf food chains around the world, especially when it comes to beef. They are aiming for the best of both worlds—quality and sustainability. The Angus MAX Beef Burger is a prime example of this, with its beef sourced from the prime cuts of Australia. This is a big part of why they teamed up with Ant Group to push sustainable practices.

Zhang Yan'er, Vice President of Supply Chain Management of McDonald's China, once said at the 2019 Fortune Global Sustainable Forum, "Sustainable development is not just a matter for a single

[10] Alipay, 2022, https://opendocs.alipay.com/b/04ougx?pathHash=6fbb968b.

company but a task that requires the joint efforts of the entire industry and upstream and downstream partners. McDonald's pays special attention to the sustainability of beef globally because the impact of beef production on the environment is particularly significant. However, the industry currently lacks a unified standard for sustainable development, and McDonald's hopes to promote the establishment of industry standards through its own efforts, covering aspects such as feed, breeding methods, and upstream production, making the entire supply chain more green and environmentally friendly."[11]

The idea of "sustainable beef" was first put on the table at the "Global Roundtable on Sustainable Beef" back in 2012, with the goal of making beef production more efficient to ease the burden on natural resources like land and water, all while keeping the nutritional value of beef high. Although this idea has gained a lot of support, it has also stirred up some debates because there are no clear standards yet. Some people question whether McDonald's is utilizing this idea to increase burger sales rather than out of concern for the environment. But McDonald's is not letting that stop them from digging deeper into sustainable beef practices. They have already said they will use sustainable beef in some of their burgers and are working with the Quaker Foundation to figure out how to measure just how sustainable their beef supply chains are. They have also put a big $4.5 million into studying how grazing affects the environment.

McDonald's once set a goal to achieve 100% sustainable beef procurement by 2016, but the actual progress did not achieve the expected results. However, McDonald's is still working hard and has committed to increasing the proportion of sustainable beef in its largest market to 85% by the end of 2020. Zhang Yan'er emphasized that the beef purchased by McDonald's must not come from farms developed by deforestation, indicating the company's strict standards on land use.[12]

McDonald's initiatives in sustainable food supply chain management work as useful models for other businesses in addition to reflecting their attention on social responsibility and environmental preservation.

[11] Finance Sina, 2019, https://finance.sina.com.cn/hy/hyjz/2019-09-06/doc-iicezueu3824815.shtml.

[12] Foodinc, 2023, https://www.foodinc.com.cn/2023/03/24/20230324001/.

McDonald's is getting closer to realizing the aim of quality for its high-quality products by working with suppliers and modern technology to create consistent sustainable standards and adopting doable steps to lessen environmental impacts.

PART III

Sustainable Financialization for Social Innovation, Decent Work and Inclusive Development

CHAPTER 4

Sustainable FinTech for Small Business, Developing Countries, and International Trade: WorldFirst, Antom

4.1 Introduction

In the process of globalization, it is not only necessary for businesses to seek economic gain, but to be concerned as well with the environment, their societies, and their governance. The Sustainable Development Goals (SDGs) put forth by the United Nations could serve to structure the practice of all other components of the society, and since enterprises are a critical element of the society, their actions are significant for the achievement of these goals. This chapter will evaluate the activities of WorldFirst and Antom in the implementation of the SDGs through the use of case studies.

In the first half of 2024, the total value of China's goods trade imports and exports reached 21.17 trillion yuan, with a new high in trade scale, a year-on-year increase of 6.1%. Among them, cross-border e-commerce has become a vital part in the development of foreign trade of China. As of July 2023, the cumulative number of enterprises in cross-border e-commerce industry has reached 31,500 in China. In the development of cross-border e-commerce, the intermediate links between buying and selling have been reduced, making it more convenient to "buy globally and sell globally." But in the wave of globalization and digitalization, payment has become a key challenge affecting the monetization of cross-border e-commerce. The reason is that as going abroad becomes the

J. Sun, *Sustainable Financialization*, Sustainable Development Goals Series, https://doi.org/10.1007/978-981-96-3580-1_4

choice of more and more Chinese companies, these people come overseas to search for new growth curves and life cycles. Faced with unfamiliar overseas environments and emerging markets where digital infrastructure needs to be improved, cross-border payments have become a headache.

The emergence of cross-border digital payment platforms has become the most important "bridge" to bridge the gap. "Doing cross-border business is as simple as doing local business," said Shi Wenyi, CEO of WorldFirst. Behind it lies countless hardships of small businesses. This chapter will discuss how WorldFirst helps small businesses sell their accounts globally, and how it has become a responsible financial services institution.

4.2 WorldFirst: Providing Convenience of Cross-Border Finance for Small Business

WorldFirst is a global provider of financial technology and international payment services. Established in London in 2004, WorldFirst is committed to creating one-stop payment and financial services, providing high-quality cross-border collection, payment, and exchange services for small businesses worldwide. After the acquisition of Ant Group in 2019, WorldFirst started providing better services for Chinese sellers. It enables customers in framing versatile product offerings along with the guarantee of financial security through its integrated global ecosystem in cross-border finance domain.

4.2.1 Global Voyage: Assisting Cross-Border E-Commerce Expansion

4.2.1.1 From Pain Points: Crafting Cross-Border E-commerce Expansion Strategies

With globalization speeding up, e-commerce companies are rushing to raise capital, move across borders, and set up sights on foreign markets. Nevertheless, two crucial challenges remain: the difficulties inherent in the mechanisms for cross-border payments and trade legislation. Different e-commerce platforms have their own requirements for the submission of materials and go through several rounds of approvals to evade duplication of efforts. In order to be able to use more than one platform, businesses must always be learning different entry requirements, sending documents, and enduring prolonged proceedings. Variances can also arise in the financial regulations and policies of different nations, and as a consequence, businesses may take a lot of time to find the market they require, making global expansion hard and time-consuming.

Moreover, from 2022 onwards, many Chinese foreign trade enterprises have had to contend with three major issues: problems with the supply chain, declining foreign trade, and profit tightening due to changes in the prices of commodities. "Goods must be sold, and money must be collected" is perhaps the greatest headache of both large and small foreign trade companies, including cross-border e-businesses.[1]

But this struggle is proportional to the size of the small business. They are unable to acquire the funds necessary for cross-border transactions in a timely manner because their small size and instability make it hard to obtain credit. Small businesses in cross-border trade also have a level of mistrust from financial institutions. Businesses talk about trust, when given what is needed for their operations deals with a higher threshold. Small enterprises do not only have to deal with moving goods across and payment systems but also solve how to transfer the cash flows in an efficient manner.

Even when these small businesses tussle with such impediments, their next hurdle includes the identification of the potential markets. Finding about the target regions and the target services for expansion is a skill set that many cross-border small businesses do not have. Their inability to conduct appropriate market research and service model development in the appropriate manner restricts their international business growth.

Acknowledging the hurdles experienced by tiny businesses in penetrating foreign markets, WorldFirst set out the "Global Voyage" program in 2019. The target of this program concerns seeking quality e-commerce cross-border platforms in order to enable small enterprises to penetrate foreign markets in a more efficient manner. This project not only eases the how to set up foreign stores and speeds up the waiting time for approval, but also extends from taxation, through logistics, to over marketing. By providing assistance at the beginning and till the undertaking of activities, this assists in lowering the barriers of international trade while advocating for the expansion of inclusive finance as well.

[1] The Economic observer, 2023, https://baijiahao.baidu.com/s?id=1759797681514417134&wfr=spider&for=pc8.

The "Global Voyage" initiative partners with 49 well-known e-commerce platforms worldwide, covering key markets in regions like Europe, North America, and Southeast Asia. It offers sellers a fast track for official partnerships, which means that sellers submit one set of application papers to quickly open stores across numerous platforms. This makes the application that was tedious, complicated, and time-consuming, simple. The project not only helps sellers save time and resources in anticipating unnecessary problems, but it also enhances the probability of the seller successfully opening a store which exposes small-scale businesses to global market.

From a financial viewpoint, WorldFirst helps companies with a simple cross-border payment which helps companies track global currency in real-time. Funds' transfer and settlement become speedy and decrease costs and time which are associated with the classical cross-border financial services. It is a useful financial management tool for small businesses that enables them to access trade proceeds from overseas quickly and conveniently, therefore enabling the businesses to grow and expand in the future.

To conclude, WorldFirst's "Global Voyage" initiative not only lowers the threshold for small businesses to enter international markets but also provides a comprehensive set of cross-border payment and operational support services that help businesses become more competitive globally. By simplifying processes, reducing costs, and improving cash flow management, this initiative offers cross-border e-commerce businesses a solid path for international expansion.

Shi Wenyi, Vice President of Ant Group's International Business Unit, said, "Since 2022, we have leveraged cross-border RMB settlement facilitation policies to launch the 'Global Voyage' initiative. By partnering with global leading cross-border e-commerce platforms and foreign trade service providers, we offer small exporters one-stop solutions for store opening, payments, and collections. In 2022 alone, Ant Group's RMB payment services for export e-commerce saw a year-on-year increase of 114%."

WorldFirst's Global Market (*Source* Internet)

David Li, a Guangdong-based cross-border e-commerce entrepreneur specializing in home and lighting products, is one of WorldFirst's long time clients. With the advent of digital cross-border financial services, he said, "it's like having a digital CFO free of charge." He then added that with the presence of a tech-savvy "digital CFO" through cross-border digital platforms, trade is made safer and more secure. Even micro-multinational enterprises need not fret about financial worries anymore. "Cross-border payments do not only have to be swift and easy, but also very much safe."[2]

4.2.1.2 By Deep Collaboration: Creating Ample Growth Opportunities for Sellers

As mentioned earlier, sellers in the early stages of cross-border trade should be focused on acquiring market understanding, operating competences, and sourcing the relevant industry resources. As a result, the organization makes a contribution to the Global Voyage not just by providing store opening services, but moreover by bringing industry resources such as periodic online and offline seller training, lessons from top-performing sellers in the industry, and collaborative marketing with

[2] Ibid.

e-commerce sites. Such partnerships provide sellers with unique support policies and incentives. This is aimed at ensuring that sellers expand fast, do not repeat recurrent operational weaknesses, and get embedded in the global markets rapidly.

WorldFirst collaborates closely with major cross-border e-commerce platforms like TikTok Shop and Amazon, offering sellers regular training sessions both online and offline. These trainings cover platform rules, operational strategies, and market analysis; all of which assist sellers to be cost-effective in market trends and research. For example, WorldFirst helps new sellers using its discount commissions through its partnership with TikTok Shop to build the market quickly through unique policies and support of the platform. To further enhance their presence in the market, WorldFirst also conducts seller meetings on a regular basis where representatives from e-commerce platforms give presentations. The sellers are then able to engage directly with the management of the platforms to get practical guidance on how to grow the market.

WorldFirst Offers One Billion Generous Subsidies for TikTok Shop Sellers (*Source* WorldFirst)

Another example is WorldFirst's collaboration with Brazil's largest e-commerce platform, B2W, which brings significant advantages to merchants looking to enter the Latin American market. Under the framework of the "Global Voyage" initiative, merchants benefit from three core advantages:

1. Flexible Payment Options and Financial Security: One key advantage of the B2W platform is its unique installment payment model. Brazilian consumers can pay in installments, while B2W pays the seller the full amount upfront. This ensures that sellers receive immediate payment, alleviating cash flow pressure that could arise from customer installment plans. For small businesses, this cash flow assurance is crucial, helping them quickly replenish inventory or expand their operations.
2. Strong Brand Reputation and Extensive Coverage: With over 2,000 physical stores and the strong brand reputation of its parent company, Lojas Americanas, B2W provides sellers with a powerful sales channel. Lojas Americanas, a long-established retail giant in Brazil, enjoys high consumer trust. By partnering with such a platform, sellers can reach online customers and gain additional exposure and sales opportunities through its offline network. This combination of online and offline channels significantly boosts product visibility and credibility for sellers.
3. Localized Operational Services and Market Support: B2W understands the Brazilian market deeply, offering localized operational services, including Portuguese-language support tailored to local consumer preferences and behaviors. Additionally, B2W promotes merchants through television ads and other media channels, increasing brand penetration and helping sellers effectively connect with local consumers. For international merchants unfamiliar with the Brazilian market, this localized support reduces the barriers to entry, allowing them to integrate into the local consumer ecosystem more quickly.

These advantages make the WorldFirst-B2W partnership a convenient gateway for businesses participating in the "Global Voyage" initiative to tap into Brazil's vast consumer market. Sellers can leverage the platform to lower entry barriers, reduce risks, and establish a foothold in the competitive global market, paving the way for sustainable business growth.

Furthermore, WorldFirst employs its international payment network to enable sellers to easily control money and quickly receive cross-border payment on platforms such as B2W. In this regard, for the Brazilian market, merchants operating in the country are able to sell and pay in Brazilian currency and receive payment in a timely manner which enhances the cash flow efficiency something small enterprises appreciate given their cash flow problem most of the time.

The WorldFirst company works in conjunction with the providers of various ecosystem services to deliver comprehensive global services such as logistics, tax, and warehousing solutions to help sellers overcome numerous challenges in cross-border operations. For example, when there are special international Amazon Seller promotions, sellers who link their accounts to WorldFirst do not have to pay any fees for withdrawals or international transfers, in addition to being able to take advantage of several cashback opportunities. Most of the time, cross-border e-commerce sellers have high withdrawal fees when depositing funds using other banks or platforms especially when dealing with large amounts. By employing a zero-fee policy, WorldFirst eliminates the need for sellers to incur unnecessary expenses when transacting large amounts, which is usually the case in cross-border payments. Such activities and benefits reduce greatly the operational expenses for sellers, thus increasing chances for many small businesses to go global.

Sellers frequently find themselves in a disadvantageous position, as they are usually charged high transaction costs. However, initiatives like WorldFirst's "Global Voyage" have benefited the sellers because they generally charge reasonable transaction fees. WorldFirst capped the transaction rate at 0.3% for all platforms, which is far superior to the existing rates in the B2C foreign trade domain. Small businesses, as a result, incur lower transaction costs and enjoy greater profit margins, since their competitiveness in the global market increases. At the same time, WorldFirst managed to cut the duration of cross-border payments from several weeks to a few minutes. In 2023, such services as fund tracking and rapid settlement were launched as well. Their purpose was to address the issues confronted by small cross-border e-commerce businesses. These services assist small enterprises in reducing operational costs, reducing perennial challenges associated with transaction costs, and managing to enter the global marketplace efficiently.

4.2.1.3 *From Cross-Border E-commerce to Inclusive Finance*

The Global Voyage of WorldFirst is not a mere innovative feat in cross-border e-commerce; it is the inherent spirit of financial inclusion. That is adding small businesses into the equation of international trade through accessible cross-border payment and finance infrastructure and services.

Reaching out to more businesses especially smaller e-commerce businesses out there. Startups and small companies usually are grappling with cash deficiency and market penetration among other factors. The "Global Voyage" initiative effectively helps these enterprises overcome such difficulties by simplifying the account opening process, offering diverse financial tools, and reducing transaction costs, allowing them to focus on business expansion and achieve sustainable growth.

By supporting small businesses, WorldFirst has significantly expanded the reach of financial services, ensuring that financial resources are no longer confined to large corporations or dominant players in mature markets. This kind of broad inclusivity means larger firms and smaller firms in various parts of the world will have access to financial resources likely giving more room for fairness in global trade. Furthermore, through its considerable partnerships with numerous global e-commerce channels, WorldFirst makes it practical for small businesses to participate in global supply networks which raises their earnings and contributes in the process towards sustainable inclusive growth of the economy.

WorldFirst's program seeks to empower the social aspect of international small businesses, something the company considers an emphasis on the 'S' of ESG. The program promotes social inclusion and economic equity by assisting micro-enterprises to surmount the confines of cross-border payments. Such a situation not only generates employment opportunities for thousands of enterprises but also ensures balanced international economic growth. Moreover, World First provides timely and reliable financial solutions that reduce inefficiencies in cash flow, increasing operational efficiency and ultimately supporting economic growth in an indirect way.

In the future, with the growth of the "Global Voyage" program, WorldFirst will continue to focus on innovative financing models in its efforts to assist more micro-enterprises in their cross-border activities.

4.2.2 WorldTrade: Safeguarding B2B Foreign Trade Payments

4.2.2.1 WorldTrade: Simple, Efficient, Secure

WorldTrade is one of the products developed by WorldFirst to provide payment and transaction protection services for global trade orders. It aims to make cross-border trade simpler, more efficient, and secure, helping merchants seamlessly connect to global business opportunities with a single click. In China, WorldTrade serves businesses engaged in

CN B2B trade from Mainland China and Hong Kong SAR, such as factories, trading companies, and integrated manufacturing and trading entities. These businesses, involved in traditional trade, general trade, or cross-border B2B commerce, can draft trade orders and complete order collections efficiently via the WorldTrade platform. Globally, it serves enterprises in all regions except those prohibited by policy, such as overseas retailers, wholesalers, and trading companies. Buyers can pay foreign trade orders to Chinese suppliers through WorldTrade and manage orders accordingly.

Public information shows that B2B foreign trade merchants are facing many challenges. Especially in the global economic environment, overseas inflationary pressures have increased, and economic growth has slowed down, which directly affects the stability of international trade. At the same time, foreign trade formats are becoming increasingly diversified, and the market's requirements for transaction speed are constantly increasing. However, traditional foreign trade settlement methods are inadequate, as they are difficult to open accounts, expensive, and slow to collect payments, and the exchange process is often accompanied by high costs and complex procedures. In addition, compliance risks are becoming increasingly prominent, and frequent bank card freezing issues are causing headaches for many businesses. Once a "frozen card" occurs, there may be a risk of fund chain breakage, which seriously affects the relationship with upstream and downstream suppliers and hinders normal trade operations.

In this context, foreign trade merchants need a new payment channel that can improve transaction efficiency and reduce operating costs, especially for the trust problem between buyers and sellers. To mitigate this problem, cross-border payment service providers like WorldFirst have appeared in the market and have come up with a comprehensive big-picture solution to the transaction process starting from placing an order to seeing it fulfilled and paid for in the deployment of a unique B2B trade service known as WorldTrade. The major objective of this service is to ensure that secure, compliant, transparent, and efficient payment solutions are provided, that merchants' transaction uncertainties and risks are reduced, and that normal trade procedures are enhanced. At the same time, it also addresses the common challenges such as late payment for goods received or late shipment of goods which are common scenarios in traditional cross-border trade.

B2B trade payments of WorldTrade improve the security and efficiency of transactions significantly. Its automated payment feature allows the system to initiate payment requests to buyers based on pre-set payment terms immediately after a trade order is established. It not only reduces the seller's workload in chasing payments but also effectively prevents some customers from deliberately delaying payments. Additionally, the payment services provided by WorldTrade support multiple payment methods, including credit cards, debit cards, and the WorldFirst global account (World Account), offering overseas buyers more flexible options. Most importantly, through WorldTrade, the process of trade fund transfers is greatly simplified, allowing for real-time settlements. Compared to the average 1–7-day settlement times typical of traditional cross-border payments, this is a significant improvement that also substantially reduces transaction costs.

Furthermore, WorldTrade also improves the transparency of transactions and helps track every order to guarantee that it is processed and completed, thereby assisting in the establishment of a form of "digital credit" between the buyer and seller. Small businesses that use WorldTrade can focus on the payment processing rates that are at 2.9% which is relatively lower than any other competitors in the market. At the same time, both users save on transaction and currency exchange costs whenever both parties use the WorldFirst global account (World Account). For instance, a business making over $10 million in yearly collections is estimated to save $10,000 yearly on transaction costs by enabling WorldTrade's services. The transaction execution and payment management in a single platform increases the efficiency of capital movement and allows for better options and cost savings to foreign trade enterprises in addition to other benefits.[3]

Finally, using its strong digital technology, there is also reasonable order management and a good system for payment processing for the B2B trade in WorldTrade. Every payment made and every order placed emanates data from different heads in one form or the other that deals with cash, information, logistics, and even customs thus enabling safe crossing of borders with funds. Integration of data does not only enhance the security of transactions but assists small businesses in export and trade

[3] Mpaypass, 2024, https://www.mpaypass.com.cn/news/202404/16145803.html.

in getting good credit history making it easy for them when accessing loans and other forms of assistance.

Also, the WorldTrade through its smart cash management function can easily deal with incoming funds to the order and use sophisticated prediction technology to recommend needful time for conversion of currencies, enabling users to achieve better rates. Such confidence in the commercial relationship is enhanced by this system which, due to the automated and transparent nature of its processes, lowers the costs of developing commercial trust between the parties intending to transact. It is anticipated that WorldTrade shall enable hundreds of thousands of B2B sellers to speedily facilitate trade settlement and improve transactional efficiency.

4.2.2.2 Africa Payment Solutions: Transparent, Compliant, and Tailored

For more than ten years, China has been the largest trading partner of African countries with a number of China companies setting base in Africa and conducting reciprocal trade. As they do so, however, there are challenges which are peculiar to the Africa region that merchants encounter while scaling their operations in the region. One of the hurdles that is experienced most frequently is that of foreign exchange control which in most cases extends the time frame of payment remittances and the efficiency of cash flow. Most of the countries in Africa have been able to control foreign currency transactions which have made it cumbersome and slow to receive payments and this adversely affects the liquidity of the merchants.

Likewise, the rapid depreciation of local currencies further aggravates the exchange rate risk. Many African nations, especially the politically or economically unstable ones, face currency risk due to erratic currency exchange rates. Such currency volatility is a considerable challenge for small businesses that have a fast turnover cash cycle.

Also, some informal financial institutions or remittance channels pose compliance issues, that stretch the risks for merchants. Lack of financial development is also a cause for concern in some parts of Africa when safety and legality of funds in cross-border transactions are in question.

In order to meet these requirements and challenges, currently, the merchants need a payment solution that is safe and secures compliance but is also effective and efficient in ensuring restitution of funds. This is where fintech companies such as WorldFirst come in to intercede in cross-border payments. They provide a customized template referred to as the

"China–Africa Trade Payment and Settlement Solution" assisting firms achieve transparent and compliant local currency payment and settlement in Africa.

WorldFirst has exclusively launched a unique China–Africa payment solution that supports local currency payments for 13 countries (*Source* WorldFirst)

This solution is designed with the African trade market in mind and mostly locals in Nigeria, Kenya, and Tanzania who use local currencies. It intends to address some of the issues that would be encountered by the China–Africa foreign trade people in cross-border remittances. The solution means that African purchasers can make payments to the Chinese seller directly in their local currency without leaving the US dollars in the WorldFirst Account with the Chinese seller; thus the payment and conversion processes have been simplified.

The solution provided by WorldFirst's Africa payment solution relieves the Chinese foreign trade enterprise's cash flow management constraints so that they can easily collectors from different African countries. Such a solution is very important as it creates a robust payment mechanism that is compliant, safe, and efficient which gives businesses the confidence they need to engage in cross-border transactions and receive payment for goods or services on time. Below is an outline of the cash flow cycle for this solution.

1. Payment Process for African Buyers

WorldFirst Africa payment solution encompasses 13 African countries enabling buyers from these regions to make payments in their local currencies. The ability to pay using the local currency is good for the solution, Salary or wages or income, as many African countries have restrictions on foreign exchange and, therefore, buyers find it difficult to procure the USD or another foreign currency. As a result, purchasers can use the national currency as the means of payment without the fuss of foreign exchange. Local payment models reduce the time lag imposed by foreign exchange and increase the willingness to pay on the part of the buyers while payments are quickly received by the sellers.

2. Role of Local Financial Institutions

Upon making purchases in Africa, buyers pay in the local currency in which case the funds first get to the local financial institutions that Partner with WorldFirst. These institutions handle the payment from buyers in local currencies and undertake the commodity's conversion to USD after a given time interval. In countries such as Nigeria and Cameroon the transfer takes approximately 1–2 business days while in other countries this can take anywhere from 2 to 5 business days.

The use of Financial Institutions locally really helps speed up the efficiency with which funds are dispersed and ensures that payments take place through legal and regulatory channels thus mitigating financial risks for both buyers and sellers. Furthermore, there are some benefits that sellers accrue because of WorldFirst's closeness to these local banks, including more robust and preferable exchange rates and lower revenue losses from switching currencies, etc. This is a notable strength given that many countries in Africa experience unprecedented currency volatility, allowing businesses to price more adequately in foreign trade.

3. Transfer of USD Funds to WorldFirst B2B Account

After the conversion is done by the local financial or banking institution, the amount is sent to the seller's WorldFirst foreign trade B2B account in United States dollars. This stage is of utmost importance, making sure that funds get transferred in a secure and reliable manner during the entire time frame of the cross-border payment but also aiding to the simplicity of the whole process. This is especially critical for sellers who can see their payments as they happen in real time and control the amount to be used for payment or withdrawals or any other fourth-party payments within their WorldFirst B2B account.

This stage of the transfer process mostly takes a day or two and facilitates that sellers expect the funds anytime soon, something that is very important for the smooth running of the business in operations, most especially the small businesses. This systematic process of cash payment flow also effectively reduces ambiguity and confusion stakes in times of foreign transactions.

4. Seller Withdrawals and Payments

Sellers can choose to withdraw the funds in their WorldFirst B2B account, or if already deposited, pay the suppliers directly. For sellers choosing to withdraw, the time taken for RMB withdrawals is one minute and the time taken for foreign currency deposits is the same day. World First B2B employs efficient fund management systems that automatically address businesses' demands for quick cash flow recovery and effective management.

Also, as WorldFirst International, they allow customers their funds to be allocated uncharged. This allows for easy movement of funds from one nation to another and from one currency to another making it easier for exporters with many foreign trade markets to optimize their cash flow management globally.

5. Fully Online Process

There is no requirement for the clients to visit the banks in order to carry out the transactions as all operations in the whole cash flow process are undertaken over the internet. All sellers also have his or her own integrated backend system whereby WorldFirst enables them to monitor the status of their transactions, manage their funds, and execute settlement activities through the system. The fact that all processes are online eliminates the chore of operational procedures and also helps in streamlining the waiting period required for offline transactions and enhances efficiency in the transactions.

Through the Africa payment solution, WorldFirst enables cross-border foreign trade merchants to easily and efficiently remit collections, assisting them in addressing common concerns in the African landscape like foreign exchange controls, exchange rate volatility, and compliance risks. WorldFirst partners with local financial institutions in Africa to make sure that funds are secured and compliant while fixing exchange rates helps in mitigating the risk of currency loss. The fact that the entire operational process is online, together with the additional capacity of making instant payments means that cross-border payments have become very easy and efficient which considerably aids Chinese foreign trade enterprises in Africa. The benefits of this solution can be classified into two attributes, namely "Compliance" and "African characteristics":

With this solution, sellers can remit funds to China or other regions without the fear of putting their funds at risk of being frozen or their accounts being compromised. WorldFirst works with qualified banks and financial institutions that are able to carry out sound regulatory activities in China and in several African countries to check and prevent the compliance of funds. This means that regardless of the size of the enterprise, they can legally and safely conduct cross-border transactions all over the world using this solution, thereby eliminating the risks that are legally and financially related to informal channels.

One of the biggest pain points for organizations in cross-border payments on a global level is the compliance part. Now in the African market, there are quite severe foreign exchange regimes and the financial regulatory framework is quite diverse and highly dynamic. WorldFirst makes it possible to increase the safety and legality of transactions with the assistance of over 60 payment licenses that they possess across the planet. Enterprises do not have to fear the ramifications of engaging in non-compliant operations, including having their accounts frozen or transactions being stalled. WorldFirst works with reputable local banks and financial institutions in Africa not only to ensure that the funds in cross-border transactions are compliant but also to assist businesses in resolving complex currency exchange and fund transfer challenges in the African market. Because of the compliance guarantees presented by WorldFirst, many foreign trade companies can be sure about entering the African market and quickly recovering their funds.

WorldFirst's Africa solution is also highly targeted towards the African market. More specifically, the ability to collect local currency enables African buyers to pay in their domestic currencies without the hassle of and costs of converting it to US dollars. The features illustrated in the image clearly indicate that after the buyer makes the payment to international features, below 1–2 working days the amount will have been converted to US dollars and transferred to a WorldFirst foreign trade business-to-business account. Enterprises are able to lock in exchange rates with a view to alleviating potential risks associated with currency movements and minimizing losses that occur through the conversion of currencies. As for the entire operational process, the buyers and sellers are also planning and executing all operations through the internet without the necessity to set foot into a bank's premises, thus enhancing the speed of cross-border transactions.

To conclude, WorldFirst remains a good cross-border payment solutions provider by curbing all regulations without compromising the regulation and all enabling factors of the African market, making it easier for companies focused to expand into Africa. This solution not only lowers the cost of changing currencies and operation legit but it also gives an operational edge when companies seek to extend their operations worldwide.

There is feedback from the customers that WorldFirst's China–Africa payment solution is quite common to the foreign trade merchants. One of the clients Mr. Liu from Xiamen remarked, "WorldFirst's China Africa solution was good for our African buyers who had problems converting US dollars for our benefit as foreign merchants. In addition, WorldFirst's speed in repatriating the funds to China is also remarkably quick: I am now a committed supporter of this WorldFirst." There were also high praises given to WorldFirst by Mr Yang from Guangzhou, "It is however crucial to bring back funds Trade Africa safely out. I have previously known the Ant Group and one of the major reasons why I registered with WorldFirst Services is because it is a subsidiary of the Ant Group which is faster but has better safety. Besides, I tested it and the overall experience, the entire cycle is alright, so I usually call to book the funds from Africa through WorldFirst."

4.2.2.3 From Cross-Border Collection to Inclusive Finance

The WorldTrade solution enables the merchants to break a lot of barriers for cross-border payments particularly those that boundary their growth like the ones that are related to the small business, foreign exchange, and capital inertia. This reduces their financial burden and also improves their competitiveness at the global scale as well as the pace of their development. The Africa payment solution is a good case of the more innovative way of addressing certain market requirements in the world. Thanks to WorldFirst's local currency settlement, African clients can make transactions with Chinese clients in their local currencies, which enables sellers to receive payments in US dollars very quickly. This greatly reduces the capital recovery period which gives suitable funding to businesses working in the complex African landscape.

With the help of these novel financial service solutions, WorldFirst assists merchants in dealing with various barriers in making cross-border payments allowing them to explore new markets especially in developing countries. Such assistance enhances not only the business advancement of merchants but also supports the sustainable growth of the global economy. Through ensuring the competitiveness of small businesses operating in international markets, WorldFirst completely fulfills its promise on an inclusive financial model for growth and development worldwide.

Moreover, these engagements also exemplify opportunities that WorldFirst provides its clients through its great concern for social responsibility. Working with its WorldTrade and Africa payment solutions, WorldFirst enables small enterprises to overcome cross-border payment difficulties, increases financial inclusiveness, while maintaining its social responsibility on ESG. In particular, such provisions enable small businesses to grow without contributing to the creation of barriers to a more open global financial order.

4.2.3 *Diversified Strategies: Serving More Developing Countries*

As a cross-border digital payment and financial services platform under Ant Group, WorldFirst is committed to global expansion. It focuses on its own commercial growth and how its services can promote social welfare and economic inclusivity. Particularly attentive to emerging markets such as Southeast Asia, the Middle East, and South America, WorldFirst looks abroad, aiming to help small businesses expand into global markets and to introduce more vibrant market brands to developing countries. By combining economic benefits with social interests, it fulfills its corporate social responsibilities. The core of this strategic plan lies in perfecting the overseas ecosystem service chain, introducing more efficient and secure cross-border payment solutions to local businesses, and supporting them in facing global economic challenges through various preferential policies and support measures. Meanwhile, it deepens partnerships around the world to build a robust cross-border payment ecosystem, aligning with the "Social" aspect of the ESG framework.

During the 2024 Canton Fair, WorldFirst displayed its five-in-one solution for cross-border payments, which includes "collection, payment, management, exchange, and lending" features and introduced its new WorldTrade product that focuses on the B2B sector. The product has greatly enhanced the processes of cross-border payments such that payment receipt challenges and the danger of having one's account frozen have been cut down which on average increases payment efficiency by over 10 times. This is critical especially for small businesses in foreign trade since they are in need of relatively cheaper and secure payment options if they are to sustain their cash flows with growing inflation and economic growth challenges in the world market.

At the same time, WorldFirst decided to organize a number of promotional activities for domestic merchants during the Spring Festival which included zero exchange transaction fees for a number of exchanges, rate cuts, promotions on the payment processing fee, and reduced annual interest in foreign trade finance. These strategies are meant to enable the small business to overcome peak stocking seasons where collections are guaranteed and cash flow is ensured, hence contributing to the growth of the foreign trade market in China.

WorldFirst with the assistance of more than 100 central ecosystem service providers seeks to develop export businesses by launching timely Universal Gravity in the year 2024. At the May conference on the ecosystem, WorldFirst announced a new initiative "Universal Gravitation", which is a full-link ecosystem service system delivering customers with integrated cash flow, information, and business processes. Over the course of "Global Financial Partners Day", the "Ecosystem finance" scheme was fully unveiled, tying together central enterprises, factor markets, and funds. Not only does it boost the service supply chain productivity, but also broadens the services to small businesses. If supported by technology from WorldFirst, innovation in products, improved customer experience and enhanced risk management will be the core focus of the financial partners. Additionally, the concept of "Ecosystem finance" contributes to the high standards in digitization and standardization of financial transaction information, enhancing the transparency of data. During the occasion, three primary awards were handed by WorldFirst to its partners for their significant contribution towards the building of the globalizing financial services and the international trade systems.

WorldFirst CEO Shi revealed that as Chinese sellers seek brand recognition, diversification, and localization abroad, WorldFirst is planning to strengthen its presence in Southeast Asia, the Middle East, and South America. Demand in these regions for mechanical products and the "new trio" (electric passenger vehicles, lithium batteries, solar cells) is growing rapidly. WorldFirst plans to address challenges in payments, finance, and ecosystem services by enhancing its overseas service chain, meeting the specific needs of local markets, and taking tangible actions to improve the quality of life for local populations.

What WorldFirst intends to achieve through market diversification is not only the international presence into the developing markets but also enabling global enterprises with more reliable and cost-effective solutions for cross-border transactions. Impacting the process of globalization, her interventions to deepen the ecosystem service chain will help improve the global payment system. At the same time, it allows small businesses as well more opportunities to grow and develop in the international business environment.

Building upon its sound technology base, WorldFirst remains capable of providing safe, compliant, transparent, and effective cross-border payment services to China's small foreign trade enterprises. By partnering with reputable global and local overseas banks, WorldFirst offers local collections in over 30 currencies on a single service which covers over 200 countries and regions with a combined total of over US$200 billion transactions processed. Also, the "EcoPay" product offered by WorldFirst and around 90 ecosystem service providers address various issues from the creation of advertising to logistics thus conveniently making procurement processes faster and cheaper. This all-in-one service model improves the ease and safety of making transactions, enabling over 1 million small enterprises to achieve the "one account sells worldwide" concept and at the same time complete the entire cycle of "collection-payment-management-exchange-lending", without any hidden costs of account opening, management and deposit, and with no risk of exchange depreciation.

WorldFirst has constantly been innovating since 2019, the year the company joined Ant Group. WorldFirst uses Ant Group's technology architecture to make enhancements to its technology for various overseas situations as the business has grown over time. As early as in 2019,

WorldFirst partnered with 1688 to provide the service of "Cross-Border Pay" such that important cross-border e-commerce sellers and purchasing agents are able to purchase currency for 1688 stock. The total transaction amount of "Cross-Border Pay" is almost 100 billion RMB. In order to assist more Chinese enterprises in going global, WorldFirst intends to cooperate with many partnership ecosystem members to find practices that help advance the international expansion of Chinese enterprises.

4.3 Antom: Creating a One-Stop Digital Payment Solution

Antom, a subsidiary of Ant International, is dedicated to providing leading payment services to global merchants, offering a one-stop industry digital payment solution that supports business growth. Currently, the services Antom provides support merchants in more than 40 countries, including mainland China and Hong Kong, Japan, Thailand, Malaysia, Indonesia, the Philippines, Vietnam, Singapore, the United States, the United Kingdom, and the European Union. The services bridge the gap between merchants and many international as well as local mainstream payment solutions which include bank cards, online banking, and e-wallets.

Antom also helps merchants foster better connections with consumers by offering them digital marketing solutions. A+ Rewards, which is M&C's digital marketing service, consists of a mini-program/SDK integrated into e-wallets and payment applications, allowing merchants all over the world to issue as many types of discounts and creative promotions as appropriate and necessary for different types of scenarios that aim to improve user acquisition and user retention.

Antom's history began when Antom was a payment service provider for the international e-commerce operations of Alibaba Group, before growing to provide such services for leading global e-commerce and digital entertainment platforms. Antom collaborated with 2C2P Global, a dominant acquiring services provider in Singapore, and formed a complete licensing structure, good institutional partners, and strong technical capabilities. This partnership has not only enhanced the technological advancement of Antom but allowed it to offer acquiring services professionally in over 40 countries supporting the business development of the enterprises across the globe.

Many enterprises now prefer cross-border and local acquiring services provided by Antom. Its service network spreads across 150 countries and regions, being able to process credit and debit card payments, online banking transfers, and e-wallets in more than 100 currencies. In China, Antom engages with partners from reputable industries such as digital entertainment, streaming, e-commerce, new energy, and tea drinks to gain more audiences outside the country. It only takes 1 integration for Enterprises to enjoy hundreds of payment networks and access 1.5 billion consumers in many economic regions of the world. In addition, Antom has also emerged as one of the official sponsors of UEFA Euro 2024, which underscores its position in the global payments market.

Joey Mattress: Quality Payment Services for Sweet Dreams Every Night[4]

Joey Mattress is an e-commerce company based in Malaysia that deals exclusively with sleep products providing direct sales to consumers to make the shopping experience convenient and easy. However, in the previous phase of the company development, Joey Mattress had several difficulties: the initial payment processor engaged not only high transaction charges which ate into the business's profit margins, but also poor service. Furthermore, it was a requirement that Joey Mattress somehow had to go through the trouble of negotiating individually, over the service of installment payment plan schemes offered by major local banks. That was also a time waster of an exercise and bore no fruits.

In order to address those problems, Joey Mattress was looking for an option that would allow making payments in a more efficient and easy way, reduce the time required to launch its goods on the market, and, if possible, suit lower the backend administrative efforts of the finance and accounting departments of the Company within reasonable transaction charges. With the incorporation of the solution from 2C2P, a subsidiary of Antom, Joey Mattress underwent a transformation in its payment services. 2C2P also assisted Joey Mattress in operating with such leading Malaysian financial institutions as Public Bank Berhad, Hong Leong Bank, and RHB Bank offering customers the opportunity to avail installment payment schemes. The services rendered by 2C2P are not only end-to-end and dependable but have also improved the payment

[4] Antom, 2024, https://www.antom.com/cn/knowledge-source/blog?id=Joey_Mattress&cate=%E5%AE%A2%E6%88%B7%E6%A1%88%E4%BE%8B.

processing times, reduced the turnaround time for refunds and streamlined reconciliation accounting processes, thereby reducing the burden on Joey Mattress' internal teams. Further, it enabled Joey Mattress to cut the transaction costs by not less than 30% leading to an increase in the Company's profits.

As Winson Chong, founder and CEO of Joey Mattress, puts it: "2C2P, a subsidiary of Antom, not only offers the most competitive rates in the industry but also provides very dependable team support. Thanks to our close collaboration, we are able to confidently advance towards a broader market."

GERALD.ph: Providing a Smoother Payment Experience for Gourmets[5]

Founded in 2016 by French expatriate Gerald Egasse, GERALD.ph is an online gourmet supermarket operating in the Philippines. Prior to switching payment service providers, GERALD.ph frequently dealt with suspected fraudulent transactions. The existing security measures were not effective, placing an additional burden on backend administrators and consuming resources that could have been used to enhance customer service. Therefore, GERALD.ph needed a payment service provider urgently with robust fraud detection capabilities to ensure transaction security and hoped for a provider that could offer prompt response services to deliver a first-class e-commerce shopping experience to its high-end clientele.

The solution provided by 2C2P helped GERALD.ph smoothly transition its payment services. Not only did 2C2P efficiently complete the system integration and coordination work with various banks, but it also equipped GERALD.ph with advanced anti-fraud detection tools. These tools ensured that while payment requests were being processed, real-time risk detection, control, and prevention were simultaneously carried out. Additionally, the solution offered by 2C2P is highly scalable. During the pandemic, when there was a surge in home-based online shopping, 2C2P maintained fast and reliable service, providing a high standard of security. GERALD.ph also leveraged 2C2P's 123 solution to implement a variety of payment methods, including ATMs, self-service terminals, and online direct debits, enriching the payment options available to its customers, which helped attract a wider customer base.

[5] Antom, 2024, https://www.antom.com/cn/knowledge-source/blog?id=GERALph&cate=%E5%AE%A2%E6%88%B7%E6%A1%88%E4%BE%8B.

Gerald Egasse, CEO of GERALD.ph, commented: "The bookkeeping process has become smoother and more efficient; we no longer encounter issues with fraudulent transactions or credit card theft. The response speed of the 2C2P team from Antom subsidiary is extremely fast, and we are very satisfied with their service."

Origin Mattress: Enhancing the Mattress Shopping Experience in Malaysia with Payment Solutions[6]

Origin Mattress is a Singapore-based manufacturer of mattress and bedding, where every product's design is done in Germany. Even though they had an online store and a 120-day free trial, their Malaysian market online store only provided credit card payments in 2020, which was a major inhibitor for sales expansion. The rapid development of Malaysia's digital/E-payment ecosystem made it necessary for Origin Mattress to look for another payment service provider to enable them to penetrate the local market and enhance their payment service systems.

In order to counter this challenge, Antom subsidiary 2C2P teamed up with Origin Mattress and offered an integrated payment solution. 2C2P paid attention and put into practice that integration with the backend of Origin's system was easy and pleasant for users. Customers can use a number of ways to pay such as e-wallets, internet banking as well as FPX that serves as an online real-time payment gateway. Also, 2C2P offered an installment payment option, improving their competitiveness in the market. Origin Mattress internal team can perform the transactions without the need to have a third party, since all transactions go through 2C2P. At the moment, untimely 70% of their cash flow goes through the 2C2P client, which increases the confidence of the origin mattress in their plans to conquer the world. As an example, Shaylynn Yu, the Business Development Manager of Origin mattress covering Singapore and Malaysia added "2C2P's solution is beneficial for us because, not only are transactions being geared for volume, but the payments modes are varied and the customers checkout seamlessly."

Ticketmelon is an online event ticketing website created specifically for event planners and is oriented towards customizing for many types of occasions using analytics. Each year, Ticketmelon offers professional ticket selling for thousands of events, including business, artistic, concert,

[6] Antom, 2024, https://www.antom.com/cn/knowledge-source/blog?id=Origin_mattress&cate=%E5%AE%A2%E6%88%B7%E6%A1%88%E4%BE%8B.

celebration, and others. In order to reach its goal of becoming the "Facebook" for the online event industry, Ticketmelon had an urgent need for a payment solution that would suit and fit in its vision.

In this regard, Antom subsidiary 2C2P provided Ticketmelon with a flexible, scalable, secure, and robust payment system which is able to manage various payment processing needs of the event planners and organizers across the range of the scale. Not only does the system come with adequate payment processing capabilities, but it can also handle peak loads that may occur during significant events. In order to guarantee even better security, 2C2P deployed a fraud detection unit that goes over potentially fraudulent transactions in a post-transaction setup. This guarantees both Ticketmelon in whichever capacity it engages and the event planners that there will not be any problems on their side during the payment process. Commenting on these developments, Thanyakorn Hovongratana, General Manager of Ticketmelon, explained: "We have practically limitless collection possibilities thanks to 2C2P. Their payment solution has enabled us to shift our focus towards business performance and scaling the organization since we do not have any pressure in terms of payment processing due to high volumes of traffic."

Antom and its subsidiary 2C2P come to the assistance of small businesses by eliminating operational and other factors presented by the mainstream financial system through innovative, simple, dependable, and secure payment solutions.

Firstly, Antom's solutions are instrumental in curtailing transaction costs. Most of the small enterprises have to incur high charges during payment processing, which cut their profit margins. For example, Joey Mattress cuts its transaction fee costs by about thirty percent because it worked with Antom. As a consequence, its financial performance improved and the firm was able to put in more effort towards product and marketing strategies. In like manner, after the implementation of 2C2P's solutions, Origin Mattress was able to manage its payment processes more efficiently and reduce operational costs as well. This mitigation of operational costs corresponds with the SDG 8 (Decent Work and Economic Growth) and assists more businesses in attaining responsible development.

Besides this, Antom has the effect of promoting consumers' financial inclusion and therefore expanding the range of the payment options available. The lack of proper payment means has made many small businesses lose potential customers. Solutions offered by Antom allow usage of different payment means like e-wallets, and online bank transfers or payment by installment. For instance, Ticketmelon through its use of 2C2P system was able to match its offerings with flexible payment means which facilitated a win–win situation for event organizers alongside consumers in terms of payments. The advantage is that it allows consumers to have a better payment experience and also attracts new customers, especially the low-income population who are often neglected by other financial services. Therefore, Antom helps to meet the objectives of SDG 10 (Reduced Inequalities) which permit more people to participate in economic affairs thereby improving the economic status of the society in general.

Moreover, Antom not only shows that making large payments to cybercriminals impossible, but also provides anti-fraud protection mechanisms for small businesses. These safety precautions help to mitigate the threat of malicious use, and consumers are more willing to accept online payments. GERALD.ph significantly lessened the frequency of swindling transactions after the integration with Antom especially due to the technology of anti-fraud systems deployed. Such safety measures enable consumers to transact on the internet safely and allow firms to concentrate on their expansion. Since the digital world develops even further, the security of payments is becoming one of the issues consumers want to be met. Antom is in line with SDG 12 (Responsible Consumption and Production) as it enables a safe payment ecosystem which further promotes safe patterns of consumption.

Antom is an example of how to help small businesses without ignoring the Environmental, Social, and Governance (ESG) factors. By improving business processes and cutting costs, Antom facilitates business development and creates more employment. For example, Origin Mattress utilized the back-office accounting system offered by 2C2P to enhance the centralization of its transaction processes, which dramatically improved its work efficiency and quality of customer service, thus enhancing its social responsibility. This cycle not only enhances corporate social responsibility but also strengthens the environmental aspect of business and addresses resource efficiency. The use of proper payment methods can help in the management of cash flow and the unnecessary wastage of company resources resulting in better business growth.

To summarize, the commitment of Antom towards fostering financial inclusion and sustainable development is commendable and reflects that it is a responsible company. Antom integrates innovation and technology to offer varied business financial solutions that are practical and enable economic growth and sustainable development in society. With the current issues that face the global economy, it is essential that examples such as those that are shown by Antom are used by other companies in order to understand how such factors as financial technology promote economic growth, social equity, and sustainable development.

4.4 Conclusion

WorldFirst provides a range of services which assist cross-border SMEs internationally, such as global collections, worldwide payments, provision of foreign currency exchange services, and supply chain finance. WorldFirst claims to have expanded its scope further providing payment services to over 200 countries across more than 40 currencies to serve over 1 million cross-border merchants by the end of 2023. It also promises fast integration into new jobs and fully resolves common operational issues.

WorldFirst's new initiative "Global Voyage" helps sellers to quickly process their products on 490 online marketplaces and enables easy access to different foreign markets. It enables sellers to expand their business without having to go through the traditional operating procedures. The WorldTrade service introduced by WorldFirst revolves around the improvement of security and efficiency quality for B2B foreign trade collection, meanwhile enabling smaller businesses to improve their credit rating due to lower costs. Other businesses aim to integrate developing nations that are economically disadvantaged into their endeavors and services.

For example, in new markets such as those in Southeast Asia, Eastern Europe, the Middle East, and South America, WorldFirst is facilitating the globalization of businesses with more reliable and cost-effective cross-border payment alternatives. They also offer a plethora of support policies as well as one-stop-shop services that settle transactions from beginning to end.

Antom provides a comprehensive industry digital payment platform, enabling more than 250 popular international and local payment options such as bank cards, online banking, and e-wallets. Antom's services are available to mainland China and Hong Kong customers as well as to the customers in Southeast Asia, Europe, and America regions and offer a variety of international and local payment options. It expands the market for businesses and makes it easier for customers. Additionally, through its A+ Rewards marketing solution, Antom enables merchants to relate and engage more seamlessly with customers, which is critical for small firms operating in developing nations.

Due to these initiatives, WorldFirst and Antom not only assist with expanding small businesses and facilitating international trade, but also aid towards the achievement of the United Nations Sustainable Development Goals (SDGs). In particular, they are involved in the achievement of the following goals: Goal 8 Decent Work and Economic Growth; Goal 9 Industry, Innovation and Infrastructure, and Goal 17 Partnerships for the Goals. WorldFirst's Global Voyage and WorldTrade services simplify the cross-border opening of stores, streamline B2B trade payments, reduce the costs of transactions, and improve the security and efficiency of transactions. Antom assists in the penetration of small businesses into new markets and the implementation of new technologies by offering a comprehensive digital payments solution for the industry and digital marketing services. Both enable the creation of additional jobs, promotion of the use of cutting-edge technologies, and contribute to the economic stability of the developing nations through far-reaching global collaboration. The expansion of enterprises and the inclusive and sustainable development of the world economy have been realized through these efforts.

CHAPTER 5

Sustainable FinTech for Social Development: Ant Farm, Ant Foundation, and Local Social Partners for Rural Development, Community Dacilities, and Humanitarian Aid

5.1 Industrial Revitalization: The Agricultural Assistance Practice of the Ant Foundation

5.1.1 *"Hundred Counties, Hundred Products": A New Model for Public Welfare Assistance*

In September 2022, during the China Farmers' Harvest Festival was launched in the Qianjiang District of Chongqing. At the event, Ant Group announced its continued increased investment in 160 national key counties for rural revitalization. This includes launching an agricultural assistance channel on Alipay, providing a free, long-term showcase window for the specialty agricultural products of these key counties. The key to rural revitalization lies in the revitalization of industries. This measure aims to further assist less developed counties in enhancing the brand awareness of their agricultural products, stimulating public consumption to support agriculture, and helping local areas to create competitive agricultural product brands, thereby promoting the prosperity of the rural economy through industrial revitalization.

Li Jianliang, General Manager of the Rural Revitalization Department of Ant Group, stated that "Most of the 160 key counties we assist have been deeply impoverished counties and are key areas for our work

J. Sun, *Sustainable Financialization*, Sustainable Development Goals Series, https://doi.org/10.1007/978-981-96-3580-1_5

in aiding rural revitalization. We hope that through the purely public welfare agricultural assistance platform of the Alipay agricultural assistance channel, combined with a series of measures such as inclusive finance, we can help these key counties achieve industrial prosperity and the enrichment of rural areas." On the day of the launch of the Alipay agricultural assistance channel, specialty agricultural products from 33 key counties for rural revitalization had already been introduced.[1]

Specialty Agricultural Products Enter Alipay's Agricultural Assistance Channel (*Source* Ant Group)

In addition to the launch of the agricultural assistance channel, Alipay also introduced a seven-day "Hundred Counties, Hundred Products" themed agricultural assistance event from the 22nd to the 28th, encouraging the public to purchase specialty agricultural products from these areas. Prior to this, from April 2021 to August 2022, Alipay had successfully held 16 "Hundred Counties, Hundred Products" themed events,

[1] Ant Group, https://www.antgroup.com/news-media/press-releases/1663826677000.

helping to sell over 10 million jin (a Chinese unit of weight, 斤) of agricultural products with total sales exceeding 70 million yuan. All sales revenue from these events goes directly to the farmers, with the Alipay platform charging no fees. Now, with the introduction of the agricultural assistance channel, specialty agricultural products from 160 key counties have a long-term, fixed exhibition and sales platform.

For instance, Ant Group has deepened its cooperation with the China Foundation for Poverty Alleviation (CFPA) and announced a series of support measures in 2022. Ant Group pledged to invest 200 million yuan for the procurement of "agricultural assistance fruits" in the central and western regions and to expand the scope and quantity of public welfare procurement through Alipay's Baba Farm. Furthermore, Ant Group also invested in hundreds of millions of traffic resources to cooperate with the CFPA to create the "One County, One Product" project, assisting selected key counties and agricultural products in achieving brand upgrades and promoting the development of local special industries.

According to the cooperation agreement, the collaboration between Ant Group and the CFPA mainly focuses on two projects: "One County, One Product" and "Public Welfare Procurement." In terms of "Public Welfare Procurement," Ant Group plans to increase the procurement of fruits from the central and western regions through Alipay's Baba Farm to support local farmers. In the "One County, One Product" project, the CFPA selects key counties and specialty agricultural products for assistance, and Alipay's Baba Farm invests a significant amount of traffic resources to jointly promote the brand building of local agricultural products. From September 22 to September 26, Alipay's Baba Farm and the CFPA jointly held a "Supporting the Harvest Festival" sales event, helping farmers from five provinces including Henan, Shanxi, Sichuan, Yunnan, and Heilongjiang to sell ecological agricultural products. It is revealed that on the first day of the event, about 830,000 jin of agricultural products were sold. Among them, the sales of red fragrant crisp pears from Sanmenxia City, Henan Province, reached 300,000 jin, equivalent to the annual output of 10,000 fruit trees; and the garlic from Qi County, Henan Province, which was backlogged due to heavy rain, also sold 60,000 jin within a day.

5.1.2 *Digital Inclusive Finance and Technological Assistance to Agriculture*

The cooperation between Ant Group and the CFPA is not limited to agricultural product sales and brand building but also includes promoting the penetration of digital inclusive financial services into rural areas. This collaboration aims to support the development of local specialty industries by providing interest-free and reduced-interest loans, as well as interest rate preferences, offering more convenient and affordable financial services to farmers and small and micro-enterprises.[2]

Ant Group Establishes Online Agricultural Assistance Channel for 160 Key Counties (*Source* Ant Group)

By the end of 2021, over 1,000 districts and counties had initiated agricultural cooperation with MyBank, a number that nearly accounts for half of the total county-level jurisdictions in the country.

[2] http://www.appbw.com/livelihood/articles-406.html.

During the Harvest Festival, MyBank provided 30 days of interest-free loan support to growers, breeders, and agricultural e-commerce operators in a hundred major agricultural product counties nationwide. Eligible users in these counties can search for "Rural Revitalization" on Alipay to apply for autumn harvest loans online via mobile phone, with a 30-day interest waiver covering the entire harvest season. For leaders and operational farmers in 160 key counties for rural revitalization, MyBank also offered annual interest reduction and interest-free loan support.

Furthermore, targeting leaders in rural revitalization, MyBank, in collaboration with the China Agricultural Science and Education Fund under the Ministry of Agriculture and Rural Affairs, provides annual interest-free loans to the "Top Ten Farmers of the Nation" selected by the Ministry, assisting them in expanding their production and operation scale and driving more farmers to increase production and income.[3]

Significant achievements have also been made in the area of technological assistance to agriculture. Satellite remote sensing technology has existed for many years, encompassing a comprehensive system of technologies for observing the Earth and celestial bodies from the ground to space. Users can obtain satellite data from remote sensing platforms and complete the reception, processing, and analysis of information through remote sensing instruments. Satellite remote sensing is widely used in agriculture, forestry, geology, oceanography, and other fields. The integration of satellite remote sensing, financial services, and agricultural development represents a unique innovation of the "Daurian Starling" project, initiated by MyBank, established by Ant Group. Specifically, MyBank utilizes spectral recognition technology from satellite remote sensing imagery to identify the types, areas, and growth conditions of crops in a given plot of land. By analyzing images, satellites can distinguish between different crops such as rice, corn, wheat, apples, etc., and in conjunction with farmers' self-certification of cultivated land, government-registered land transfer data, as well as climate, geographical location, industry prosperity, and other factors, it uses dozens of risk control models to estimate output and value, thereby providing farmers with credit limits and reasonable repayment periods.

[3] Ant Group, https://www.antgroup.com/news-media/press-releases/1663826677000.

Currently, the "Daurian Starling" typically updates satellite imagery and recognition results in real-time every 5 to 7 days, monitoring the growth of crops and determining the stages such as seedling, jointing, or harvest periods, thereby analyzing the different funding needs of farmers during transplanting, spraying, topdressing, and harvesting periods.

Historically, farmers have faced high risks in agricultural production, processing, and distribution due to the lack of collateral and guarantees, compounded by low levels of rural informatization, making it difficult for credit officers to conduct due diligence, monitor loan usage, and collect debts. Financial institutions have struggled to accurately assess the creditworthiness of farmers, resulting in persistent difficulties in obtaining loans for farmers in the industry.

"The 'Daurian Starling' provides a crucial assessment dimension for the agricultural loan business of MyBank," said Kong Yue, head of the rural finance department at MyBank, in an interview with Yicai. "The 'Daurian Starling' is not MyBank's first technological exploration in the field of rural finance. Since 2016, as one of the first private pilot banks in the country, MyBank has been promoting inclusive finance in county areas, focusing on serving small- and micro-enterprises and agricultural customers.[4]

MyBank runs on what they call the "310" model: you can apply in just 3 minutes, get your money in 1 second, and it is all done without any human touch. They have swapped old-school bank branches for cloud tech and use big data to manage risks, skipping the usual complex credit checks. This transformation in their transactions has made their services both cheap and efficient. According to their yearly reports, MyBank's bad loan rate was a low 1.52% in 2020, which is significantly lower than the average level of listed banks during the same period.

Kong Yue and his team faced the challenge that the data risk control model, which performed well in the field of small- and micro-enterprise loans, did not go smoothly in the agricultural finance sector. The "310" model is a universally applicable loan model that does not make fine industry distinctions. As a result, the inherent problems in the agricultural finance sector quickly emerged—many farmers do not frequently use the internet, lack a foundation in agricultural data, and it is difficult to

[4] Yicai, 2021, https://www.yicai.com/news/101272473.html#:~:text=%E2%80%9C%E5%A4%A7%E5%B1%B1%E9%9B%80%E2%80%9D%E6%8A%80%E6%9C%AF%E7%8A%B9%E5%A6%82%E5%9C%A8.

construct user portraits, making risk control difficult. Kong Yue recalled that due to the many uncertain factors in the agricultural industry, the loan amounts initially provided by MyBank often could not fully meet the production needs of farmers. Therefore, only by deeply understanding the agricultural industry and truly reflecting the production and operation status of farmers can the rural credit problem be effectively solved. In order to cover as many farmers as possible, in July 2019, Kong Yue and his team chose to take the planting industry as the breakthrough point and ultimately chose satellite remote sensing technology as the technological innovation for this project.

"The reason why remote sensing technology became our final choice is that it naturally records the production and operation process of farmers," emphasized Kong Yue. The historical information traceback function is unique to remote sensing, which means that remote sensing can not only record the present but also reflect the crop growth of historical years. The continuity of the production process is also one of the important factors in MyBank's risk control system.

Satellite remote sensing technology (*Source* Mybank)

The technology has been fully commercially used in the field of rural finance, becoming the "new key" to solving the problem of difficult loans for farmers. As spring goes and autumn comes, with the ripening of oranges and tangerines, "Daurian Starling" begins to benefit fruit farmers. In Yugan County, Jiangxi Province, farmer Yang Zuobo became the first person to obtain contactless loans from MyBank through satellite remote sensing technology. Without having to shuttle back and forth between banks and the fields and waiting anxiously, Yang Zuobo simply opened the Alipay mini-program on his mobile phone, circled his plot on the map, and got a loan within a few minutes. Peng Qingqing, a citrus farmer from Xinyu, Jiangxi, is also a beneficiary of the project. She circled her orchard on her mobile phone, and MyBank predicted the output by analyzing satellite images. Peng Qingqing quickly obtained a loan of 220,000 yuan, solving her capital turnover problem.

"The citrus will soon enter the harvest season this year. With the support of this fund, I am more confident in the bumper harvest!" said Peng Qingqing.[5]

By the end of 2022, MyBank had signed cooperation covering 1,230 counties nationwide, involving about 50% of the county-level jurisdictions involved in agriculture in the country. Now "Daurian Starling" has been applied in 31 provinces, autonomous regions, and municipalities directly under the Central Government in China, covering 15 agricultural product industries nationwide, providing pure credit, mortgage-free financial services for more than 1.5 million planters.[6]

[5] Ant Group, https://www.antgroup.com/banner-detail/001.

[6] Ant Group, https://www.antgroup.com/esg/digital-hp-help-agriculture-details/help-tab-2.

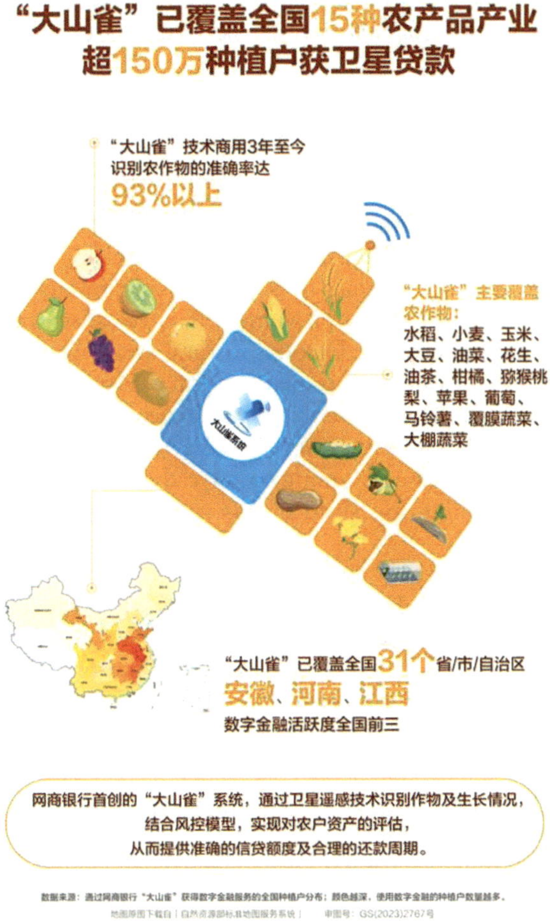

The Three-Year Impact of MYbank’s “Daurian Starling” Technology in Commercial Use (*Source* Sina)

In November 2021, the International Data Corporation (IDC) included MYbank’s Daurian Starling satellite remote sensing risk control technology in its report “Practice and Exploration of Remote Banking

Construction," highlighting its enhancement of various innovative technologies based on the needs of specific business scenarios.[7] This has improved the digital and intelligent level of various business links and is recognized as a leading practice in the field of remote banking and rural finance.

5.1.3 *Benefiting Disadvantaged Groups: Four Core Projects of the Special Fund*

In 2022, the Zhejiang Ant Public Welfare Foundation and Ant Group jointly established the "Zhejiang Ant Public Welfare Foundation Ant Manor Special Fund," aiming to support rural development through various means, particularly by providing effective assistance and support for the needs of disadvantaged groups in rural areas. At its inception, the special fund set up four core projects: "Ant Manor Open Cooperation Plan," "Go for the Dream · Disability Assistance Project," "Baby Health Plan," and "Ant Manor Happy New Year's Eve Dinner."[8] These projects target different social issues, aiming to solve the practical problems faced by rural areas through diversified means and improve the quality of life for rural residents.

In March 2023, the "Go for the Dream · Disability Assistance Project" and the "Baby Health Plan" were separated due to their significant progress in the fields of health and disability assistance, becoming independently operated projects. These two projects have been further developed and improved on their original basis, providing more professional health support and services to rural populations and creating better living conditions and opportunities for development for people with disabilities. Additionally, the "Ant Manor Happy New Year's Eve Dinner" project has successfully completed its mission and was concluded. During this period, the "Ant Manor Open Cooperation Plan," as one of the key projects of the special fund, continued to receive sustained support. This plan, through cooperation with more charitable organizations, not only strengthened support for rural revitalization but also promoted the effective allocation and utilization of charitable resources. With the continuous upgrading of rules, the cooperation plan is more

[7] IDC, 2021, https://www.idc.com/research/viewtoc.jsp?containerId=CHC47257621.

[8] Ant Foundation, https://www.antfoundation.org.cn/content/530.

focused on rural revitalization and support for vulnerable groups and has also introduced credibility assessment mechanisms such as organizational ratings to ensure that more outstanding institutions can participate in the project and jointly promote sustainable rural development.

5.1.3.1 Ant Manor Open Cooperation Plan

In order to support the positive development of more social organizations and public welfare projects, especially those committed to helping vulnerable groups and promoting rural revitalization, the Zhejiang Ant Public Welfare Foundation, in conjunction with the Alipay Public Welfare Platform and Ant Manor, jointly launched the "Ant Manor Open Cooperation Plan." This plan aims to promote the revitalization and development of rural areas through cooperation with outstanding public welfare projects.

The special fund has initiated thematic activities for different social issues during several important public welfare festivals, hoping to attract more public attention and support for groups in rural areas who are easily returned to poverty. For instance, on significant days such as International Women's Day (March 8th), World Autism Awareness Day (April 2nd), World Red Cross and Red Crescent Day (May 8th), Global Accessibility Awareness Day (May 21st), and International Children's Day (June 1st), the special fund organizes various activities, including innovative interactions, promotional videos, and graphic displays, to convey charitable concepts and the importance of issues to users, thereby stimulating public participation and attracting the active involvement of tens of millions of users.

By the end of 2022, 24 public welfare organizations had participated in the "Ant Manor Open Cooperation Plan," supporting 46 rural public welfare projects and providing substantial help and services to more than 500,000 disadvantaged groups in rural areas. These projects cover multiple aspects, from educational support to health services, and infrastructure improvement, comprehensively enhancing the quality of life for rural residents.

In 2023, the special fund's management committee took the lead in upgrading the open cooperation rules. The main focus of the rules upgrade was on rural revitalization and support for vulnerable groups, providing more opportunities for more developmental projects; credibility evaluation such as organizational ratings was added to provide opportunities for more excellent institutions.

5.1.3.2 *Go for the Dream—Disability Assistance Project*

To ensure the basic living needs of people with disabilities and promote their comprehensive development, under the guidance of the Rehabilitation Department of the China Disabled Persons' Federation, the Zhejiang Ant Public Welfare Foundation and the China Disabled Persons' Welfare Foundation have closely cooperated. They have upgraded and optimized the existing prosthetic fitting assistance project to ensure that more rural people with physical disabilities can receive high-quality prosthetic fitting services and corresponding rehabilitation training. This cooperation not only provides necessary human care for rural people with disabilities, solves their practical difficulties, but also creates a more equal and inclusive social environment for them.

By the end of 2022, the Zhejiang Ant Public Welfare Foundation had donated a total of 2 million yuan to the China Disabled Persons' Welfare Foundation to support this important public welfare project. The investment of this fund has enabled 56 people with physical disabilities living in rural areas to successfully fit new prostheses and receive necessary rehabilitation training. Under the technical guidance of the China Rehabilitation Research Center, these people with disabilities have not only regained basic mobility but also significantly improved their quality of life.

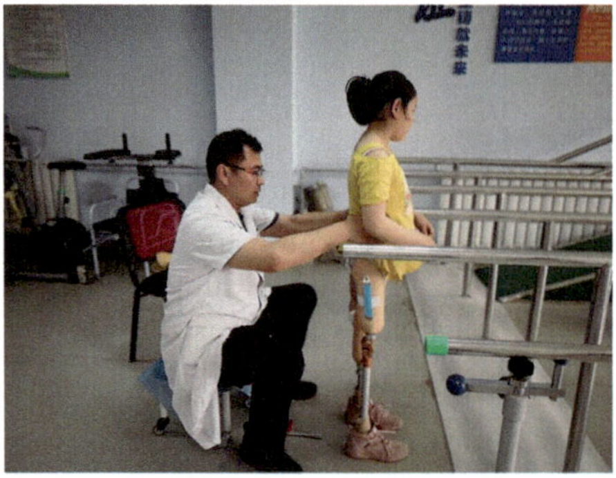

Prostheses Fitting and Rehabilitation Training for Rural Individuals with Physical Disabilities (*Source* Ant Group)

5.1.3.3 Baby Health Plan

In response to the issue of family poverty caused by major illnesses or accidental disabilities in children from low-income families in underdeveloped areas, the Zhejiang Ant Public Welfare Foundation, in collaboration with the China Foundation for Rural Development and China National Petroleum Corporation, launched the "Baby Health Plan" in September 2022. They provide these children with public welfare health insurance to alleviate the financial burden caused by illness or accidents and reduce the risk of poverty resulting from such incidents.

The project targets children from low-income families in underdeveloped areas, aged 0–14, by providing comprehensive protection including hospitalization medical care, specific diseases, and accidents, thereby reducing the incidence of children in rural low-income families returning to poverty due to illness or accidents, and helping to improve the health level of children in underdeveloped areas.

For rural low-income families, in addition to bearing the cost of illness, caring for a sick child can lead to a reduction in family labor and a decrease in income. It is quite common for families to return to poverty after their children suffer from major illnesses. Wang Xiaojing, Secretary-General of the co-initiating Zhejiang Ant Public Welfare Foundation, stated that currently, there are few public welfare projects that provide support for the living, nursing, and post-illness rehabilitation costs for families with serious illnesses, and these invisible costs can be a huge burden that a difficult family finds hard to bear. The "Baby Health Plan" project was born out of this need. Based on the "Baby Health Plan" project, insured low-income families' children can receive up to 100,000 yuan in medical insurance + up to 90,000 yuan in specific disease insurance and accident insurance, reducing the financial burden on families seeking medical treatment.

The project initially covers 12 counties in 6 provinces, including Jishishan Bonan Autonomous County of Gansu Province, Julu County of Hebei Province, Taiqian County and Fan County of Henan Province, Xishui County of Guizhou Province, Hengfeng County of Jiangxi Province, Nileke County, Tori County, Barkol Kazakh Autonomous County, Jimunai County, Qinghe County, and Chabukqiar County of Xinjiang.

The project fully leverages the advantages of internet insurance, providing protection for rural children through online insurance, policy inquiry, and online claims. In the event of a claim, uploading medical vouchers online can achieve rapid review and rapid claims, with the claim to account completed within 3–5 working days. The project adopts an "Internet + Public Welfare + Insurance" mechanism, applying blockchain technology to achieve full coverage, full tracking, and full transparency.

As co-initiators, China National Petroleum Corporation and Ant Public Welfare Foundation provide financial support for the project, and Ant Group will also make full use of its platforms such as Alipay Public Welfare and Ant Manor to drive public attention and support for the basic protection of people in difficult situations in rural areas.

In 2022, the "Baby Health Plan" successfully benefited 13 cities (prefectures) in 9 provinces (autonomous regions, municipalities directly under the Central Government) including Xinjiang, Inner Mongolia, Sichuan, and Guizhou, as well as 16 underdeveloped counties (districts). The project provided comprehensive health protection for more than 131,100 children aged 0 to 14 from low-income families. Among them, the Zhejiang Ant Public Welfare Foundation specifically donated funds to provide insurance protection for about 108,300 children.

5.1.3.4 Ant Manor Happy New Year's Eve Dinner

In response to the social issues of population aging and rural left-behind elderly, the Zhejiang Ant Public Welfare Foundation cooperated with the China Aging Development Foundation to jointly implement the "Ant Manor Happy New Year's Eve Dinner" project. By providing a warm New Year's Eve dinner experience for rural elderly people, the project aims to provide care and support for rural left-behind elderly and other elderly people in need, ensuring that they feel the warmth and care of society during the traditional festival. At the end of 2022, the Zhejiang Ant Public Welfare Foundation donated 1,818,400 yuan to the China Aging Development Foundation for the joint "Ant Manor—Happy New Year's Eve Dinner" public welfare activity. This activity is not only aimed at allowing the elderly to enjoy delicious food during the Spring Festival but more importantly, to let them feel the care and respect from society, enhancing their sense of happiness and belonging.

In the first year of the launch, the "Ant Manor Happy New Year's Eve Dinner" public welfare project successfully benefited 44 counties (including 15 national key counties for rural revitalization and 13 provincial key counties for rural revitalization), benefiting more than 20,000 rural elderly people including left-behind elderly, difficult elderly, retired veterans, etc., sending them local characteristic New Year's Eve dinners to share the strong flavor of the New Year together.

In Mabian Yi Autonomous County, Sichuan, local elderly people dressed in ethnic costumes participated in the New Year's Eve dinner activity, and the scene also alternated with the writing of couplets and sending "blessing" characters, creating a strong atmosphere for the New Year. In Guizhou, young volunteers chatted with the elderly while making glutinous rice balls. In Leishan County, Guizhou, the execution team of the Happy New Year's Eve Dinner carefully prepared 18 dishes, including special dishes such as pork rib hot pot and preserved vegetable braised pork, and also customized easy-to-chew and digestible dishes such as noodle soup according to the physical condition of the elderly. So far, more than 300 elderly people in Leishan County have participated in related activities, including community widowed elderly, empty nest elderly, disabled elderly, families without children, and special family members. Elderly people with limited mobility also enjoyed home delivery services. In Sinan County, Guizhou, the cultural traditions of local ethnic minorities were also integrated into the New Year's Eve dinner activity. People followed the most solemn eating and drinking method of the Tujia people, splicing together piece by piece of the eight immortal tables, and filling the entire Anhua Ancient Street. 66-year-old Li surnamed elderly said, "I really like this lively place. So many people eating together can make up for the loneliness and regret of not having relatives around."[9]

This activity has also expanded its influence. It has been reported by local media in various provinces, cities, and counties in China, and has also been widely disseminated on social media platforms such as People's Daily, China News Service, Tencent, Douyin, and Weibo, causing a widespread impact and resonance in society. With "Happy New Year's Eve Dinner" as the keyword, Baidu search shows nearly 100,000 pieces of information. Local governments and social organizations regard this activity as an important opportunity to inherit the tradition of respecting and loving the elderly, and to guide social trends, vigorously publicizing and even including it in local government work. The local people, especially the elderly, spread the word, forming a social effect. In terms of

[9] China Ageing Development Foundation, 2023, http://www.cadf.org.cn/ueditor/php/upload/file/20230315/1678865425615791.pdf#:~:text=%E5%BC%98%E6%89%AC%E4%B8%AD%E5%8D%8E%E6%B0%91%E6%97%8F%E6%95%AC%E8%80%81%E3%80%81.

effect, this public welfare attempt has not only increased the whole society's attention to the elderly group but also laid a good foundation for the subsequent public welfare projects.

New Year's Eve Dinner Event Scene (*Source* China Ageing Development Foundation)

5.2 Community Development: Public Welfare Shops and Construction of Urban Infrastructure

5.2.1 Public Welfare Shop Support Project

MyBank and the Ant Foundation have teamed up to start a project called the "Public Welfare Shop Support Project" back in 2022. This initiative is all about giving a boost to small shops that keep up the good work and service in their communities, especially when it comes to looking after those who need a helping hand.

The project primarily supports three categories of public welfare shops: Package A shops, disability assistance shops, and cancer-fighting shops. Package A shops provide affordable or complimentary services to individuals facing hardships, offering essential support to those in need through their efforts. Disability assistance shops are dedicated to assisting individuals with disabilities, offering specialized services and support to help them better integrate into society. Cancer-fighting shops offer support to patients with severe illnesses, such as supplying nutritional meals or other forms of assistance to help lighten the financial load on patients.

As of May 30, 2024, the Public Welfare Shop Alliance has supported a total of 1,500 public welfare shops across the country, covering 30 provinces, autonomous regions, and municipalities. These shops have benefited more than 7.2 million recipients by providing various forms of assistance and support. To further expand the impact of this project, MyBank also announced the "111" plan for 2024: to support 1,000 public welfare shops throughout the year, cumulatively join forces with 100 business partners, establish benchmark public welfare shops in 10 provinces and cities, and participate in urban public welfare day activities. In addition, MyBank has also launched the TikTok Talent Public Welfare Shop Exploration Group, planning that a team composed of TikTok talents will visit these public welfare shops in the future, using the power of social media to further enhance public awareness and support for these shops.[10]

5.2.2 *Blue Sports Corner*

On September 21, 2022, under the guidance of the Hangzhou Asian Games Organizing Committee, the first "Asian Games Blue Sports Corner" funded and built by the Ant Foundation was officially opened on the rooftop of the Dongshan Nong Community Party Construction Service Center in Hangzhou. This project transformed the original idle space, making it a new place for community residents to enjoy the fun of sports. In just three months, the Blue Sports Corner in Dongshan Nong Community has provided a venue for more than 20 events and 18 community groups. Notably, badminton world champion Wang Lin also brought a "champion class" to the community residents. According to statistics, more than 8,500 people have enjoyed the "Asian Games dividend" here, experiencing the joy of sports.

The public welfare project aims to use idle spaces in communities, such as rooftops of residential buildings, vacant plots, and street corners, to help streets and communities build multifunctional, flexible sports venues, allowing community families, especially teenagers, to exercise physically conveniently and safely near their homes.

At the same time, another "Blue Sports Corner" has been put into use in Xinhua Road Street, Shanghai. This venue is designed with a child-friendly theme, focusing on the design of sports and entertainment spaces for children and teenagers, providing a good place for leisure and entertainment for more than 3,000 residents in the surrounding community.

[10] https://new.qq.com/rain/a/20230616A069PL00.

The "Blue Sports Corner" project is designed to infuse the joy of sports into the everyday lives of local residents. It facilitates this into practice through a variety of approaches, including transforming spaces, equipping them with sports gear, and organizing community events. The goal is to promote national fitness and give residents a chance to experience the excitement of the Asian Games firsthand. These revamped venues not only enhance the cultural and athletic life of the community but also create more opportunities for residents to engage in sports. By encouraging participation in physical activities, the project aims to boost residents' fitness levels and overall quality of life.

"Blue Sports Corner" transformed from idle space in the community

5.3 Emergency Relief: From Combating Pandemics to Resisting Natural Disasters

5.3.1 *Combating the COVID-19 Pandemic*

In early 2022, Xi'an, Shaanxi, China faced a severe challenge in fighting the pandemic. On January 1 of that year, the Ant Foundation donated 10 million yuan to the Xi'an Red Cross for the special purpose of purchasing and distributing anti-epidemic medical and livelihood materials, as well as recruiting and supporting volunteer teams. This fund effectively alleviated the tension of local medical resources, ensuring sufficient protective equipment for frontline anti-epidemic personnel, while also ensuring the basic living needs of the public, such as the supply of key materials like food and medicine.

In February of the same year, when Hong Kong was hit by the fifth wave of the pandemic, the Ant Foundation once again extended its aid, donating 10 million Hong Kong dollars to specifically support the replenishment of protective materials and living security for medical staff. These materials were quickly delivered to the "epidemic walker base" through an efficient logistics system and were relayed by the "Eighteen Districts Anti-epidemic Love Team"—they went deep into the community, delivering anti-epidemic materials door to door to the hands of the infected, home-isolated individuals, and the elderly living alone, greatly alleviating the plight of vulnerable groups.

Hong Kong "Eighteen Districts Anti-Epidemic Love Team" Activity Record

The Ant Foundation focuses not only on direct anti-epidemic assistance but also on the needs of frontline civilian volunteers. In May 2022, to continue promoting the spirit of volunteer service, the Ant Foundation donated 1 million yuan to fully support the "Volunteer Service Care Action" for epidemic prevention and control, jointly initiated by the Central Propaganda Department and the Central Civilization Office. The Ant Foundation, in collaboration with the China Volunteer Service Foundation, precisely connected with volunteer service organizations and volunteers on the frontlines of the epidemic in Jilin Province. The donated funds were used to purchase high-quality epidemic prevention supplies, such as masks, protective clothing, disinfectants, etc., to ensure the personal safety and health of volunteers during their missions. At the same time, specially customized volunteer service care action vests were provided to give volunteers a unified identity.

Rural areas, with their vast territories, large populations, and weak foundations, are key areas and weak links in epidemic prevention and control, with the rural elderly being a key focus of the Ant Foundation. Based on the research results in December 2022, at the beginning of

2023, the Ant Foundation, in conjunction with Ant Group, donated 500 oxygen generators and 50,000 nasal oxygen tubes to public welfare institutions for the elderly in 160 national key counties for rural revitalization, helping to alleviate the shortage of oxygen generators in these institutions.

In addition, psychological assistance is also one of the key areas of the foundation's work. At the beginning of 2020, in the face of the widespread panic and psychological pressure caused by the pneumonia epidemic triggered by the new coronavirus, the Ant Foundation (then known as the Alipay Public Welfare Foundation) joined hands with authoritative institutions such as the Chinese Psychological Society, including the Institute of Psychology of the Chinese Academy of Sciences, the China Association of Social Work, the School of Psychological and Cognitive Sciences of Peking University, and the Beijing Qiao'ai Foundation, to launch a special psychological assistance action against the COVID-19 epidemic. Users can simply search for the keyword "psychological assistance" through the convenient search function of the Alipay APP to immediately access professional psychological counseling services and obtain necessary emotional support and psychological guidance.

5.3.1.1 Resisting Natural Disasters

During the Henan extreme rainstorm disaster in 2021, the Ant Foundation urgently activated its emergency mechanism, quickly raising and donating tens of millions of yuan in funds and rescue materials, including emergency rescue equipment such as rubber boats and kayaks, as well as daily necessities such as food, drinking water, and tents. These were quickly delivered to the disaster area through partners, providing timely and effective assistance to the affected people.

The Ant Foundation also leveraged its technological CCF-Ant Green Computing Special Research Fund prowess in a big way, partnering with top tech firms to harness the latest in big data and cloud computing. They lent a hand in getting communications back up and rebuilding what was lost in the disaster zone. By quickly setting up temporary networks, crunching disaster-related numbers, and making sure aid was sent where it was needed most, they showed just how powerful technology can be in a crisis.

In addition to the rescue operations, the Ant Foundation also paid attention to the importance of post-disaster reconstruction and long-term development. They not only established a special fund to support the repair of infrastructure and the reconstruction of educational projects in the disaster area but also actively promoted economic recovery and sustainable development projects, helping the affected areas to gradually emerge from difficulties and rebuild their beautiful homes. For example, the Ant Foundation donated 20 million yuan in the later stage of the Henan disaster relief and invested technology and talent to support relevant institutions in carrying out the construction of digital disaster prevention and mitigation capabilities. The China Internet Development Foundation and Zhejiang Alipay Public Welfare Foundation jointly launched the "Digital Disaster Prevention and Mitigation Capacity Building" public welfare project, which was implemented by the China Internet Development Foundation.[11]

The Digital Disaster Prevention and Mitigation Integrated Service Platform is committed to uniting social emergency forces, creating a public service ecosystem and a professional service ecosystem in the field of disaster prevention and mitigation, and building a smooth and effective collaborative bridge between the government, industry, social forces, and the public. This is an open platform, adopting a co-construction, co-management, and sharing internet-based operation model. The establishment and operation of the platform have provided the field of disaster prevention and mitigation with an internet platform tool facing the society, which can more efficiently unite various emergency forces to participate in disaster prevention and rescue work; it can more efficiently promote the promotion of emergency professional products and services and promote the healthy development of the entire emergency industry.

5.4 Deep Industry-Academia-Research Collaboration: CCF—Ant Green Computing Special Research Fund

On September 30, 2022, the "CCF-Ant Green Computing Special Research Fund" was officially launched. This is the first "green computing" themed research fund introduced by the Chinese technology industry, aimed at supporting outstanding global scholars in conducting cutting-edge green computing technology research oriented towards real industrial scenarios. The fund provides technological reserves to promote

[11] Thepaper, https://www.thepaper.cn/newsDetail_forward_24577444.

the achievement of carbon neutrality goals through a collaborative industry-academia-research model.

CCF—Ant Green computing special research Fund (*Source* Ant Group)

The CCF-Ant Green Computing Special Research Fund is part of the CCF-Ant Research Fund (initiated in 2020). The "CCF-Ant Research Fund" was jointly initiated by the China Computer Federation (CCF) and Ant Group in 2020. The fund builds a collaboration and academic exchange platform for global university scholars, connecting industrial practice issues with academic research issues, and supports scholars in conducting cutting-edge research that combines with industry. Over the four years of operation, the CCF-Ant Research Fund has attracted widespread attention from global scholars, with a total fund support exceeding 30 million, attracting nearly a thousand experts and scholars to apply, and supporting over a hundred scientific research projects. Research work has been carried out comprehensively in fields such as artificial intelligence, software engineering, blockchain, data security, databases, and the Internet of Things, with many research results being applied in Ant Group's business scenarios.[12]

In April 2021, Ant Research was established. Ant Research is committed to conducting useful and imaginative scientific research. Facing the digital and intelligent future, aiming at the forefront of

[12] CCF, 2024, https://www.ccf.org.cn/Collaboration/Enterprise_Fund/News/my/2024-06-12/823715.shtml#:~:text=%E2%80%9CCCF-%E8%9A%82%E8%9A%81%E7%A7%91%E7%A0%94%E5%9F%BA.

global technology, promoting key core technologies, and facilitating deep integration of "industry, academia, research, and application."

In 2022, under the framework of the "CCF-Ant Research Fund," Ant Group, in collaboration with the CCF, newly launched the "CCF-Ant Green Computing Special Research Fund." The fund aims to build an industry-academia cooperation platform, connecting industrial practice issues with academic research issues, and supports outstanding scholars at home and abroad in conducting cutting-edge research that combines with industry. The plan is to invest more than 6 million in research funds within one year, launching 32 research topics covering multiple technical directions, including green machine learning, green applications, and green engineering systems/clusters.

As we look ahead to trends like the metaverse and 5G, China's need for computing power is skyrocketing, and so is its energy use. That is also why backing the carbon neutrality initiative with tech is getting more and more crucial. People may use data and AI brains far more efficiently by implementing green computing technology and staying up-to-date with innovations. This paves the way for a better, cleaner digital economy that benefits society as a whole while also being environmentally friendly.

Ant Group has long been concerned with green computing power and algorithms and hopes to promote technological development through an open and collaborative model. As one of the first domestic technology companies to research and practice "green computing" technology, Ant began exploring green technology to improve resource utilization in 2019 and officially announced its carbon neutrality goal in March 2021. This year, this "green computing" technology has been certified by Guojin, confirming that Ant reduced 27,727.48 tons of carbon dioxide equivalent through the use of this technology in 2021. Currently, Ant's green computing technology is being opened to the industry and society through open source, free patent opening, and commercialization, accelerating the improvement of the green computing industry ecosystem.[13]

Since its start, the "CCF-Ant Research Fund" has been a big booster for science. In just two years, it is backed 73 projects with over 20 million dollars, focusing on cutting-edge topics like AI, data privacy, blockchain, Internet of Things, and databases training. A big moment was the birth of China's first "CCF-Ant Privacy Computing Special Research Fund" on

[13] Ant Group, https://www.antgroup.com/news-media/press-releases/1664529606000.

July 4, 2022. On top of that, Ant Group has teamed up with more than ten top universities worldwide including Tsinghua University, Zhejiang University, Shanghai Jiao Tong University, and even the University of California and Berkeley to set up labs for cooperation on breakthrough research.

5.5 Conclusion

This chapter delves into how Ant Group has used financial technology to drive sustainability in different areas of society. With their "Hundred Counties, Hundred Products" project, they have upped the profile of local farm goods and given them a steady sales spot on Alipay's Agriculture Channel, which is a big boost for rural economies. In the digital finance space, Mybank's "Daurian Starling" satellite technology for lending has made financial services more accessible and tailored for local farmers, tackling the challenge of getting loans, especially in rural areas. When it comes to crises like COVID-19 and natural disasters, Ant Group has shown up with quick and proactive moves, proving how important companies can be in tough times. For community life, projects like the "Public Welfare Shop Support" and "Blue Sports Corner" have upgraded local facilities and made life more convenient for local residents. Last but not least, the "CCF-Ant Green Computing Special Research Fund" has pushed for close connections between industries, universities, and research labs, all aimed at backing technology that helps reach carbon neutrality.

CHAPTER 6

Sustainable FinTech for Underbanked/ Unbanked Social Groups: MY Bank in China, Anext Bank in Singapore, and Bettr in Thailand

6.1 Inclusive Finance: The Practice of MY Bank

In today's global economy, MSMEs, solo entrepreneurs, and farmers are the backbone. They are the powerhouses of economic growth and the bedrock of innovation and social stability. Yet, these vital players often struggle to secure funding, which can cap their growth and, in some cases, doom many a promising venture before it gets off the ground. For ages, the challenge of securing finance has been a thorn in the side of MSMEs and small businesses, due to factors like information gaps, high transaction costs, elevated default risks, and a mismatch between the financial offerings and their actual needs.

Nevertheless, the financial technology revolution is shifting this landscape. Cutting-edge tools and technology, from big data analytics to satellite imaging, from AI to cloud computing, are transforming how small businesses and merchants access finance. These technologies are making the financial ecosystem more inclusive, efficient, and sustainable. By going digital, fintech can significantly cut down on transaction costs, level the playing field when it comes to information, and thus enhance the reach and relevance of financial services.

Since its establishment in June 2015, MY Bank has been committed to a development strategy focused on serving micro and small clients

J. Sun, *Sustainable Financialization*, Sustainable Development Goals Series, https://doi.org/10.1007/978-981-96-3580-1_6

and the rural market, actively promoting inclusive financial services, and developing small and micro credit businesses. Leveraging payment and consumption data provided by the Alibaba platform, MY Bank utilizes big data risk control technology to transform customer behavior into credit points, enabling micro and small clients to obtain loans without the need for guarantees or collateral. At the same time, MY Bank's credit approval process follows the "310" model, which stands for "3 minutes for application, 1 second for loan disbursement, and 0 human review," completing the entire process online and meeting the characteristics of small and micro credit needs, which are "short, small, frequent, and urgent."

By the end of 2019, MY Bank had cumulatively issued loans of 70.03 billion yuan, serving 20.87 million small and micro-enterprises and individual operators, with an average balance of only 31,000 yuan per account. According to calculations by iResearch, the number of small and micro-enterprises and individual industrial and commercial households in China reached 121.319 million in 2019. Based on this, by the end of 2019, MY Bank had served over 17% of the nation's small and micro-enterprises. In addition, according to MY Bank's annual report, in 2019, MY Bank provided loan limits for more than half of the small shops nationwide, with an average growth rate of transactions reaching 35% for various small shops with a daily turnover of less than 30,000 yuan.

6.1.1 Supply Chain Challenges for Small- and Micro-Enterprises

The first issue is the rapid change in the market environment, where financing services for upstream and downstream small and micro entities in the supply chain cannot keep up. With the development of digital commerce, significant changes have occurred in the business community: the first is the digitalization of the entire business process; the second is the increasing role of brands throughout the process, with brands playing a larger and more controlling role in merchants, business operations, and economic production activities; the third is channel deepening; and the fourth is the personalized consumption and emerging population demands that give rise to a quick response model requiring an agile supply chain to match.

The second issue is the different financing difficulties faced by different entities and scenarios within the supply chain of brand enterprises. A close analysis of the upstream and downstream small and micro entities in the brand merchant supply chain shows that suppliers, dealers, and terminal retailers all suffer from the problems of difficult and expensive financing in different scenarios. Their development is constrained by capital, which also hinders the development of brand enterprises.

In the procurement and ordering scenario, dealers are in a weak position between brand merchants and wholesalers or terminal retailers. In actual operations, when dealers purchase and order from brand merchants, they need to prepay the purchase amount, and brand merchants produce and deliver based on order scheduling. When selling to downstream retailers, there is usually a settlement period with larger retailers.

In the operating equipment credit sale scenario, the traditional credit sale scenario involves three groups: brand merchants/manufacturers/dealers, terminal businesses, and traditional financial institutions. The small goal of small and micro businesses to purchase equipment and expand operations through traditional financial channels is often difficult to achieve. With economic fluctuations, dealers face greater pressure to recover payments. For new customers, dealers are generally unwilling to grant credit sales, which limits their sales scale.

In the industrial payment settlement scenario—which mainly refers to the fund settlement issues in the brand distribution channels, franchise channels, and wholesale and retail markets—brand distribution channels have many pain points in fund management due to their long depth, wide coverage, and involvement of many entities. In the post-pandemic era, the survival rate of franchise stores has also decreased significantly. For small- and micro-enterprises themselves, there are also fund management difficulties.

In the bond, winning bid, and contract scenarios, for government procurement, there is often a period of half a month to 30 days from winning the bid to signing the contract, during which almost all enterprises have about 50% of their capital gap.

In addition to this, there are other scenarios in the brand merchant supply chain, such as terminal distribution collection, flexible employment, and franchise chain operations. The pain point in the terminal distribution scenario is the long account period of downstream retail stores, a large amount of pressure on capital, and difficulty in collection and accounting; the pain point in the flexible employment scenario is the difficulty for employers to recruit and find money; in the franchise chain scenario, the cost of brand franchise stores is relatively high, with large initial investments, making it difficult for franchisees to expand and operate stores.

6.1.1.1 The "Wild Goose System" Boosts Supply Chain Financing

In response to the aforementioned challenges faced by MSMEs, MY Bank has developed the "Wild Goose System," a digital credit and fund management tool designed to address the pain points in various scenarios within the supply chains of brand enterprises. The system has developed a range of digital financial products, including Contract Loan, Purchase Loan, Franchisee Loan, Salary Loan, MY Bank subsidy, Bill Payment, Cloud Funding, and Recovery Treasure, aiming to meet the full range of financial needs for MSMEs in the supply chain of core brands, from upstream to downstream. This includes financing for supply, purchase ordering, distribution recovery, salary and franchise payments, penetrating financial services to the capillaries, and revitalizing the supply chain from the main line to the branch lines. Compared to traditional supply chain financial services, the Wild Goose System leverages digital risk control advantages and a micro-service gene to introduce the "310" service philosophy to the supply chain finance sector for the first time, providing pure credit services to core brand supply chains and terminal operators, thereby improving the supply chain financial service experience.

Purchase Loan Alleviates Dealers' Working Capital Difficulties

For the supply chain's purchase ordering scenario, MY Bank has collaborated with brand merchants to develop the Purchase Loan product, offering customized pure credit ordering/purchase loans for their dealers. This addresses the issues of difficult purchasing, slow turnover, and low credit limits for small and micro-operators, providing financial support for brand merchants to expand the breadth of their supply chain. Unlike traditional ordering models, in the Purchase Loan model, dealers apply for funds to purchase goods first and repay after selling the goods, effectively expanding the scale of purchases.

The inclusive nature is supported by the core technology of the Wild Goose System: first, identification and verification technology provide key credit dimensions for micro and small enterprises. Second, large-scale graph computing, and data processing technology identify enterprise relationships, creating a comprehensive database of corporate credit data. Additionally, MY Bank uses graph computing and data processing technology to restore the supply chain trade relationship network, thereby predicting credit risks and implementing risk prevention strategies such as credit enhancement, limit increase, targeted payment, and anti-fraud and anti-cash laundering for users.

At the same time, during the operation, MY Bank maintains full and in-depth interaction and cooperation with brand merchants before, during, and after the loan, which helps small and micro dealers solve working capital difficulties while controlling risks.

Installment Payments Help MSMEs Easily Purchase Operating Equipment

In the unfamiliar supply chain credit sale scenario, Installment Payments provide a service where buyers can pay in installments, and sellers can receive payments in advance.

Installment Payments provide information online, enabling efficient real-time online credit granting; they support flexible borrowing and repayment, which can relatively alleviate the financial pressure on MSMEs; moreover, Installment Payments are essentially secured transactions, not loans, with funds paid through MY Bank directly to dealers/brand merchants and cannot be diverted for other uses. Traditional financing leases or loans are aimed at enterprises, while Installment Payments support individuals and industrial and commercial households, with a credit approval rate of over 70% under the logic of big data risk control.[1]

Supplier Digital Financial Service Solutions Help Suppliers Flexibly Recover Funds Online

Supplier Digital Financial Service Solutions focus on the capital needs of suppliers and apply to upstream scenarios of core enterprises or government procurement, providing a comprehensive financial service package for procurement suppliers throughout the entire procurement cycle. Currently, the Wild Goose System supports central, state-owned enterprises, and government upstream procurement supply chain scenarios, and will gradually be extended to high-quality private enterprises and listed company scenarios with continuous experience accumulation. The current supplier digital financial services include four products: Supply Loan, Bid Loan, Contract Loan, and Bid Bond.

[1] NetEase, 2022, https://www.163.com/dy/article/H8AHLTRA0518IJ4K.html.

Cloud Funding Unblocks the Supply Chain and Empowers Brand Merchants to Reduce Costs and Increase Efficiency

Based on the brand supply chain scenario, Cloud Funding provides brand customers with a complete set of online, low-cost, and easy reconciliation services for transaction splitting, procurement payment, fund allocation, and fund management, helping to achieve the compliance and comprehensiveness of brand enterprise fund management, and reducing costs and increasing efficiency throughout the chain. Compared with other products in the Wild Goose System, Cloud Funding serves a longer chain, spanning more supply chain nodes from core brand enterprises to small and micro businesses.

The current Cloud Funding service customers mainly include three categories: brand distribution channels, franchise chain enterprises, and wholesale market integrated retail enterprises. Within the radiation range or industry chain of these customers, there are a large number of MSMEs. MY Bank hopes to cooperate with these core brand enterprises to help MSMEs also enjoy the online convenience brought by digital upgrading or reduce the complexity of operations for MSMEs.

After three years of practice, the Wild Goose System has linked more than 500 brand cooperations, starting from the fast-moving consumer goods industry and gradually covering 18 industries; the loan availability rate of upstream and downstream supply chain customers has been greatly improved, reaching 80%, without the need for core enterprise guarantees; dealers who have used loan services have seen an average increase of 20% in their purchase volume, improving the financial blood supply of the brand supply chain "capillaries" and truly realizing financial assistance for business development. MY Bank's 2021 annual report shows that about 70% of micro users' average loan duration is within three months, the interest expenditure of more than 70% of single loans is not higher than 100 yuan, and the overall interest rate of inclusive micro loans has continued to decline for four years.

6.1.2 *Identifying Rural Financial Pain Points*

China's seventh national population census sheds light on a significant figure—around 500 million individuals reside in China's rural regions, with nearly 70% of this demographic being financially underserved, creating a capital shortfall of an estimated 30 trillion yuan. The inherent uncertainties and risks in agricultural endeavors have consistently posed a challenge, leaving rural financial services without a robust set of solutions.

Agricultural operations are inherently risky, often described as "eating by the weather," with production yields and quality heavily influenced by the climate and environment. Additionally, farming operations are characterized by fragmentation. Most farmers work on a small-scale without mechanization or automation, leading to high production costs, unstable output quality, weak risk resistance, and high operational risks. Moreover, most farmers lack credit information and collateral assets: according to the "China Inclusive Finance Indicator Analysis Report (2019)," by the end of 2019, the national financial credit information database included 1.02 billion natural persons, of which 834 million urban residents had credit files, with a coverage rate of 92.7%; in contrast, only 186 million of the 500 million rural residents had credit files, with a coverage rate of just 37.2%. Furthermore, not all farmers hold collateralizable assets.

This leads to difficulties in rural areas in obtaining support from traditional financial institutions. Traditional financial institutions' credit services mainly rely on credit information and collateral. However, individual farmers are scattered, remote, small-scale, and have small loan demands. Therefore, the cost for banks and rural credit cooperatives to assess borrowers' actual production capacity, land management, genuine loan needs, and repayment ability is high. For banks with more branches and higher efficiency, the cost of credit review for farmers is about 300 to 500 yuan per household, while for banks with fewer branches, this cost can even reach thousands of yuan. However, for a small-scale farmer to borrow 20,000 yuan for three months, the interest return is only about 300 yuan. Traditional financial institutions' operational efficiency and risk control capabilities are insufficient to meet the fragmented, low-yield capital needs of rural operations.

This situation leaves farmers without the financial backing they need to scale up their operations, lease the necessary machinery and equipment, or upgrade their production technology. As a result, they struggle with low production efficiency and subpar product quality. This perpetuates a cycle where agricultural progress moves at a snail's pace. Over the past decade, the issue of severe land abandonment in China has escalated, affecting a staggering 300 million mu. This not only squanders valuable land resources but also poses a threat to the nation's food security.

6.1.2.1 *MY Bank's Exploration of Rural Finance*

At the end of 2016, Ant Group launched a rural finance strategy, relying on the Rural Finance Department of MY Bank to officially enter the agricultural and rural fields, attempting to use existing resources to solve the problems of difficult and expensive rural financing and to serve the comprehensive needs of "three rural" users, promoting inclusive finance to take root in rural areas.

Unlike traditional financial institutions, MY Bank assesses the borrowers' repayment ability as a risk control assessment, trying to enter the rural financial market through manual approval risk control methods, and launched the first generation of rural financial products: The Rural Finance Department of MY Bank initially based on the borrowers' authorized online transactions, payments, fees, and credit data on external trading platforms, combined with offline manual recommendations from external rural e-commerce operators, to assess and grant credit to farmers.

However, after the launch of the first generation of manual rural finance, more problems than expected arose, mainly for two reasons: first, there is a lack of dynamic supervision means, and manual reviews are difficult to real-time re-enact the state of borrowing and lending; second, there is a lack of various credible data required for credit.

MY Bank's solution to this problem is the village and town government. The village and town governments are deeply involved in the daily production and life of farmers. In the second phase of rural finance business, MY Bank began to try to use public data from county governments to portray the real production and operation conditions of farmers. There are more than 1,800 counties involved in agriculture nationwide, and the county governments with actual data statistical capabilities and data storage capabilities hold key data such as land rights confirmation and agricultural insurance. Based on this data, MY Bank can use technology as a medium to form credit assets from rural transactions, logistics, payments, and other information, and establish a comprehensive credit assessment system in a secure environment.

Today, MY Bank has reached strategic cooperation with more than 850 counties and districts involved in agriculture, covering 28 provinces, autonomous regions, and municipalities directly under the Central Government.

After integrating county data, MY Bank can generate a relatively comprehensive data portrait of the actual long-term local life and agricultural production of farmers. With the increase of input data, the risk control model is trained and optimized. With the strong computing support of the cloud computing platform, MY Bank has achieved the "310" credit service model: 3 minutes online application, 1 second fund arrival, and the entire process is zero manual intervention.

The introduction of county public data in rural financial services is a turning point for MY Bank's rural digital inclusive finance business. Online data accumulation and model breaking have finally entered a positive cycle.

6.1.2.2 Case Study: The Financing Process of a Farmer in Jiangxi, China

During the busy farming seasons of June and July 2018, Yingfeng Liu, a major landholder in Shizicun, Huangshagang Town, Gao'an City, Jiangxi Province, managed over 300 mu of land. That year, she purchased a batch of agricultural machinery and equipment, along with necessary agricultural supplies and labor costs, and was still short of tens of thousands of yuan to get through the harvest season. "Money indeed runs out as it's used; no matter how you calculate, there's always a gap at the end. ... I had inquired at the bank before, but without any collateral, I couldn't get a loan. I heard that the Rural Credit Cooperatives offered small credit loans without collateral, and they told me that based on the land transfer area, I could get an interest-free loan of 2 million yuan. However, when I took out the land transfer contract, I found that it was in my father's name, not mine." This situation is very common for young people like Yingfeng Liu: after gaining some work experience in industries and internet e-commerce in first- and second-tier cities, they want to return to their hometowns to become "large-scale growers" or "small farm owners," while also taking care of their elderly parents who are nearby. However, they often lack conventional collateral and are usually not local residents, thus unable to obtain land transfers or enjoy government subsidies for their own land transfers. As a result, they are likely to use the identity of their parents or relatives to obtain land transfers and cooperate with their families in planting or breeding. Nevertheless, for financial institutions, borrowers who are parents with local household registration and are over retirement age can bring some difficulties to business operations.

"The borrowing entity must be consistent with the identity of the land transferor, and since the transfer contract was signed in my father's name, only he could borrow. But he is already 65 years old, over 60, and the

Rural Credit Cooperatives consider him not capable of production and operation, not meeting the loan conditions, so the Rural Credit Cooperatives are also not an option." With the busy farming season approaching and funds not being smoothly available, Yingfeng Liu had to look for other ways.

Before returning to her hometown, Yingfeng Liu had worked in online sales and was familiar with emerging industries such as e-commerce and the internet. While making an online payment and worrying about funds, Yingfeng Liu noticed a push message from the MY Bank and tried to borrow a credit loan from it. After spending a minute or two completing the authorization, Yingfeng Liu tried to apply for credit. She thought the approval process would take several days, but she was surprised to find it completed in just a few seconds, with a credit limit of nearly 40,000 yuan. Yingfeng Liu needed 20,000 yuan to refuel the harvester, so she applied for a loan of 20,000 yuan, and the funds were almost immediately available. In this way, with the help of the MY Bank, Yingfeng Liu solved her urgent financial problem.

After the rice harvest was over and the sales were collected, Yingfeng Liu quickly repaid the loan. Because it was a flexible borrowing and repayment model, and the actual loan period was only a few days, Yingfeng Liu only paid interest costs of thirty to forty yuan to get through this busy farming season. "The first time I used the credit loan from the MY Bank, I felt it was convenient and cheap. If I borrowed 20,000 yuan from a bank, although the annual interest rate seems relatively low, I would borrow it for three months or half a year, and the interest might be three to four hundred yuan. This is different; I can repay it immediately after receiving the payment for selling rice, and the actual interest rate is much lower."

Since it can match Yingfeng Liu's short-term funding needs during the busy farming season and the characteristic of quick repayment after sales collection, the credit loan from the MY Bank has become Yingfeng Liu's financial helper during the busy farming season every year. "My credit limit is 40,000 yuan, but I can borrow and repay it monthly, so the actual total amount I borrow each year is one to two hundred thousand yuan." As she continues to borrow and repay, Yingfeng Liu's credit is also accumulating.

At the same time, Yingfeng Liu's production and management capabilities have also increased. By 2019, Yingfeng Liu had managed nearly 600 mu of farmland, and her funding needs also increased. One day, she saw a message that the Internet Merchant Loan could increase the loan limit. With a try-it-out attitude, Yingfeng Liu followed the prompts on her mobile phone and circled her cultivated land on the satellite map. After a few seconds, she found that her credit limit had been successfully increased to 400,000 yuan. Behind this operation, the MY Bank actually used satellite remote sensing technology to identify the cultivated land area, crop varieties, etc., of her planting and increased the loan amount accordingly.

The increase in the loan limit, coupled with the characteristics of flexible borrowing and repayment and interest calculation based on the actual usage days, has made the funding issue no longer the main factor restricting Yingfeng Liu's production. "Now, planting, management, and technical work have formed a fixed routine, and it is relatively easy to operate 1,500 mu of land."

6.1.2.3 The "Daurian Starling" Project Brings Funds to Rural Areas

The core purpose of the "Daurian Starling" project is to return to the borrower itself, identify their loan needs, judge whether they are good operators, and thereby better promote the inclusive development of rural agriculture with funds.

Farmers' core assets are land and crops. At the beginning, the project team focused on identifying the actual cultivated land and crop information of farmers. However, after practical experience, it was found that flat cameras could not photograph the entire view of the farmland at all. To identify the farmland, there should be a bird's-eye view and a panoramic view.

The project team first thought of drones: by installing a camera on the drone, and flying around, they could obtain farmland information. However, the cost of drones is too high, with an average identification cost of 120–150 yuan per mu of land; moreover, if spraying pesticides from the air while taking pictures of the land, the area of the country where drones spray pesticides is less than one-thousandth of the total. Coupled with the complexity of farmers operating drones, it is difficult to ensure that usable and stable images can be obtained, so the path of drones is also not feasible.

When the project team encountered difficulties, a project team member suddenly proposed, "Why don't we try satellites?" This idea excited the entire team because satellite remote sensing is a "super drone" set up in space. In July 2019, the MY Bank officially launched the "Daurian Starling" project, hoping to use satellite remote sensing technology to obtain effective, sustainable, and dynamically updated data on farmers' core assets with the authorization of farmers, superimpose algorithmic technology capabilities, quickly and uniformly identify the asset situation of farmers across the country, and expand the scale of credit.

Satellite remote sensing technology has been mature for a long time, but it has never been applied in the credit field because of the complex trade-offs between shooting frequency, precision, coverage range, and identification cost. The accuracy of domestic commercial satellites can reach 0.5 m (i.e., the unit pixel point is 0.5 m × 0.5 m), but the number of satellites is not much, and for land conditions, it can only be updated once every half a month, and the cost is relatively high. Whether it is pre-loan assessment or loan management, the MY Bank needs to obtain high-time-effective feedback on land conditions and timely understand the growth situation of crops. The data information returned by domestic satellites is not enough to meet the needs of the MY Bank.

In order to save time and cost, the MY Bank decided to use high-time-effective, low-cost, low-precision satellites instead of low-time-effective, high cost, high-precision satellites, and organized a team to train crop recognition models. Rice and corn look green from the images returned by satellite remote sensing, but their economic value is different. How to use technology to finely distinguish the crops on the land, and how to ensure the recognition accuracy under the influence of factors such as clouds, terrain, climate, and sunlight, the MY Bank spent 8 months and finally developed an AI image recognition technology for crops, plots, and crop growth trends based on satellite remote sensing data and spectrum, establishing 28 satellite recognition models, which can not only judge the quality of the plot and the crops planted on it but also judge the disaster range and situation when the crops are affected by disasters, achieving a breakthrough attempt of satellite remote sensing technology in the credit field.

Farmers can enjoy simple operations and fast loan disbursement through this. When farmers need loans or want to increase credit limits, they only need to open the mobile software, walk around the cultivated land, circle out the land they actually plant, or directly circle out the cultivated land in the satellite map manually, and provide complete authorization as required. Within 1 second, the risk control system of the MY Bank's backend will evaluate the credit level and risk of the farmer based on the internal crop recognition model and other multi-dimensional data, provide credit access for farmers who meet the standards, and display the credit limit for farmers who have obtained access. Farmers can then withdraw and use funds through the online payment platform.

The MY Bank uses satellite data from the sky and land circle information from farmers, combined with multiple data sources such as public land data from county governments, to jointly confirm that the borrower is operating a certain crop on a certain piece of cultivated land. "Every time we identify a source of repayment, we add a part of the credit for the borrower. For example, we learn that the borrower also runs a small store locally, or has some e-commerce sales, fast-moving consumer goods distribution, we will add these repayment capabilities one by one." After excluding other liabilities of farmers obtained from the credit system, the comprehensive credit limit of the MY Bank for this farmer is formed.

Using satellite remote sensing technology to identify farmers' cultivated land and crops is a successful start. In the future, it may be combined with soil moisture analysis, greenhouse, economic crop analysis, etc., for refined iteration according to credit needs; it may also combine satellite remote sensing technology with thermal sensing, sonar, and other technologies to serve farmers with sheds, aquaculture, etc., to explore other sources of repayment capabilities for farmers.

The "Daurian Starling" project was born on the flexible management mechanism of the MY Bank. In addition to the innovative iteration of risk control technology, the combination with other technologies is also more likely. This gives the MY Bank a differentiated advantage in financial service capabilities.

As of August 2021, the "Daurian Starling" project has obtained 13 national patents and has been selected into the People's Bank of China's innovation regulatory sandbox. After several iterations, the crop recognition accuracy rate has reached 93%, "opening a door to trace the dynamics of agricultural production using new technology." At present, the satellite remote sensing credit technology of the "Daurian Starling" project has covered one-third of the county-level regions involved in agriculture across the country, and 1.6 million large grain farmers across the country have obtained contactless loan limits through the satellite remote sensing technology applied by the MY Bank.

Based on strong technology and strict risk control, the non-performing loan rate of the MY Bank's agricultural loans is even lower than the overall non-performing level of the MY Bank's loan business. However, the road to exploring technology-enabled inclusive finance is still not easy, and the integration of rural development and digital finance will provide more practices for evaluation in the future.

6.2 MSME-Oriented: Anext Digital Bank's Innovative Practices

Anext Bank, a digital wholesale bank based in Singapore and overseen by the Monetary Authority of Singapore, is a fully owned subsidiary of Ant Group. It zeroes in on serving local and Southeast Asian MSMEs in Singapore, especially those in the cross-border trade arena. Besides, it offers these businesses cutting-edge and secure digital financial services to help the local enterprises embrace digital transformation and expand on the global stage.

In terms of financial product layout, Anext Bank primarily covers the following three areas of business: First, the launch of its inaugural product, the Anext Business Account. This product allows non-local companies to remotely open accounts. As of now, companies registered in Singapore, regardless of whether their main business operations are conducted in Singapore, can log on to the bank's website at www.Anext.com.sg to register and achieve "one-click account opening"; Second, import and export trade financing services for small- and micro-enterprises. On June 6, 2022, the day Anext Bank announced its opening, it signed a two-year memorandum of understanding with Proxtera, a Singaporean small and medium enterprise (SME) service agency. Proxtera is an online platform jointly initiated by the Monetary Authority of Singapore (MAS) and the Infocomm Media Development Authority (IMDA) and currently serves 400,000 SMEs in Asia and Africa. The bank will be the first to provide financial services to buyers and sellers on the Proxtera network, offering timely financial support for MSMEs to participate in foreign trade imports and exports; Third, global innovation in mobile payments and account management for small and medium-sized merchants.

6.2.1 Targeting Market Pain Points: Meeting the Needs of MSMEs

During the trial operation of Anext Bank, its CEO, Toh Su Mei, recalled a period early in her career when a former colleague made a data entry error in the banking records of an MSME. "This seemingly minor mistake ultimately affected the SME's request and ability to obtain credit facilities to finance his business operations. Since I first witnessed how the right financing solutions and banking service quality can make or break a business, I have not looked back. This is even more evident for SMEs", said Toh Su Mei, in an interview with Asian Banking and Finance.

Currently, traditional banks are finding it tough to keep up with the financial demands of SMEs in Asia. Toh Su Mei sees the crux of the problem as the swiftly shifting requirements of these businesses—they are navigating dynamic markets, embracing new business models, and their financial needs are diversifying at a rapid pace.

As a digital bank, Anext is stepping up to address this variety of needs by rolling out financial services on a grand scale. It is trying to make finance more accessible and straightforward for SMEs across the board, not just the big players. Anext is changing in tandem with the business world's transition to hybrid and digital-first models. As Deloitte's research indicates, small businesses command over 85% of the cross-border e-commerce market in the Asia–Pacific. In this regard, Anext Bank is aiming to cater to this segment of the SME community that is often left underserved or unserved.

Toh Su Mei stated that, essentially, SMEs that have shifted their business models to hybrid and digital-first models will anticipate and meet their financing needs through embedded financing—essentially enabling them to access financing from where they conduct their business digitally. Furthermore, there is potential for cooperation between digital banks, fintech firms, and traditional banks. As SMEs expand, the two sectors can work in tandem to offer a wide range of services.

Toh Su Mei also mentioned in the interview that Anext Bank wants to make banking for SMEs an easy and beneficial experience because they know that business owners do not have time to "deal with bank accounts" or figure out cross-border foreign exchange, interest rates, etc. With this in mind, Anext Bank introduced the Anext Business Account, a simplified business account that has no startup or maintenance fees, to assist with operating expenses.

This business account comes with the convenience of remote onboarding, so that international SMEs are not required to go through the full account opening rigmarole. This feature is a game-changer for many foreign business owners. As a result, Anext Bank is able to dive deeper into understanding the unique needs of various SME roles, which inspires them to roll out features that really hit the mark.

Anext Bank's next priority is to innovate with partners. Adopting an open and collaborative approach, it works hand in hand with industry partners and public sector institutions to address industry pain points on a large scale by leveraging each partner's unique position, knowledge, and subject matter expertise.

6.2.2 *Facing the Future: Innovation and Challenges Coexist*

In August 2022, Anext Digital Bank unveiled details of its new financing product for SMEs—the Anext Business Loan. The loan offers a maximum amount of $300,000 and a minimum of $5,000, with a term of 5 years and an interest rate as low as 6.8% (Effective Interest Rate, EIR), without any management fees or collateral requirements.

Compared to traditional bank business loans, the Anext Business Loan has three distinct advantages for MSMEs. First is the paperless application process. Emphasizing financial inclusion, the digital bank allows applications for loans under $30,000 without any documentation, reducing the need for bank statements compared to traditional credit assessment processes. In fact, for SME customers who have bank accounts with three local banks, their applications also require no documentation. Second is the inclusive loan amount. Most traditional banks are reluctant to accept business loans below $30,000 due to the maintenance and monitoring required, which can exceed the loan amount itself in terms of the cost of processing by credit officers. Anext Digital Bank can serve customers with smaller needs by digitizing processes and credit assessments, significantly reducing costs. Lastly, the interest rate is as low as 6.8% annually (EIR), well below the industry average of approximately 7.25–8.25% under government-funded enterprise financing programs, and the absence of management fees saves MSMEs up to 1% in costs.

The launch of Anext Bank's loans undoubtedly brings convenience to MSME financing. Anext Bank reported in mid-2024 that its total customer base doubled over the past 12 months, and the total value of cross-border transactions facilitated increased fivefold year-on-year. Moreover, 69% of Anext Bank's customers are micro-enterprises. When broken down by product category, 56% of these micro-enterprises are loan customers, and 80% are deposit customers.

Since kicking off in June 2022, Anext Bank has been on a mission to empower MSMEs, with a special focus on those on the rise. Right from the get-go, Anext Bank has been all about fast-tracking financial inclusion with technology at its core, making financial services a breeze for MSMEs. They have teamed up with industry players through the Anext Program for Industry Specialists (APIs), bringing financing services straight to their platform and cutting down on the usual banking red tape. For instance, joining forces with e-commerce platforms has made it easier for SMEs to secure loans right where they operate. A 2023 survey by Anext Bank found that 70% of Singaporean SMEs were keen on getting financial services through non-traditional banking channels, spurring Anext Bank to step up with more adaptable and cost-effective financing choices. These initiatives have been a boon for Anext Bank's customer base, which swelled significantly. By the end of 2023, micro-enterprises made up a whopping 68% of Anext Bank's clientele, with 31% being foreign entrepreneurs based in Singapore.[2]

Entering 2024, Anext Bank saw some impressive gains in expanding both regionally and globally. The bank boasted a threefold increase in its customer base over the last year, and a whopping sixfold surge in cross-border transaction volumes, showing its strong support for MSMEs going international. There's also been a notable uptick in cross-border transactions by Anext Bank's customers in key sectors like wholesale and retail, professional services, and ICT. By May 31, 2024, Anext Bank counted 31% of its customers as entrepreneurs from 78 different countries.

Anext Bank's innovative moves have led to quick wins and opened up a wider array of financial services for MSMEs. In March 2023, Ant Group pumped in an investment of $188 million into Anext Bank, and then another $148 million in March 2024, bringing the total investment to around $502.61 million. This financial boost underscores Ant Group's backing of Anext Bank and its digital finance strategy in Southeast Asia.

Despite this financial boost, Anext Bank is not without its financial hurdles. In 2022, the bank's losses rose to $19.72 million, a sharp increase from the year before, with revenues at only $2.31 million. By the close of the fiscal year on December 31, 2023, Anext Bank's total revenue had climbed to 24.5 million Singapore dollars (about $18.11 million), marking significant growth. Nevertheless, without full disclosure of the bank's loss situation, it is expected that a thorough assessment of Anext

[2] The Digital Banker, 2023, https://thedigitalbanker.com/exclusive-interview-with-toh-su-mei-chief-executive-officer-anext-bank-through-apis-we-work-with-industry-partners-to-serve-their-customers-financing-needs-directly-on-their-platforms-to/.

Bank's profitability and strength requires a longer view of its market performance.[3]

Anext Bank's ongoing growth and financial health are a mirror image of Singapore's digital banking scene, where numerous banks lean on investments from their parent companies while also scoring some notable successes, particularly in lending. Despite financial hurdles and the ups and downs of the global market, the gains that digital banks like Anext are reaping are expected to continue alongside the challenges they face.

6.3 Embedded Finance Provides Digital Loans for Micro and Small Enterprises: Bettr in Thailand

Bettr, known as a digital lending initiative from Ant International, sets its sights on offering microloans to MSMEs. Yang Peng, the CEO of Ant International, has stated, "Bettr is on a mission to simplify and streamline financing for MSMEs, who are set to be the new engines of trade in both regional and international markets." According to estimates by the International Finance Corporation (IFC), around 65 million businesses in developing countries—constituting about 40% of all formal small- and micro-enterprises—experience an annual funding shortfall of up to $5.2 trillion.

Bettr's first major partnership is with Asset World Corporation (AWC), a leading lifestyle and real estate group in Thailand and part of the prominent TCC Group, a major Thai property conglomerate. Through this collaboration, Bettr delivers smooth and user-friendly digital loan services to micro and small businesses within the AWC network via integrated financial solutions. The goal is to boost financial inclusion for MSMEs and to raise the bar for Thailand's food wholesale sector with dependable financial support. The partnership's first phase includes Phenix, which is AWC's food wholesale hub. With a grand opening on July 2, 2024, Phenix is situated in the heart of Bangkok and is known for being Bangkok's largest food and leisure destination, featuring a pioneering business model that marries food wholesaling with retail. Phenix connects food producers with buyers in both physical and digital marketplaces,

[3] 7.5 Degree, 2024, https://www.baijing.cn/article/49077.

creating a unique blend of entertainment, dining, and lifestyle experiences. Following this initial collaboration, Bettr and AWC plan to extend their partnership to include merchants in AWC's hospitality, retail, wholesale, and other business domains.

Cooperation between Ant International and Thailand Food Industry (*Source* AWC)

Bettr's initial team-up with AWC zeros in on AWC's flagship project, Phenix, which calls Pratunam home as the world's largest food wholesale hub. This alliance is poised to be a win for sellers, buyers, and AWC's own suppliers who ply their trade on the Phenix platform, both in the digital and physical realms, and it is set to trickle down benefits to merchants across AWC' hospitality, retail, wholesale, and commercial sectors.

The July 2024 launch of Phenix aligns with the Thai government's strategic push to put the country's food and tourism sectors on the global map, with the goal of boosting the international prestige of Thai cuisine as a form of cultural diplomacy. Phenix is on a mission to draw food entrepreneurs, culinary experts, and influential personalities from across the globe to cement Thailand's status as a premier "food ecosystem destination." The platform extends an open invitation to global food manufacturers to exhibit their offerings at Phenix, engage with prospective clients, and tap into a wealth of collaboration opportunities within the food industry.

Phenix offers a comprehensive wholesale food center, providing a range of products from food and ingredients to kitchen equipment, all offered at bulk prices, and sourced directly from manufacturers, ensuring the best possible prices for business owners and customers. With over 200 restaurants and cafes, Phenix is located in the bustling Pratunam district, offering a large co-living area of over 20,000 square meters, providing a multifunctional space for wholesalers to connect, collaborate, and conduct business. It collaborates with leading global food industry players, such as Koelnmesse, the organizer of globally renowned food exhibitions like Anuga and Thaifex, and the Zhejiang China Commodity City Group Co., Ltd., the world's largest grocery wholesale market operator located in Yiwu, China. Phenix connects wholesale food businesses, resellers, and Thai entrepreneurs, providing them with an integrated platform for domestic and international trade, while also targeting regional export markets. Additionally, Phenix actively embraces the digital economy by launching Thailand's first and only B2B/B2B2C online wholesale platform, seamlessly integrating physical and digital channels (Omni Channel). They aim to create a user-friendly, secure, 24/7 available platform connecting buyers and sellers, supporting wholesale businesses in the digital economy.[4]

This collaboration between AWC and Ant International directly connects to the ASEAN Trade Center (ATF), the largest urban food wholesale market in Southeast Asia. The 800 wholesalers in ATF will be the first to use WorldFirst, facilitating more convenient and secure import and export trade with China and 14 other countries and regions. Leveraging WorldFirst, a one-stop digital payment and financial services platform serving global cross-border MSMEs, AWC's e-commerce platform PhenixBox will also enable Thai local merchants to "buy and sell globally with one account." PhenixBox, an Omni Channel platform under Asset World Corporation (AWC), connects offline to online transaction activities. The collaboration between PhenixBox and Bettr marks Bettr's first embedded finance collaboration in Thailand, aiming to provide digital loan services for MSMEs on AWC's platforms. Located in Pratunam, Phenix is the world's largest food wholesale center with the largest food lounge. Through dependable financial solutions, Bettr will provide alternative digital financial solutions for food entrepreneurs involved in the Phenix project, fostering the international growth of Thailand's food wholesale sector.

[4] Phenix, https://www.phenixbox.com/.

Wallapa Traisorat, CEO and President of AWC, stated that this collaboration will benefit MSME sellers, buyers, and AWC suppliers conducting business online and offline on the Phenix platform, eventually expanding to other AWC projects. In the future, AWC merchants may also achieve cross-border direct procurement on 1688.com through WorldFirst, allowing Thai consumers to access more cost-effective "Made in China" products.[5]

At the same time, AWC's major hotels and commercial retail networks will also integrate with Ant's Alipay+ cross-border payment and marketing solution network, facilitating tourists from Mainland China, Hong Kong, Malaysia, South Korea, and other countries and regions to use their domestic e-wallets to scan and pay at AWC's chain hotels and shopping centers across Thailand, enjoying preferential exchange rates and other benefits. Alipay+ will also provide AWC merchants with a wealth of digital marketing tools for cross-border business.

Compared to directly issuing digital loans, Ant's newly launched Bettr focuses on embedded financial services, integrating financial services or products into non-financial platforms. In other words, Bettr and AWC have established an embedded financial partnership to first provide digital loan services to MSMEs under AWC.

Through creative financial and technology solutions, Bettr's services will assist Thai MSMEs and entrepreneurs in becoming globally recognized. Furthermore, Ant International's Alipay+ cross-border mobile payment and digital technology solutions, along with its one-stop digital payment and financial services platform for international business, WorldFirst, will contribute to the development of Thailand's digital economy, strengthen cross-border business ties, and encourage sustainable international expansion for Thai entrepreneurs.

Right now, Bettr is still feeling its way through uncharted territory. While Bettr is geared towards boosting financial inclusion and giving a leg up to MSMEs, there are hurdles to clear for this partnership to be a success. Navigating the international market could mean tackling issues like keeping loans affordable, managing risks, and widening access to services in developing countries like Thailand. Moreover, getting people on board with digital financial services, especially in areas where cash is king, could be an uphill battle. Compliance is another layer of

[5] Xing Zhou, 2023, https://www.dsb.cn/224868.html.

complexity. Bettr has to adhere to Thailand's Personal Data Protection Act (PDPA), the country's inaugural comprehensive legislation that governs the handling of personal data, including its collection, use, sharing, and international transfer, with far-reaching implications. Bettr must ensure they are in lockstep with PDPA when it comes to managing personal data, from gathering to safeguarding it. On top of that, Thailand's new Royal Decree on the Operation of Digital Platform Service Businesses that Require Prior Notification (2022) mandates that digital platforms, like Bettr, must report their activities to the Electronic Transactions Development Agency (ETDA) before a set deadline.

In summary, the partnership between Bettr and AWC marks a significant milestone for Thailand in the realms of fintech and food distribution. By offering groundbreaking financial solutions and a solid trading platform, this joint effort is set to spur the growth of Thai MSMEs, sharpen their global market competitiveness, and fuel the Thai economy's ongoing prosperity and expansion. As this collaboration grows and strengthens, it is reasonable to anticipate that Thailand will play a more pivotal role in the international food industry, with Bettr's digital lending services emerging as a crucial catalyst in this journey.

6.4 Conclusion

Riding the wave of the digital economy, the funding challenges faced by MSMEs, individual merchants, and farmers have come under the global spotlight. These players are the heartbeat of economic growth and the foundation of innovation and social stability. Yet, they grapple with financing hurdles due to information gaps, high transaction costs, elevated default risks, and a mismatch between the financial products available and their actual needs. The financial technology revolution, especially with technologies like big data, AI, and cloud computing, is redefining the financing landscape for MSMEs and making financial services more accessible and relevant.

This chapter takes a close look at the strategies of MY Bank in China, Anext Bank in Singapore, and Bettr in Thailand, illustrating how fintech is extending sustainable financial services to groups that have been traditionally underserved. MY Bank satisfies the microfinance needs with big data risk control and a "310" credit service model. Anext Bank targets MSMEs, particularly those in cross-border trade, with digital financial services that support their digital transformation and global expansion.

Following that, Bettr's practice has been introduced. Bettr offers smooth and convenient digital loan services for MSMEs in Thailand through embedded financial solutions.

The above approaches have not only accelerated the digital economic transition but also bolstered banks' capacity. By harnessing technological innovation and collaboration, they have made financial services more inclusive, reduced transaction costs, and bridged information asymmetry, thus providing more effective financing support for MSMEs. Moreover, these examples show how to tackle cross-border business challenges within a compliant framework, such as keeping loans affordable, managing risks, and broadening service reach.

Despite the many challenges, fintech holds immense potential to empower MSMEs, boost cross-border trade, and drive the digital economic transition. With ongoing technological progress and innovative collaboration models, there is every reason to believe that fintech will keep fueling global economic development, especially by playing a pivotal role in nurturing MSME growth and enhancing their competitive edge on the world stage.

CHAPTER 7

Sustainable Fintech for Decent Work and Economic Growth: 10 x 1000 Tech for Inclusion

In the world of fintech, the drive for inclusive growth has taken center stage as a key priority for boosting the global economy. The 10 × 1000 Tech for Inclusion program (10 × 1000), a brainchild of Ant Group in partnership with the International Finance Corporation (IFC), represents a giant stride in this quest. Launched in 2018, this visionary initiative has been pioneering in grooming tech leaders who are not just the vanguard of innovation but also the engines of economic change in emerging markets. As we approach 2023, the program has made impressive progress, having successfully equipped a diverse group of 6,243 learners from more than 100 countries and regions with the necessary certifications. This chapter takes a deep dive into the profound influence of the 10 × 1000, exploring its cutting-edge educational model, the vibrant community it has nurtured, and its pivotal role in sculpting a sustainable future for fintech that promotes quality employment and economic prosperity.

J. Sun, *Sustainable Financialization*, Sustainable Development Goals Series, https://doi.org/10.1007/978-981-96-3580-1_7

7.1 Empowering Through Education: An Introduction to the 10 × 1000 Tech for Inclusion Program

Ant Group's ambition to drive inclusive growth in the field of financial technology is to cultivate at least 1,000 emerging talents and technical leaders in this domain annually over a decade, positioning them as catalysts for digital economic expansion. The inclusive fintech project called "10 × 1000 Tech for Inclusion" (10 × 1000) was jointly introduced by Alipay and the International Finance Corporation, a part of the World Bank Group, on October 13, 2018, during the World Bank Annual Meeting. The vision of this program is to establish a global and open learning platform driven by industry and partner collaboration, dedicated to nurturing technological leaders in emerging markets and providing opportunities for learners to drive digital economic growth. Since its inception in 2018, the program has pledged to train 1,000 technological leaders annually for emerging markets over ten years. By focusing on knowledge sharing and exchange in the field of financial technology, particularly drawing on China's extensive experience in fintech, the initiative aims to promote sustainable economic development in these countries and regions, bridge the digital divide, and bring tangible benefits to the local areas.

"Talent is the most crucial element in technological innovation. Only by changing people's mindsets and enhancing their skills can we truly enter the digital economy era. What emerging markets urgently need are tech talents. The Alibaba Digital Economy aims to cultivate future tech leaders for emerging markets," stated Jack Ma, founder of Alibaba Group, on the significance of the 10 × 1000 project.

7.1.1 *Integration of Curriculum and Community*

In the 10 × 1000 curriculum, participants gain key insights from the mindset, knowledge, and skill modules of the instructors. Each course is designed to be scalable and is modified and enriched according to the needs of learners.

The 10 × 1000 Flex model allows learners to complete courses entirely at their own pace online. The Flex program is designed to provide learners with the mindset, knowledge, and skills to become drivers of digital economic growth. The curriculum includes foundational courses, expert programs, mini-series, and more. Among them, the Digital Financial Services Mini-Series is a masterclass for current and future industry leaders—it offers cutting-edge insights from thought leaders in key areas of digital financial services, including best practices for the digital transformation of banks and traditional financial institutions, how fintech companies can collaborate with fin Integration of Curriculum and Community financial institutions to leverage each other's strengths, real insights into introducing inclusive digital financial services to frontier communities with limited infrastructure, and in-depth market trends and case studies on the rapidly growing vertical of digital wealth management.

The 10 × 1000 Capstone program, on the other hand, is a short-term, intensive experience program. It offers an immersive learning experience over five weeks, including video courses, live Q&A, interactive discussions, and team Capstone projects. The "Fintech Leadership Program" included is specifically designed for senior leaders, providing executives with an opportunity to exchange ideas around key themes of technology, strategy, and leadership. Learn from and interact with expert instructors and industry peers. The content covers a comprehensive overview of the impact of digital technology on inclusive finance, major trends in inclusive finance in Asia, and how top players design business frameworks around inclusivity; an introduction to the key technologies behind fintech (blockchain, artificial intelligence, security, internet of things, cloud) and the latest applications of blockchain and artificial intelligence in financial services; and best practices on how to use digital technology to solve business problems and achieve scalability.

Compared to the curriculum, what sets 10 × 1000 apart is its community building. The 10 × 1000 community is a place for learners to share ideas, seek solutions, and engage in discussions. By hosting exclusive community events, it provides learners worldwide with opportunities to interact with experts and connect with other professionals. The community encourages all learners to participate in its construction, hoping to create a more inclusive and sustainable future through mutual support. Mr. Jason Pau, the project leader of 10 × 1000, stated, "Deepening the understanding of emerging technology applications, cross-regional sharing, and local networking are the top priorities for 10 × 1000 learners. 10 × 1000 recognizes the potential of cross-national and cross-regional knowledge sharing and focuses on connecting and cultivating a vibrant global fintech community composed of learners, experts, and partners to foster learning. In 2021, 17 expert instructors and 31 leading global organizations, tech investors, and industry organizations joined the community, becoming our close partners, and together we seek learners for 10 × 1000."[1]

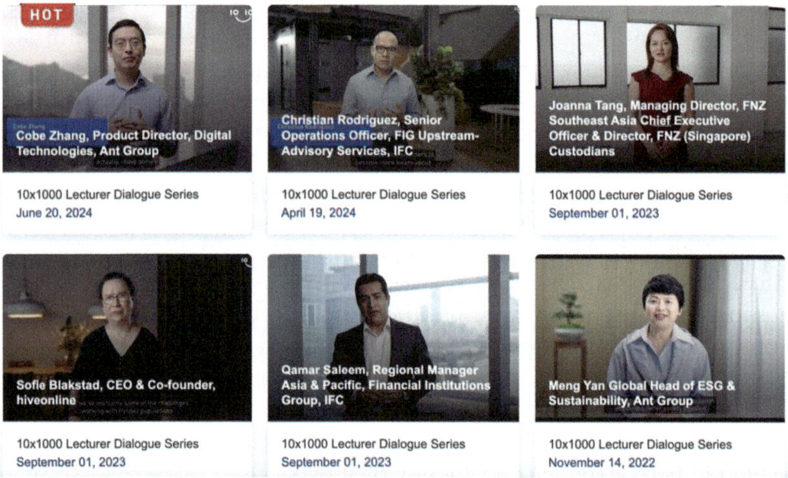

Experts Insights of 10 × 1000 Platform (*Source* 10 × 1000 Tech for Inclusion)

7.1.2 Stage Overview: Progress of the 10 × 1000 Program in 2021 as an Example

In 2021, the online course of 10 × 1000 wrapped up with a remarkable 1067 participants hailing from a diverse pool of 66 countries and regions. A significant chunk, nearly 80%, of these tech enthusiasts were from the burgeoning markets of South and Southeast Asia, with hotspots like Bangladesh, Indonesia, and the Philippines leading the pack. Adding to the diversity, 23% of the cohort represented small and medium-sized enterprises (SMEs), showcasing a broad appeal of this program.

The initiative has garnered global recognition, with 31 organizations spanning national, international, and regional scopes coming on

[1] Business Wire, 2022, https://www.businesswire.com/news/home/20220331006106/en/10x1000-Tech-for-Inclusion-Enhances-Fintech-Learning-Programs-to-Continue-Bridging-the-Global-Digital-Skills-Gap.

board as partners. This esteemed group includes heavyweights like the World Economic Forum, the United Nations World Food Program, and the United Nations Economic Commission for Africa, highlighting the program's far-reaching impact and commitment to fostering inclusive tech growth worldwide.

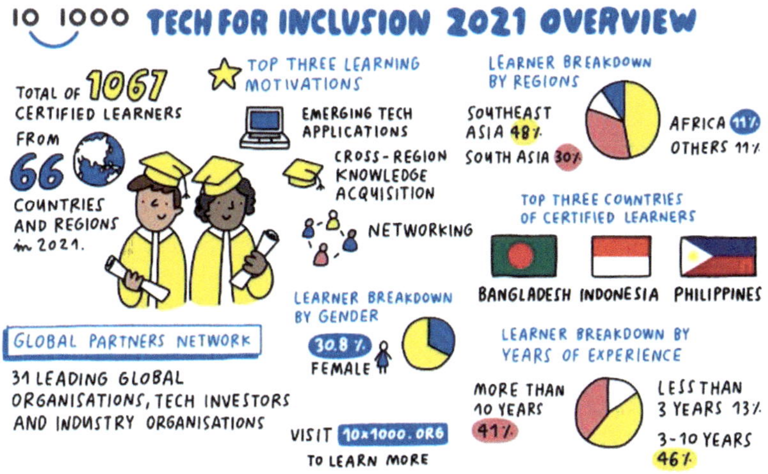

10 × 1000 Tech for Inclusion 2021 Overview (*Source* Business Wire)

The practicality of the program's courses has been widely praised by participants, with 88.5% of the participants stating that the training was positively helpful and enlightening for their daily work and career development. Angela Lum-neh, founder of Steer Analytics from Cameroon, Africa, participated in the program in 2021. She said, "Unlike other online courses, this program has a large community that brings together over 1000 talents and leaders from various industries worldwide. In addition to the knowledge gained from the program, I was able to connect with peers and members in the same field through the community, which was very rewarding."[2]

Emmanuel Elieza Mrok Kibanga, founder and CEO of ClickPay from Tanzania, Africa, also expressed his satisfaction in an interview, saying, "Although the training period was short, it was more practical than

[2] https://www.antgroup.com/news-media/press-releases/1648718782000.

a three-year bachelor's degree program. I would sincerely recommend friends to join this program."

"In 2021, we received an enthusiastic response and participation from over 1000 participants worldwide, setting a historical record. Especially notable was the active involvement from South and Southeast Asia," said Jason Pau, the project leader. "With the strong support of 17 expert instructors from 10 countries, we were able to develop a rich curriculum and promote industry experience exchange across countries and regions." He added, "In 2022, the 10 × 1000 program will also introduce a more in-depth Fintech Expert Program, a French version of the basic course, and studies in green Fintech, hoping to cultivate Fintech technical talents for more countries and regions." Specifically, these plans include the Fintech Expert Program, Green Fintech Mini-Series, and the French Fintech Foundation Project, among others. They will better meet the needs and reach more global learners, enhancing participants' understanding of various topics such as green Fintech and ESG.

MO Tarek, the entrepreneurial spirit hailing from Egypt and educated in Finland, took to social media to extend his warm congratulations to the 1067 graduates of the 10 × 1000 Plan. He is now a resident of Malaysia, one of the world's leading countries in waste import. Tarek's involvement in the 10 × 1000 initiative has equipped him with profound insights into waste management and sustainable economic growth.

"China's dominance in electric vehicle production, accounting for half of the global market, and its meticulous legislation to curb waste import, serve as an exemplary model for Southeast Asia," Tarek proclaimed. His experience from the 10 × 1000 Plan in Hangzhou has not only been a learning journey but also a catalyst for change. Upon returning to Kuala Lumpur, he integrated the knowledge gained into his startup, tackling the critical last mile challenge in electronic waste recycling in Malaysia. Tarek's story is a testament to the transformative power of the 10 × 1000 program, turning ideas into actionable solutions for a greener tomorrow.

MO Tarek attended the offline course of the 10 × 1000 in Hangzhou in 2019 (*Source* China Daily)

7.2 Case Stories: 10 × 1000 Empowering Participants

7.2.1 Participant Palwasha Faizan: The Emerging Female Leader

Yang Dongyang, the operations head of the 10 × 1000 program, noted that women in the digital economy sector are increasingly making their presence known in Southeast and South Asia. For instance, in Indonesia, 60% of SME managers are women. "These once-hidden 'alphas' are stepping into the limelight. We want to help them gain more opportunities," he said.

At Pakistan's established bank, Habib Bank, 28-year-old Palwasha Faizan is not only the only woman in the Digital Products and Platform Development Department but also its head. For her, such challenges are not new. At the age of four, Palwasha's father gave her a microphone to practice speaking in front of a room full of male family members. She always remembers her father's words, "In our country, in most situations, you are likely to be the only woman in the room."

Currently, the established bank is attempting to transform into a tech company that embraces innovation, connecting ordinary people who were previously excluded from financial services to the country's financial system and providing financial support to more people without collateral. Palwasha's proposal to offer unsecured credit loans for users has gained much approval, inspired by a training program called "10 × 1000" she recently attended. Instructors from China's internet banks shared their

310 loan model: apply in 3 min, disburse in 1 s, with 0 human intervention. By the end of 2021, this internet bank had provided loans to over 45 million Chinese SME owners and individual businesses.[3]

In 2021, she was nominated by her bank to participate in the "10 × 1000" training program. Palwasha was excited that the program not only allowed her to learn from top figures in the digital economy from around the world, such as the founder of India's mobile payment company Paytm, Vijay, and the CEO of Indonesia's mobile payment company DANA, Vincent, but also enabled her to efficiently share learning materials and discuss issues with other participants online through a simple WhatsApp study group, without leaving her home.

After joining the "10 × 1000" program, Palwasha found that 30% of the members were women. She said this made her feel less alone and unsupported.

Having benefited greatly from the program, Palwasha was particularly interested in the 310 model of online microloans taught by Chinese instructors from the MyBank internet bank, which gave her new insights into promoting financial inclusion. "To promote inclusive finance in Pakistan, it is necessary to use technology to reduce risks and costs while ensuring profits, only then is it feasible and sustainable," she said.[4]

Talking about the two-month project study, the word Palwasha used most was "inspiration." Vincent was the instructor who left the deepest impression on Palwasha, not only because of his sharing but also because of the country he is from. "The development of Indonesia's digital economy is of great reference to Pakistan. DANA's expansion strategy and problem-solving methods have been very inspiring to me," Palwasha said.

Currently, the challenge Palwasha faces is that traditional financial institutions in Pakistan, including Habib Bank, have realized the necessity and urgency of digitalization, but user acceptance remains relatively low, with cash still being the preferred choice for ordinary people. Palwasha carefully studied Alipay and MyBank, and she and her colleagues began to try

[3] https://news.sina.com.cn/c/2022-06-28/doc-imizmscu9124246.shtml#:~:text=%E9%A9%AC%E6%9D%A5%E8%A5%BF%E4%BA%9AERTH%E5%85%AC%E5%8F%B8.

[4] https://www.thepaper.cn/newsDetail_forward_18977048.

to promote the digitalization of government subsidies, agricultural loans, and social security payments on a small scale, guiding more and more ordinary people to use mobile payments through these rigid demands, so that they "feel its convenience."

7.2.2 Participant Utari Octavianty: The Young Pragmatic Innovator for Inclusive Fishermen

Utari Octavianty grew up in a fishing village, where her mother made a living selling fishing gear. This background fueled her strong desire to improve the challenging living conditions of local fishermen. In 2015, she and her university classmates decided to establish an e-commerce platform to help local fishermen bypass intermediaries and sell their catch directly to customers worldwide. This initiative resulted in the creation of Aruna.

Aruna was the first Indonesian firm dedicated to using cutting-edge technology to link regional fishermen with larger markets. After its establishment, Aruna quickly attracted a large number of fishermen, with 2.7 million already using the site and securing investment rounds upon rounds. Thanks to their efforts, the fishermen's incomes increased by 3 to 12 times. Aruna also started concentrating on empowering women in the coastal region, which helped them earn a gold prize at the 2018 Asean ICT Awards.

Utari was not one to rest on her laurels; she had her sights set on a bigger impact for the fishing community, undeterred by the daunting challenges of scaling up. In 2019, she took a bold step by joining the "Southeast Asia Social Innovation Challenge", a prestigious event co-hosted by Alipay, the National University of Singapore, and the World Bank's IFC. Her determination and innovation shone through as she emerged victorious, securing her place in the "10 × 1000" training program.[5]

In 2019, Utari and her entrepreneurial partners traveled to Hangzhou to attend the 10 × 1000 Tech for Inclusion program training class, co-hosted by Alipay and the IFC of the World Bank. Mentors provided face-to-face consultations on her business model and offered numerous suggestions. This valuable learning experience significantly broadened Utari's horizons.

[5] Ibid.

Utari stated that her trip to Hangzhou was highly rewarding. Before her visit, she had never imagined a city that could so comprehensively support electronic payments, underpinned by a value system of inclusivity and the power of technology. This was her most profound realization after learning about the growth trajectories of companies like Alipay during the training program.[6]

Utari hopes to bring this value system and technological power back to Indonesia to improve the lives of more fishermen. Upon returning to her country, she upgraded the Aruna platform, integrating online and offline elements, and encompassing various stages such as fishing, fish product processing, quality control, and technical training, creating a new market that is distinctly different from traditional ones. The results have been impressive—even during the most severe period of the pandemic in the first half of 2020, Aruna's revenue increased by an astonishing 86 times. In addition, **Utari** has been awarded the 2023 FORTUNE Indonesia 40 Under 40 for her capability to create opportunities and utilize technology for positive impacts on the Indonesian fisheries and marine industry.

Aruna's Co-Founder and Chief Sustainability Officer, Utari

7.2.3 Mentor Vincent: The Inspiring Sharer of Insights

As a short-term training program lasting 4 to 6 weeks, the 10 × 1000 curriculum is designed with flexibility, practicality, and conciseness in mind. Each mentor's course is segmented into sections of just a few minutes each, allowing participants to learn anytime, anywhere, utilizing fragmented time slots. Instead of delving into complex theories and grand

[6] https://www.sohu.com/a/341474426_474131.

principles, mentors focus on presenting the most up-to-date case studies and sharing experiences that offer the most practical insights.

These leaders from China and the financial technology and internet enterprises of Southeast and South Asia understand the urgent need for participants to keep their knowledge structures and perspectives aligned with the rapid pace of the digital era. They continuously break boundaries through mutual inspiration and influence, sparking various micro-innovations.[7]

In 2021, Vincent, the founder of the Indonesian mobile payment platform DANA, was voted the "Most Popular Teacher" by the participants. Vincent Iswara, the CEO of the Indonesian DANA wallet and one of the mentors of 10 × 1000, stated, "Indonesia's digital economy is in a golden age, full of endless opportunities. More than ever, Indonesia needs more tech talent to drive the development of the digital economy. I am honored to participate in this project and contribute to cultivating future talents."

Vincent Iswaratioso, CEO of Indonesian company DANA and mentor of the 10 × 1000 program

During his lectures, Vincent not only personally designs course materials but also uses his entrepreneurial story as evidence. For instance, when discussing the idea that "one person walks fast, but a group walks far," Vincent recalled the early days of DANA when resources were

7 https://news.sina.com.cn/c/2022-06-28/doc-imizmscu9124246.shtml#:~:text=%E9%A9%AC%E6%9D%A5%E8%A5%BF%E4%BA%9AERTH%E5%85%AC%E5%8F%B8.

very limited, and they began to consider moving forward with partners. DANA's earliest partner was the largest cinema chain in Indonesia, which held an 80% market share locally. This partnership allowed DANA to enter the online ticketing scenario and became an instant hit when launched.

Many participants have expressed that after the training, they still often revisit Vincent's videos to help them "find inspiration to solve real-world problems."

This Indonesian entrepreneur, who schedules his time down to the minute, says he also gains a great sense of achievement from this process, as he is also a beneficiary of the exchange: "The sharing of knowledge and experience accelerates the development of the digital economy ecosystem in Southeast and South Asia."[8]

7.3 Deepening Global Cooperation: 10 × 1000 in Hong Kong and Indonesia

In 2021, the 10 × 1000 program certified 1067 participants from 66 countries and regions, helping local entrepreneurs and professionals enhance their fintech mindset, knowledge, and skills to better promote digital economic growth.

The 10 × 1000 Tech for Inclusion initiative has attracted an increasing number of governments and institutions committed to advancing global inclusive development. Thirty-one international and regional organizations, including the United Nations Economic Commission for Africa, the United Nations World Food Programme, and the World Economic Forum, have become partners of the 10 × 1000 program, joining forces to recruit participants and develop communities. Through 10 × 1000's digital experience sharing, which empowers individuals rather than merely providing fish, the seeds of inclusive development are taking root in more places globally, further aiding the digital upgrade of local economies and societies.[9]

[8] https://news.sina.com.cn/c/2022-06-28/doc-imizmscu9124246.shtml#:~:text=%E9%A9%AC%E6%9D%A5%E8%A5%BF%E4%BA%9AERTH%E5%85%AC%E5%8F%B8.

[9] https://economy.gmw.cn/2022-11/02/content_36133268.htm#:~:text=11%E6%9C%881%E6%97%A5%EF%BC%8C%E5%9C%A8%E9%A6%99%E6%B8%AF%E9%87%91.

7.3.1 *Boosting the Development of Hong Kong's Tech Talent*

During Hong Kong Fintech Week on November 1, 2022, the Hong Kong Investment Promotion Agency (InvestHK) announced a strategic partnership with 10 × 1000 Tech for Inclusion, a worldwide fintech talent training platform. Through this partnership, 160 professionals from Hong Kong's digital ecosystem had either enrolled or finished the 10 × 1000 training program at that point.

The alliance is said to encompass recruiting homegrown Hong Kong talent for the 10 × 1000 program and forging new collaborative endeavors. The 10 × 1000 program is also part of the Global Fast Track initiative, a brainchild of InvestHK and the Monetary Authority, and is deeply involved in Hong Kong's fintech scene. The goal is to nurture local talent, quicken the pace of digital transformation, back entrepreneurship, and invigorate the digital economy's growth. As a result, the program is poised to cultivate local fintech experts, elevate the caliber of micro, small, and medium-sized enterprises (MSMEs), boost women's involvement, and propel Hong Kong's digital economic progress.[10]

Since forming a strategic alliance with InvestHK, the 10 × 1000 Tech for Inclusion program has further entrenched its role within the local fintech ecosystem. The program has successively joined forces with the FinTech Applied Research Academy of the Chinese University of Hong Kong (CUHK) and the Business School of the Hong Kong University of Science and Technology (HKUST). These partnerships aim to bolster the training and development of local fintech talent. The 10 × 1000 program crafts specialized course modules, conducts seminars and lectures, invites industry veterans to share their insights, and offers internships and career development opportunities for students and professionals. These collaborations not only transform academic research into actionable industry solutions but also grant students access to the latest fintech trends, effectively enhancing their technical skills and industry acumen, thereby contributing to the evolution of Hong Kong's fintech industry. This initiative aligns with the talent policies and measures to bolster Hong Kong's competitive edge as outlined in the first policy address by the Chief Executive of the Hong Kong Special Administrative Region, John Lee.

[10] https://www.antgroup.com/press-releases/1667367366000.

CUHK and HKUST Sign Memorandum of Understanding with Ant Group (*Source* CUHK and HKUST)

When it comes to fueling fintech innovation, the 10 × 1000 initiative is right on the money with Hong Kong's aspirations, offering robust backing for nurturing homegrown fintech prowess. CHU Ho Kan, a finance whiz and MBA aspirant at the Hong Kong University of Science and Technology, dove deep into the inclusive essence and vast prospects of fintech through his involvement with the 10 × 1000 initiative. He spotlighted "the BASIC framework—encompassing blockchain, artificial intelligence, security, the Internet of Things, and computing power—as the cornerstone for fintech's core components."

Prof. TAM Kar Yan, Dean of the Business School at HKUST, underscored fintech's game-changing role in finance and shared his enthusiasm for the partnership with Ant Group. This collaboration is set to equip students with hands-on fintech wisdom and expertise.

Chris, a 27-year-old scholar at CUHK, played down the idea of launching a startup in the next decade, underscoring the value of work experience. He remarked, "Post-graduation, I am keen on staying in Hong Kong and diving into finance or fintech, possibly data analysis. With Hong Kong as the gateway to Mainland China's opportunities, the future looks promising. The Greater Bay Area is also on my radar, given my family's roots in Guangzhou and the ease of shuttling between the cities." He also noted the synergies among powerhouses like Shenzhen, Hong Kong, and Guangzhou, expressing bullish optimism about the Greater Bay Area's growth trajectory.

Jennifer TAN, the Executive Vice President of Strategy Development and Government Affairs for Ant Group's GBA domain, also highlighted the 10 × 1000 program's essential collaboration with Hong Kong's academic institutions. By year-end, the 10 × 1000 alumni network had expanded to 2,808 graduates from 92 countries and regions. Over 40 industry experts and academics from some 20 locales lent their input to the curriculum's design and evolution. Their collective insights and experience have sharpened the learners' perspectives, knowledge base, and skill sets, positioning them as trailblazers in the digital economy's new wave.

In November 2022, ACCA Hong Kong also signed a memorandum of understanding with Ant Group to jointly promote the development of fintech talent. ACCA Hong Kong members can gain pre-approval and free registration to participate in the three-week fintech crash course through the 10 × 1000 program. After completing the course, members can continue to access the latest fintech development trends from 10 × 1000 and join the 10 × 1000 alumni network, which consists of fintech and digital economy professionals from around the world. This shared and supportive model, based on the digital economy and internet technology, not only deepens the integration of the accounting industry with fintech but also provides more possibilities for this international financial center.

Moreover, the 10 × 1000 program's foray into Hong Kong is keen on bolstering women's involvement in fintech, aiming to foster a more diverse and inclusive sector. Through the creation of scholarships, the orchestration of women's leadership seminars, and a suite of other motivational tactics, this program is encouraging more women to step into the fintech arena. These efforts are pivotal in dismantling gender barriers and championing equality within the industry.

Mr. Leung Hon-kwan, Head of Fintech at InvestHK, stated that "according to feedback from fintech companies and investors, greater support for talent cultivation, especially at the middle management level, is very valuable to our fintech ecosystem in Hong Kong. InvestHK is pleased to partner with 10 × 1000 to help our fintech community adopt this influential talent program more extensively."[11]

In 2022, a total of 1,741 learners completed the fintech foundation course or fintech expert course under 10 × 1000, with the number of graduates increasing by more than 63% compared to 2021. Among them, 115 graduates were local Hong Kong individuals from various organizations.[12]

[11] ChinaDailyHK, 2022, https://www.chinadailyhk.com/hk/article/297838.

[12] CUHK, 2023, https://www.erg.cuhk.edu.hk/erg/node/2551#:~:text=The%20Chinese%20University%20of%20Hong%20Kong%20(CUHK)%20Engineering.

Jason Pau, Senior Director and Head of the 10 × 1000 program at Ant Group, said, "We are honored to collaborate with InvestHK to achieve our shared goals of cultivating fintech talent in Hong Kong and promoting digital economic growth. Human resources are Hong Kong's competitive advantage, and in the digital era, local talent in Hong Kong has tremendous potential to further drive Hong Kong's development. In this regard, 10 × 1000 Tech for Inclusion will work with our partners to bring more world-class fintech training projects and opportunities to local participants."[13]

7.3.2 Fostering Indonesia's Digital Economy Innovation Momentum

On May 24, 2022, in Jakarta, Indonesia, the Indonesian Ministry of Communication and Information Technology ("Kominfo") announced on its official website a partnership with the fintech training platform 10 × 1000 Tech for Inclusion to assist the Indonesian government in achieving its goal of cultivating 60,000 local digital talents to stimulate the rapid development of the digital economy. This marks the introduction of an "Indonesian version" of the China-initiated 10 × 1000 Tech for Inclusion program,[14] indicating that the Chinese fintech talent cultivation model is gradually expanding into the Southeast Asian region, injecting new vitality into the local digital economy.

The Indonesian Ministry of Communication and Information is responsible for executing government affairs in the field of communication and information technology, assisting the President in managing the government. Its main tasks and functions include formulating national policies, implementing policies, and technical policies in the fields of communication and information, including postal services, telecommunications, broadcasting, information and communication technology, multimedia services, and information dissemination. It is understood that the specific implementation of this cooperation will be advanced by the Digital Entrepreneurship Academy (DEA) under Kominfo. The DEA aims to cultivate and nurture talents to accelerate digital transformation

[13] https://economy.gmw.cn/2022-11/02/content_36133268.htm#:~:text=11%E6%9C%881%E6%97%A5%EF%BC%8C%E5%9C%A8%E9%A6%99%E6%B8%AF%E9%87%91.

[14] https://tech.china.com.cn/roll/20220524/387831.shtml#:~:text=5%E6%9C%8824%E6%97%A5%EF%BC%8C%E5%9C%A8%E5%8D%B0%E5%B0%BC%E9%9B%85.

in the entrepreneurial field and improve the digital economy, with a goal to train 60,000 people in 2022.[15]

Adhering to its vision of cultivating 1,000 emerging talents and tech leaders over ten years, in 2021, 1067 participants from 66 countries and regions completed the "Flex" Fintech Foundation Program and the "Fintech Leadership Program", with over 150 participants from Indonesia, making it the second-largest recipient of participants globally. Subsequently, in 2022, the "Indonesian version" of the 10 × 1000 training courses was launched in collaboration. Jason Pau, the project lead for 10 × 1000, said, "We are deeply encouraged by the positive response from Indonesia last year and are delighted to partner with Kominfo to further expand our impact in Indonesia, enabling more local entrepreneurs and professionals to use Flex to enhance their fintech mindset, knowledge, and skills, and better drive digital economic growth. Technology permeates almost every aspect of our lives, and we hope this collaboration will help Indonesia bridge its digital skills gap while enhancing and improving the lives of local communities. Together with Kominfo and other local partners, we look forward to contributing more to Indonesia's digital development goals."[16]

The Digital Entrepreneurship Academy's collaboration with the 10 × 1000 program is laser-focused on boosting women's involvement in fintech and beefing up support for MSMEs, all to ensure that training programs are as inclusive and universal as they can be. Femmy Eka Kartika Putri, a rep from the Ministry of Human Development and Culture, has noted, "the '10 × 1000' fintech training courses have a substantial female presence, with 60% of MSMEs in Indonesia in the hands of women, contributing a whopping 61.07% to Indonesia's GDP." The 10 × 1000 Tech for Inclusion "Flex" Fintech Foundation program is designed to inspire female entrepreneurs to supercharge their businesses with digital technology. This training is anticipated to help women entrepreneurs and MSMEs navigate challenges like limited entrepreneurial exposure and

[15] https://www.businesswire.com/news/home/20220523006072/en/Kominfo-Partners-with-10x1000-Tech-for-Inclusion-to-Bring-%E2%80%9CFlex%E2%80%9D-Fintech-Foundation-Program-to-Indonesia#:~:text=JAKARTA,%20Indonesia--%20(%20BUSINESS%20WIRE%20)--The%20Ministry%20of%20Communication%20and.

[16] https://www.businesswire.com/news/home/20220523006072/en/Kominfo-Partners-with-10x1000-Tech-for-Inclusion-to-Bring-"Flex"-Fintech-Foundation-Program-to-Indonesia#:~:text=JAKARTA,%20Indonesia--%20(%20BUSINESS%20WIRE%20)--The%20Ministry%20of%20Communication%20and.

digital tech know-how, as well as funding hurdles. These initiatives are not just a win for Indonesia's digital economy growth but also a stride towards achieving the United Nations Sustainable Development Goals (SDGs).

7.4 Inclusive Tech Careers in China: Diverse Practices by Different Enterprises

7.4.1 HUAWEI Women Developers

In fact, it is not just Ant Group that is stepping up to the plate for sustainable development goals; other Chinese enterprises are also actively seeking ways to make a difference. Huawei, for instance, took a significant step in March 2021 with the launch of the HUAWEI Women Developers (HWD) program, a global initiative designed to empower women developers to craft applications and tools with world-changing potential. This program opens doors for women developers by offering them opportunities and platforms for career advancement and professional skill enhancement, thereby encouraging more women to dive into the realm of technology innovation.

The initiative is accessible to female developers worldwide through the Huawei Developer Alliance's official website, where they can access courses and training focused on technological innovation and career development. Experts from various fields are also invited to share the latest technologies and tools, providing a rich learning experience. Huawei has gone a step further by establishing a dedicated community for women developers on the Developer Alliance platform and organizing a series of online and offline events to foster connection and collaboration.

Furthermore, the Huawei Shining Star program provides exclusive incentives for women developers, supporting and encouraging their innovation and entrepreneurship. Female developers of potential and high-quality applications have the opportunity to receive project-related promotional support and invitations to official Huawei events.

Huawei has carried out women's digital technology empowerment projects in several countries, including Ireland, Argentina, Bangladesh, Kenya, and South Africa. In programs such as the "Seeds for the Future" ICT training, the participation of female students has exceeded 30%.[17]

[17] Huawei, 2021, https://www.huawei.com/cn/news/2021/3/huawei-launched-huawei-women-developers-program.

HWD 2021 in Guangzhou (*Source* HUAWEI)

7.4.2 Hang Lung Future Women Leaders Program

Initiated and hosted by Hang Lung Properties in 2022, the Hang Lung Future Women Leaders Program is a nationwide community investment initiative co-organized by the Hong Kong Youth Association's Leadership Academy and the Women's Federation of Shanghai's Xuhui and Jing'an Districts. The program aims to support the diverse development of female college students across the country through mentorship, leadership training, and internship opportunities, witnessing the emergence of new female forces.

In recent years, the pressing issue of youth unemployment has hit the headlines, with college students grappling with significant job worries. The entrenched gender roles and limited career path perceptions have only intensified the employment challenges faced by female college students. They are raring to break free from these constraints and uncover a broader spectrum of opportunities for their own and women's career advancement. In this context, the Hang Lung Future Women Leaders Program has been a breath of fresh air, garnering immediate acclaim among female college students from leading universities as soon as it was introduced.

The first phase of the project was launched in both Shanghai and Hong Kong. The Hang Lung Future Women Leaders Program began its first phase in these two cities, employing 40 elite women from various industries as mentors. Through six months of mentorship, multi-dimensional training, and community practice activities, the program aimed to enhance the personal and professional skills of 200 female college students nationwide, encouraging a new generation of women to leverage their strengths and improve their leadership. Outstanding students also had the opportunity to receive internships provided by Hang Lung and Future Women Leaders Program mentors. At the same time, the project also built a more efficient and broader communication platform for young women in both Shanghai and Hong Kong through online dialogues, taking practical actions to build a sustainable future.

Close to 180 female college students from all corners of the nation joined the inaugural Hang Lung Future Women Leaders Program, with a remarkable 75% of them being pioneers in their family's third generation to pursue higher education. Throughout the program's six-month journey, these students clocked over 6,300 h in leadership training and mentorship, with an impressive 90% completion rate. At the capstone graduation ceremony, top-performing students from mainland China and Hong Kong engaged in a dynamic dialogue, swapping insights and reflections from their transformative experiences.

The Hang Lung Future Women Leaders Program, a national community investment project, has unearthed and stimulated the personal and professional development potential of participants through six months of mentorship by 43 public welfare mentors, over 2,800 h of guidance, and nearly 3,500 h of comprehensive leadership training courses. Additionally, with the guidance of mentors and the support of the Hang Lung Volunteer Team, the participants completed the practical implementation of 31 sustainable development projects, benefiting over 4,500 community members and promoting social welfare. Some outstanding students were also provided with two-month internship opportunities at Hang Lung and companies of Future Women Leaders Program's mentors and partners, further accumulating practical experience for the transition from campus to career.[18]

[18] Thepaper, 2023, https://www.thepaper.cn/newsDetail_forward_23505644.

Graduation Ceremony of the Hang Lung Future Women Leaders Program in 2023 (*Source* Hang Lung)

7.5 Conclusion

Since its inception in 2018, the 10 × 1000 Tech for Inclusion initiative has emerged as a powerhouse in fostering inclusive growth within the global fintech landscape. In tandem with the IFC and Ant Group, the initiative has transcended mere knowledge dissemination in fintech. It has effectively groomed a significant cadre of tech leaders in emerging markets, who are now instrumental in propelling the digital economic surge in their respective nations and regions.

The achievements of this program can be traced back to its pioneering educational framework and unwavering community backing. 10 × 1000's Flex and Capstone components furnish participants with a pliable learning trajectory and a deep dive educational journey, while the meticulously curated community serves as a springboard for continuous learning and dialogue. This approach has not only heightened participant engagement but also sharpened their capacity to implement their newfound knowledge in real-world scenarios.

Moreover, it is important that the 10 × 1000 initiative's spotlight on female leaders has made a commendable impact on gender parity and women's empowerment. By extending specialized training and backing for women, the program has been instrumental in dismantling gender barriers, paving the way for more women to not only enter but also persist in the tech arena.

On a global scale, the 10 × 1000 program has forged alliances with a multitude of international and regional entities. These collaborations have broadened the horizons of this initiative and opened up new avenues for nurturing fintech talent worldwide. Notably, the adaptability of 10 × 1000 has been underscored by its collaborations in Hong Kong and Indonesia, which cater to the distinct talent development needs of various geographies.

As we look to the future, 10 × 1000 is set to persist in amplifying its global footprint and cultivating fresh talent in fintech. As the program progresses and scales, it will persist in its mission to support the United Nations Sustainable Development Goals (SDGs), particularly in the realms of quality education (SDG 4), decent work and economic growth (SDG 8), and reducing inequalities (SDG 10). Anchored in the SDGs that prioritize human capital, the 10 × 1000 program equips future industry leaders with top-tier educational resources closely aligned with industry needs (Goal 4—Quality Education), empowering learners across the globe to unlock more opportunities in the digital era (Goal 8—Decent Work and Economic Growth). With a substantial participant base from developing countries in South and Southeast Asia, the initiative is also pivotal in leveling the playing field within the fintech industry (Goal 10—Reduced Inequalities).

10 × 1000 and the SDGs (*Source* 10 × 1000 Tech for Inclusion)

By the end of 2023, the 10 × 1000 Tech for Inclusion program had successfully certified 6,243 learners from over 100 countries and regions, with participants hailing from more than 2,000 businesses, of which 80%

were SMEs. Notably, female learners accounted for 49% of the total, emerging as a formidable force in the growth of the digital economy.

As this initiative forges ahead, Ant Group is gunning to shape no less than 10,000 tech leaders from 2018 to 2028, igniting the spark in the next crop of industry trailblazers and, in the process, boosting global talent cultivation and fueling the expansion of digital economy. This ambitious move by Ant Group is not just a showcase of what corporations can achieve in driving global sustainable development, but also a clarion call to the business world, inspiring other enterprises to step up and join forces in crafting a more inclusive and sustainable future.

PART IV

Sustainable Financialization for Inclusive Partnership and Good Governance

CHAPTER 8

Sustainable FinTech for Inclusive Partnership: Ant Village, Cyber Mulan for Female Entrepreneurs in China and Indonesia, Wind Rider for Girls' Football and Blue Vest for the Elderly Population in China

8.1 Introduction

ESG has emerged in recent years as a critical metric for evaluating companies, with an increasing number of investors paying attention to the social and environmental impacts of businesses, not just their financial performance. The ESG conception originates in the framework of the Principles for Responsible Investment (PRI), which further evolved into the investment philosophy of considering sustainable and social development in numerous economic activities as its core promise.

In analyzing the major components that affect the overall ESG performance of a company, it is fair to argue that the S (social) factor plays a crucial role especially when it comes to social capital and human opportunities. To improve its ESG score and encourage sustainable development strategies Ant Group's mission is to build a platform that is fair and equitable to all. In the last few years, it was possible to observe Ant Group pursuing MNC IPO through its active participation in corporate public welfare projects. Corporate social responsibility is a term widely used

J. Sun, *Sustainable Financialization*, Sustainable Development Goals Series, https://doi.org/10.1007/978-981-96-3580-1_8

to cover the responsibility of an organization towards society without neglecting profit maximization. Most commonly, such activities include philanthropy, employee volunteering, environmental initiatives, community and stakeholder engagement, and the use of products or services to make social change.

Sustainability Report 2023 of Ant Group indicates that by the end of 2023, employees had spontaneously formed 39 charitable groups. These groups engaged in a variety of social causes, such as community volunteering, museum guiding, non-visual photography, emergency response at events, environmental conservation, remote education support, and family-oriented philanthropic activities. They created eight volunteer practice bases and also undertook 1,302 charitable events in the course of the year. The number of volunteers is about 8742 people, that's 39.7 percent more than in the year ten preceding that. More than three hundred families of owners of Ant Group included their children and assumed an active role in more than 200 acts of charity.

This article will focus on the performance of Ant Group in the latest ranking of the S component of ESG using public welfare specifically the Wind Rider, Cyber Mulan, and Blue Vest activities held by the Zhejiang Ant Public Welfare Foundation as illustrations.

8.2 Three Major Philanthropic Programs of Ant Group

In the context of accelerating global digitalization, different groups face significant gaps in access to digital resources, information skills, and equitable development opportunities, particularly women, the elderly, and residents of remote areas. These groups encounter the real challenge of a "digital gap" due to technological barriers, information isolation, and social biases. In China and many other parts of the world, however, girls residing in rural areas also have a tough time integrating into such activities as technology and sports due to insufficient resources and support. The elderly are no exception as they are largely left stranded in an intelligent society as well, find it difficult to enjoy internet-enabled services, and suffer from a host of shortcomings related to, inter alia, information fraud, and socialization. Female businesspersons also have their share of the story, in that they do not have the requisite opportunities in a career and entrepreneurship development as a result of gender discrimination, lack of social support, and professional networks. These issues not only prevent the growth and development of these groups but also, in one way or the other, impede the growth and inclusivity features of the society.

Diversity, equity, and Inclusion, also known as DEI, have been catalysts of the wellbeing and stimulus coding of culture within the multifaceted aspects of organization structure. These considerations foster sustainable social development. In order to tackle these issues, Ant Group has established an integrated framework of addressing diversity, equity, and inclusion in its social welfare strategy which focuses on assisting disadvantaged groups in the digitalization efforts through the provision of relevant platforms and resource networks. Ant Group's public welfare projects cover areas such as digital inclusivity, rural revitalization, community development, and emergency disaster relief, adhering to the principle of "Technology for Good" by providing technical support, resource empowerment, and care for diverse vulnerable groups in society.

In April 2019, Ant Group managed to register the Zhejiang Ant Public Welfare Foundation with the Zhejiang Provincial Civil Affairs Department, and in 2021, it was given a rating of 5A. The foundation is dedicated to bridging the digital divide and promoting social equity and progress. Its public welfare projects mainly focus on five areas: digital inclusivity, community development, rural revitalization, emergency disaster relief, and exploration and innovation. Through precise project deployment and social influence, the Zhejiang Ant Public Welfare Foundation has provided support for public welfare projects in 98 key rural counties across 32 provinces, with a total donation amounting to 998 million yuan.

In terms of digital inclusivity, the Ant Public Welfare Foundation has launched a series of significant projects. In this article, three of its public welfare projects that address inclusive digital development will be discussed: Wind Rider, Cyber Mulan, and Blue Vest. Wind Rider focuses on rural women's football, helping rural girls engage in sports, build confidence, and develop teamwork skills through funding for competitions and facility construction. Cyber Mulan aims to enhance women's entrepreneurial and career development by leveraging digital technology to improve their economic independence, providing resources like entrepreneurship training and mentoring courses to empower women with greater competitiveness in the tech industry and help them gain more autonomy in the digital age. Blue Vest aims to narrow the adaptation gap for the elderly in a digital society, ensuring they can access safer digital services in areas like mobile payments and information retrieval.

These projects are not merely philanthropic activities; they are guarantees and empowerments of development opportunities for specific groups in the digital age. By integrating digital technology and platform resources into these public welfare projects, Ant Group provides multi-level support ranging from infrastructure to skills training for rural girls, female entrepreneurs, and the elderly through both technical and caring approaches. This helps these segments to acquire skills to cope up with the age of computerization and have increased self-esteem, more chances, and ownership of their lives and professions. Such enduring social support also impacts the beneficiaries' identity in a positive manner and stimulates the internal and external development of caring and innovative corporate culture contributing critical forces towards well-rounded social development.

8.2.1 Wind Rider for Girls' Football

In 1999, when the U.S. women's football team won the World Cup in a dramatic penalty shootout, the Chinese team, despite finishing as runners-up, still garnered significant attention. Unfortunately, this short-term fame did not catalyze long-term support as in the coming years, interest and establishment of women's football gradually disappeared in the view of the public. In the next 20 years, the women's game of football went through a long journey of finding and establishing itself. The 2004 Athens Olympics marked a turning point when the Chinese women's team suffered a 0–8 defeat to Germany in the group stage, signaling the end of its golden era. Social and media interests declined over the years and so did any hopes of women's football which started to fade in the domestic women's football scene.

In the ensuing ten years, Chinese women's football faced numerous challenges. The once-prominent players dispersed, domestic leagues received little attention, and media coverage was scarce. A glaring example of this lack of attention occurred in 2016 when a ticket to a Shanghai Shenhua men's game granted fans free admission to a women's game, which served as a dreadful example of this neglect. This arrangement was intended to boost attendance but highlighted the severe undervaluation of women's football in the market. Women's matches not only had low ticket prices but also relied on the men's game to attract viewership, effectively rendering them as "ancillary." In comparison, men's matches had high prices for tickets, great exposure in the media, and decent support on the ground, hence the differences in their growth.

This low level of attention led to a scarcity of funding and resources. While there is a considerable amount of money spent on promotions, sponsorships, and bonuses in men's football, the amount of financial assistance to women's football could be said to be negligible. In China's premier women's football league, teams frequently look for but fail to solicit corporate sponsorships and some get by with little assistance from the local government's allowances. Many women players receive a few thousand yuan a month, unlike men who get multimillion salaries. After some time, however, some female players are already retired but are unemployed and have not had the opportunity to accumulate enough income where they have to look for high-paying jobs.

The weak youth training system is a direct result of this low level of attention. Unlike the well-established youth training systems in men's football leagues, women's football lacks sufficient grassroots investment and systematic training, leading to missed development opportunities for many promising players. There are relatively few football schools in China, and even fewer that cater specifically for girls, as there are limited facilities and resources available, so many people believe that girls should not play football at all. Even at the professional level, many clubs have been forced to cut their women's youth training teams or scale them down because of funding issues, and this has limited the amount of support available for the senior teams.

On an international level, Chinese women's football also faces a lack of external support and exchange. As there is a low focus towards women's matches, clubs do not have the means to fund international exchanges or invite foreign instruct. This scenario is a sharp contrast to the case in nations like the US and France which boast of strong women's football since a number of top international players and coaches are engaged in the leagues. The matches are widely covered by the mainstream media and generate viewership and thus sponsorship and investment for the teams which creates a positive cycle. In contrast, the lack of funding and social interest in Chinese women's football hampers its ability to form such resource advantages, leading to stagnation in players' tactical development.

Overall, Chinese women's football has long struggled in an environment lacking social support and attention. Improving its development requires comprehensive investment in policies, markets, and societal support. With these supports, women's football's access to equal resources and opportunities will be ensured as men's football.

It wasn't until the 2019 Women's World Cup that people's interest in women's football was reignited. This time, the Chinese women's football had an unwavering hope as well as unprecedented backing. The Ant Foundation started fundraising with the intention of providing substantial funds worth around 1 billion yuan specifically for players' welfare and training facilities. The Wind Rider did not have an advertising sponsorship which was the interesting part of the entire program because Alipay refrained from naming rights and all commercial rights, which was non-partisan to promote the development of Chinese women's football.

The origins of the "Wind Rider" can be traced back to 2017 when Ant Group sponsored the women's football team in Qiongzhong, Hainan, known as the Qiongzhong Women's Football Team. In 2019, the Ant Foundation initiated the "Wind Rider—Rural School Girls' Football Support Project" (also known as "Wind Rider Campus"), with the initial plan to sponsor 100 girls' football teams in rural schools over five years. The project aimed to provide funding and sports training resources to empower school-based girls' football programs, helping rural girls gain confidence, develop character, and access diverse developmental opportunities through football. By 2022, under the efforts of the Ant Foundation and the China Youth Development Foundation, the program had directly funded 60 girls' football teams in primary schools across 23 provinces. Additionally, 200 rural schools benefited indirectly from the initiative. The project collectively assisted 4,129 rural girls playing football. Among the 783 "Wind Rider Girls" who have graduated, 283 continued playing football after entering junior high school, 119 gained admissions to better junior high schools due to their involvement in football, and 30 entered sports schools or youth training centers for professional football training.

"Wind Rider" supports rural campus girls' football (*Source* Internet)

In rural women's football, a lack of training equipment has been a significant barrier. To counter such issues, the Wind Rider initiative supplies materials that are absorbed. In 2021, Alipay, along with the Chinese Football Association and the Chinese women's national football team, implemented the "Rose Dream Stadium" in order to create real football fields for girls who live in Far Away places of China and enable them to realize their football dreaming career. The project, which was made possible by the funding of the Ant Public Welfare Foundation, was undertaken as the Chinese women's football team was also out on international tours, mobilizing the goodwill of millions of online fans to come and help build provisions for football pitches for girls in the rural areas. The initiative not only articulates the fighting spirit of the Chinese women's football team but also motivates more girls to follow them and go for their own football dreaming career while doing their very best in order to achieve better things in life. Thus far, the project has built seven Rose Dream stadiums in areas such as Yunnan, Sichuan, and Hunan.

The construction of the Rose Dream Stadium began in the remote Shilin County of Yunnan, an area surrounded by mountains where children's football dreams have been hindered by a lack of infrastructure. In the absence of a suitable training base, these kids were forced to use the rough dirt hills as their practice fields. However, this was improved by the construction of the Rose Dream Stadium. A manufactured football pitch was built by Ant Public Welfare Foundation for the girls' team at Shilin, complete with artificial grass and necessary utilities.

Another Rose Dream Stadium was built in the central primary school of Heyue Town, Shuangfeng County, Loudi, Hunan. This rural school, located in a mountainous area, established a girls' football team two years ago but had to train on a dirt slope next to the school due to the lack of a formal field. The school's coach stated that the children even with those conditions training, still had a love for the game. The Chinese Football Association and Alipay acted quickly to ensure the construction of the Rose Dream Stadium for the central primary school in Heyue Town after they understood the requirements of the school, with the goal to allow these young girls to train and develop properly. Likewise, Alipay collaborated with the women's football team of China in an effort titled "Planting Seeds for the Future of Chinese Women's Football" which was used across the internet prior to Women's World Cup and aimed at raising the awareness of the society on the growing need for attention to be paid towards rural women's football. In the campaign, users could help the women's team by solving puzzles in the Alipay membership channel for which they earned the right to symbolically "plant" a patch of digital turf in the Rose Dream Stadium, in efforts to create more fields for women players in inaccessible regions. This concept of virtual engagement was really appealing as 300,000+ people sent their support and wishes towards rural women's football.

Boya Wang, age 10, voiced her joy after using the new field and said that with the new stadium, they know they are safely playing without the need to be concerned about the uneven surface and weeds, hence this helped them train better. Today, the Rose Dream Stadium has been the goal-post for the dreams of rural girl football and new stadiums are taking place in Yunnan, Sichuan, Hunan, and many more. Every field has the laughter and sweat of children in preparation and is a manifestation of the society's love and support for women's football in rural areas. Through the Rose Dream Stadium, the Ant Public Welfare Foundation not only helps these rural girls fulfill their dreams of playing football but brings them hope for new things, refreshing the women's football in rural areas.

The Alipay activity interface for "I Plant a Patch of Turf for the Rose Dream Stadium" (*Source* Alipay)

Rural women's football has long lacked opportunities for high-level competition, and the Wind Rider initiative aims to create these opportunities. In 2023, under the guidance of the Hangzhou Asian Games Organizing Committee, the China Youth Development Foundation, the Zhejiang Provincial Committee of the Communist Youth League, and the Zhejiang Ant Public Welfare Foundation jointly hosted the "Finding 2022 Asian Games Dreams | Rural Campus Women's Football Dream Fulfillment Action—2023 First Wind Rider League Finals." As the first national rural campus women's football league, the Wind Rider League drew in 35 rural campus teams from all over the country. After 71 matches held in three major areas, six teams from Guizhou, Xinjiang, Jiangxi, Hubei, Anhui, and Yunnan managed to grab victory flags and made their way to the Huanglong Sports Center, the host for these teams to shine before the Asian Games football matches.

The 2023 First Wind Rider League Finals successfully concluded at the Huanglong Stadium

12-year-old Yang Meng from Xuanwei, Yunnan, earned the title of "Best Player" due to her outstanding performance. For her, this competition was not merely an athletic contest but a milestone in her biography. All the feeds and shots she executed on the field were all done with the assurance and the ardor in her heart. She said that this occasion was the first in her life to be on board a plane and the first time to take part in a national competition where she experienced the notion of competitiveness and the strength that comes with teamwork between different regions of the country.

During the post-match activities, Yang Meng had the chance to meet her idol, Chinese women's football star Wang Shuang, making her dream come true. Wang not only encouraged Yang Meng but also shared some practical football tips, which greatly inspired her. She recalled, "I gave it my all on the field because I wanted more people to see our hard work and dreams." This experience strengthened Yang Meng's determination to pursue her football dreams, and she hopes to become an excellent player like Wang Shuang in the future, representing her country in international competitions.

Chinese women's football star Wang Shuang made a special trip back from Europe to cheer for the participating rural girls' teams at the competition (*Source* Wind Rider)

Through the Wind Rider League, Yang Meng and her teammates not only showcased their skills, but also learned to cooperate and built significant relationships with others which boosted their self-esteem. They then understood that crossing the ball over to the other side or kicking to where the goal is located is not merely a game, but a lifestyle with countless opportunities.

In October 2021, a milestone for the "Wind Rider Girls" was marked when, Wang Jiayue, one of the participants from the first batch of schools selected by the Wind Rider was seen on the large screen at the San Siro Stadium in Milan, Italy. In the recent May finals, she was noted for Winning the championship and being Named the Best player in the Southwest region of the Wind Rider League. Wang Jiayue also hoisted a similar trophy that was held by Euro 2020's top goal scorer, Cristiano Ronaldo. Her touching and inspiring tale of escaping the mountains with football gripped fans around the world during the UEFA Nations League final.

As the Chinese women's football team continues to grow and develop into a more competitive side on the international stage, the Ant Public Welfare Foundation has come up with new provisions to enhance its support for overseas activities. Together with Alipay and the Chinese Football Association, the foundation built an overseas base for the women's team to prepare them for stronger competitions in the run-up to the Asian Games and World Cup and in the process hone their competitive skills. This initiative also aims to expand the influence of women's football, helping to build a positive public image and role models for the sport.

The story of Wang Jiayue, a "Wind Rider Girl" from Bijie, Guizhou, was featured on the big screen at the UEFA Nations League final (*Source* Wind Rider)

The Wind Rider initiative not only provides material support for rural women's football but also plays a significant role in promoting sustainable social development and achieving corporate social responsibility (ESG) goals. According to the social dimension (S) of ESG, the Wind Rider initiative empowers rural girls by improving educational and sports opportunities, helping them overcome gender and regional limitations. This initiative directly aligns with the United Nations Sustainable Development Goals (SDGs), specifically Goal 5: "Gender Equality" and Goal 4: "Quality Education," by providing equal educational and growth opportunities for rural girls through sports.

The ongoing investment in the Wind Rider initiative has improved sports infrastructure in remote areas, such as the construction of the "Rose Dream Stadium," which directly creates training and development spaces for rural girls. Moreover, the league not only offers high-level competition opportunities but also encourages greater social participation, increasing awareness of rural women's football. These efforts contribute to reducing gender inequality, promoting social inclusion, and conveying positive social values through sports culture.

In the long term, the Wind Rider initiative not only helps a generation of girls pursue their football dreams but also fosters a social culture that values equality, inclusion, and diversity. This philanthropic practice, mediated by sports, not only meets the social responsibility requirements of ESG but also promotes ongoing societal attention to issues of gender equality and youth development.

8.2.2 Cyber Mulan for Female Entrepreneurs

Globally, women lead approximately one-third of enterprises. In developing and transitional economies, however, businesses owned by women are typically small and often operate in the informal economy. It was estimated that women comprised 56% of those who were not included in the digital economy in 2021. Feminine entrepreneurs face more challenges than their male counterparts when it comes to entering the business space or convincing investors. The poorest population of 783 million people, of which nearly 600 million are women, was identified in the Global Gender Gap Report of information available in December last year as released by the World Economic Forum last year. In the age group of 15–64 years, the female employment ratio is 55 whereas the figure stands at 78 for males—a gap of 23 percentage points. Women continue to be underrepresented in almost all emerging industries, particularly in the data and artificial intelligence industry where 26% are represented as males and in cloud computing industries with the figure being only 12%.

The "AI Bean" initiative, in particular, was undertaken in 2019 with the purpose of looking for ways of enhancing the digital space for women. In an event organized by Alibaba for its poverty alleviation campaign on July 16, 2020, Alibaba and Ant Group announced the Cyber Mulan initiative. The plan intends to help 50 million women worldwide in five years by improving their skill sets to make them more contributive and competitive in the digital sphere. In 2021, the Cyber Mulan Employment Special Training Fund was launched and devoted to furthering the employment of women through digital techniques. The Cyber Mulan initiative is one of Ant Group's much broad initiatives aimed at the development and promotion of women's entrepreneurship and employment with two features of security and accomplishment as fundamental principles behind its design.

The program offers multi-layered support from skills acquisition to business acumen development and access to finance and practical tools and technical advice to women. In a bid to increase the share of women in high-tech jobs, Cyber Mulan adapts some of its positions like artificial intelligence data annotators to women. This role reduces operations and communication complexities by offering specialized AI solutions that enable women in remote parts of the world to work online from their homes. It has generated many local jobs for women and has greatly increased their earnings in places such as northern Shaanxi province in China.

Additionally, Cyber Mulan offers systematic resource support for female entrepreneurs. Compared to their male counterparts, women often face greater barriers to market access and funding. Cyber Mulan's answer to this is the provision of training grants and small low-interest loans which ease the financial burden on female entrepreneurs. For example, Ant Group has provided zero-interest loans to women-owned businesses as well as startup subsidies. At the same time, the program's special fund supports women at all stages of business development, providing assistance from market research to advertising and thereby facilitating women in a dynamic environment.

"Security" serves as the foundation for women's upward mobility and includes the right and opportunity for women to access educational and medical resources. Over the years, the Ant Foundation has consistently supported the "Go for It Mulan"—a women's public welfare insurance project, collaborating with partners like the China Women's Development Foundation. Aligning with the national healthcare and education security realities, the project donates "education + medical" public welfare insurance to women living in poverty in underdeveloped counties, reducing the incidence of returning to poverty due to illness or education costs and aiding in improving the educational and medical standards in less developed regions. By the end of 2023, the "Go for It Mulan" project had benefited more than 4.6 million women across 57 counties in 17 provinces nationwide.

Decent jobs and satisfactory incomes are essential avenues for helping women realize their self-worth and enhance their sense of achievement. In many rural areas, female laborers bear significant responsibilities for their families and local economic development, yet they often face a shortage of non-agricultural employment opportunities. In this regard, the Ant Public Welfare Foundation, in unison with Ant Group, initiated the Cyber Mulan Employment Training Program. The program intends to seek out new types of work suitable for women in rural settings, train them on the skills which will be in demand, improve their digital skills, and develop ways of searching for jobs that are relevant in these underdeveloped regions. The purpose is to enable rural women to work as AI trainers, online customer care agents, management of rural guesthouses, and community services allowing for sustainable employment and development within their communities.

Through partnerships with various social organizations, the program has supported thousands of rural women in acquiring digital skills and achieving stable incomes in their hometowns. In regions like Gansu and Shaanxi, the prospects for many women to earn over 3,000 yuan monthly from jobs have resolved some of the "left-behind" problems brought about by labor migration while fostering rural advancement. By moving out of the four walls of their homes, these rural women not only have economic self-sufficiency but also improve their family status and self-identity.

Cyber Mulan also focuses on comprehensive skill enhancement through programs like the "Mulan Housekeeper Training Course," which offers integrated training in service quality, specialized operations, and digital skills, enabling rural women to play a more significant role in the tourism industry. Additionally, the program provides loans and resource support to low-income women, encouraging entrepreneurship and facilitating the marketization of local specialty products, such as intangible cultural heritage crafts, further supporting women's economic development and self-actualization.

When the "AI Bean" initiative was implemented in Jishixian County, Gansu Province, Zhang Juan, born in 1999, became an AI trainer. Now, she is the main breadwinner of her family, able to work locally while earning a relatively high income. It allows her to stay close to her family and gives her the confidence to plan for a better future. In

2023, Zhang Juan participated in the World Artificial Intelligence Conference as an AI professional. By the end of 2023, the "AI Bean" initiative had supported the establishment of 17 county-level digital employment centers in Shaanxi, Gansu, and other regions, facilitating local employment for over 5,800 people. Approximately 70% of these employees are women, making several of these centers the largest employers in their respective areas.

With the growth of rural tourism, homestays have become a new hotspot. In 2022, the Ant Foundation began exploring a training program for homestay managers targeting rural women, in collaboration with the China Rural Revitalization Foundation. Based on past practices, in 2023, the Market Management Department of the Ministry of Culture and Tourism, along with the Ant Foundation, launched the Cyber Mulan Homestay Manager Training Program, implemented jointly with the China Rural Revitalization Foundation. The program offers free training to homestay service providers, primarily rural women, with classes divided into "Mulan General Classes" and "Digital Enhancement Classes." These classes provide different types of training aimed at enhancing the policy awareness, basic service skills, specialized service skills, and digital literacy of "Mulan Managers." This initiative aims to cultivate a group of more competitive homestay managers equipped for the digital age, helping rural women seize employment opportunities brought by tourism and increase their income. By the end of 2023, nearly 4,000 rural homestay managers had been trained across 14 provinces, including Gansu, Sichuan, and Guizhou.

Data released by MYBank suggests that 47% of their micro-enterprise female clients apply for credit on five occasions or more. In light of the great need for male and female employment and entrepreneurial activities, MYBank, which is an online bank for micro-enterprises, has extended its facilities to over 17 million females since it started in June 2015. Every year over 5 million female micro-entrepreneurs are provided with business-based financial solutions like deposits and saving services, wealth management, and loans. In July 2020, MYBank launched the Cyber Mulan Entrepreneurship Initiative with the aim of providing financial assistance and business support to female entrepreneurs. People like Guo Juan and Niu Qinghua got interest-free loans amounting to 200,000 yuan for one year through this initiative.

In 2023, the Cyber Mulan Entrepreneurship Initiative was further upgraded, including loan interest reductions for one million women and providing 100 annual interest-free loan slots specifically for women. As a representative of female entrepreneurs served by MYBank, Guo Juan shared her entrepreneurial journey at the conference. After marrying and moving to Gansu, Guo Juan started learning to weave handmade hemp shoes from scratch with her mother-in-law. Her invention of "waterproof hemp shoes" made her Taobao shop an instant hit. In 2018, she had the idea to open physical stores in various scenic areas to boost sales of the handmade shoes. However, setting up these stores required deposits ranging from 50,000 to 100,000 yuan. With the help of a loan from MYBank, she opened her physical store. Over the years, the offline store's revenue has surpassed that of her online shop. Throughout her entrepreneurial journey spanning over a decade, Guo Juan has sold local hemp shoes across the country and even to Southeast Asia. Additionally, because of her efforts to re-employ women in her community, she received a 200,000-yuan interest-free business loan from the Cyber Mulan Entrepreneurship Initiative to purchase silk and land.

In conclusion, Cyber Mulan succeeds at uplifting women with the help of providing digital skills training, resources, and opportunities to earn online. The program aims at imparting new work skills to women in the productive forces of the digital economy, such as artificial intelligence, online customer support, and rural tourism. The other most important option is for underdeveloped, rural areas—jobs like AI data labeling, or online customer service provide rural women with work opportunities in their vicinity and make them more competitive in the digital economy.

The support to women with acquiring digital skills and access to entrepreneurial tools by Cyber Mulan takes them a notch higher concerning income and standard of living which in turn enhances their economic independence. By acquiring and improving digital skills and receiving employment assistance, women are removed out of poverty, and at the same time, these changes enable them to adjust their roles and identities in the family and in society. This approach not only boosts women's sense of achievement but also encourages more women to engage in emerging technologies and economic fields, expanding their career possibilities. As a flagship initiative for corporate social responsibility, Cyber Mulan actively responds to sustainable development goals such as "gender equality" and "decent work and economic growth."

Cyber Mulan has made important contributions to the "Social" and "Governance" aspects of ESG (Environmental, Social, Governance). The program helps to ensure gender equity and empowers more vulnerable groups by providing them with relevant digital skills and learning how to work online. These initiatives increase the competitiveness and professional qualifications of women in rural areas and enable them to join such sectors as AI, customer service, and rural tourism thus improving the economic status of their families and enhancing their status in the communities. By integrating goals like gender equality and social equity into its strategic planning, Cyber Mulan demonstrates the company's commitment to diversity and inclusion, establishing a brand image that reflects greater social responsibility. The promotion of Cyber Mulan not only helps the female community but contributes to a healthy social value of the company and builds a responsible image of governance among its stakeholders such as the government, social institutions, and the consumers.

8.2.3 Blue Vest for the Elderly Population

Digital and intelligent services have profoundly transformed daily life. However, this shift has posed adaptation challenges for many elderly individuals, who increasingly face the reality of a digital divide in their everyday lives. Older people experience many problems in the information society due to lower digital competences. They can be able to adjust to electronic payments, smart self-care systems, online services, or information retrieval systems even with the help of accessible collections of such resources. Because of this situation, both the state and different social sectors are making efforts to promote age-friendly information solutions. But even eased computer usage among the elderly will not help more than a few users: more efforts should be placed on information skills training and education for older persons so that they can cope with the demands of the information age.

Specifically, one of the main challenges elderly people face in the information era is the use of smart devices. Many smartphones, tablets, and various software applications are not designed with older users in mind, making complex operation processes and small fonts obstacles to their adaptation. For instance, in the realm of electronic payments, many public services and merchant payment systems have since adopted mechanisms that are more favorable to the youthful population, opting for the use of QR code scanning or mobile payments and leaving older people who are not mobile payment users disadvantaged. Therefore, this has created problems of "accessibility" to other essential activities such as shopping, healthcare services, and transportation.

Moreover, information security has emerged as a significant risk for elderly individuals during the digital transformation. Due to a limited awareness of identifying online fraud and protecting personal information, they are particularly vulnerable to scams. Telecom fraud has become rampant over the past few years against older people as criminals capitalized on these individuals' gullibility to bilk them through fake investments and fake health product advertisements leading to very big financial losses to these victims. Being low-tech literate, use of elderly does not have the capacity to perceive fraudulent information and conspiracies aimed at scamming them greatly hence the sanitization and training of the elderly on information security and scam prevention strategies is essential and needed immediately.

What's more, as a result of the rising utilization of digital services, there has been a gradual decline or complete loss of numerous offline services such as those in banking, health, and public service. With the transition of the majority of the services to the online realm, older adults become used to this service model with great difficulty. In healthcare, for example, a number of hospitals started using the system of booking appointments via the Internet, for the elderly patients who are not familiar with how to operate the devices or have no devices, this becomes a challenge. Similar patterns can be observed in public transport where some cities abandoned ticket sales by a person for scanning of QR code making transport trips a nightmare for the elderly.

To address the issues of elderly individuals in rural areas who are hesitant to use smartphones due to fear, lack of knowledge, and insufficient guidance, the Nandu Big Data Institute and the Aging Society Governance Strategy Research Center of Nankai University, in collaboration with the China Aging Industry Association, conducted a survey and released the "Survey Report on Digital Literacy Differences Among Urban and Rural Elderly" (hereinafter referred to as the "Report"). The report reveals that nearly 30% of surveyed rural elderly people are "afraid" to use smartphones, citing three main reasons: "worried about bothering others," "afraid of being scammed," and "concerned about safety." Nearly 40% of the surveyed population stated they can only operate three features on the smartphone while around 22.77% can operate more than ten features. Moreover, more than 40% of the rural elderly respondents stated that they do not engage in mobile payments and cash is the preferred method of 66% of them.

Accompanying this fear is a tendency to become "addicted" to mobile devices. The report highlights that over 80% of surveyed rural elderly individuals have had their lives affected by excessive smartphone use, with nearly 60% stating they often spend more than three hours watching short videos. Furthermore, 40% of rural elderly respondents frequently forget their tasks due to mobile phone distractions.

On the other hand, retired people living in rural areas are at a clear disadvantage when it comes to exposure to the negative effects of the Internet, such as speculative rumors, phishing scams, aggressive advertising problematic for consumers, network marketing schemes, and rebate scams. In rural areas, nearly 60% of the respondents acknowledged the rumors and thought there was some truth to them, while a similar claim among urban elderly was about 40%. In the given context, almost 60% of the surveyed population expressed the need to design an age-matched tier-based digital literacy program for the elderly.

In summary, elderly individuals are facing various inconveniences and challenges due to the digital divide. We can offer educational programs and improve their informational literacy which will allow them to embrace a digital life in a better manner and help them in achieving better social engagement. Making progress with the provision of access to age-friendly information is not only a matter of social obligation but also helps in enhancing equity in treatment and safeguarding of the rights of aged individuals within the information society, which in turn leads to balanced development across society.

In 2020, the Blue Vest initiative was launched under the guidance of the National Anti-Fraud Center, the Department of Aging Health Services of the National Health Commission, and the Anti-Fraud Center of the Ministry of Industry and Information Technology. Initiated by Ant Group and the Ant Foundation, the public welfare campaign involves community lectures, public exhibitions, and outreach via mobile caravans. Its aim is to assist seniors who are hesitant or unable to use smartphones, helping bridge the digital gap and allowing them to enjoy digital life. The initiative conducts "Financial Fraud Prevention" seminars, sets up "Digital Support for Seniors" volunteer stations, and organizes "Community Living Festivals" in multiple cities across the nation to contribute to senior-friendly digital support and safe fraud prevention efforts.

The Blue Vest assists elderly people in preventing scams (*Source* Internet)

The Blue Vest campaign's reach increased even further from 2020 to 2023 when it covered 25 provinces and a hundred plus cities nationwide conducting more than 30,000 events and reaching out to over 390,000 people. The initiative adopted a new approach to its outreach program by organizing a "Blue Vest Caravan" which resolved the last mile of the elderly people's access to technology by making sure that a digital outreach is conducted for both urban and rural areas It also began promoting the establishment of online services in a developed inclusive manner. In 2021, the "Blue Vest" mini-program which is a supportive volunteer platform uplifted most seniors as it was a video tutorial that educated seniors on app usage and fraud prevention.

Starting in 2021, Alipay introduced a dedicated hotline, "Warm Hotline," for users aged 65 and above, enabling them to quickly connect with human operators. Su Mengqi, a customer service representative for Warm Hotline, noted that while simple queries might take just a few minutes for regular users, consultations with seniors aged 65 and above typically lasted over 30 min. Accustomed to fast-paced interactions, Su adapted to the slower pace required by the Warm Hotline. In 2022, Ant Group launched the "Digital Elderly Assistance Lab" under the Blue Vest, introducing VR anti-fraud machines that demonstrate various scam techniques through interactive VR experiences. It helped both young and old understand common telecom fraud tactics. The effort also formed five fixed elderly lecture teams, creating a "Silver Bells Mutual Assistance Model," where seniors help each other, fostering mutual understanding and allowing seniors to rediscover their own values.

In 2023, the Blue Vest saw a "Service+" upgrade, signifying both enhancement and connectivity. It aims to integrate more partners and volunteers to offer holistic services addressing the four aspects of seniors' lives: health, food, housing, and travel. For example, it renovated rest corners for the elderly in restaurants and community bookstores; the nation's first anti-fraud restaurant has already opened in Kunming. Regarding health, the effort collaborates with hospitals and Alipay's health services to conduct regular medical science popularization, guide digital medical visits, and invite medical professionals to debunk myths and educate on health topics within communities, helping seniors identify fraudulent health products.

The success of the Blue Vest lies in its innovative promotional methods, the creation of a seamless offline and online dual-platform, and a diverse volunteer corps, including the unique "Silver Bells Mutual Assistance Model." Through this way, Ant Group explored a new model of "technology + service" philanthropy. It demonstrates how to assist the elderly in bridging the digital divide, offering a reference point for solving entrenched problems in people's lives and addressing new issues arising from development processes. Bridging the digital divide benefits everyone, contributing to societal well-being. Everyone can put on a "blue vest" to help the elderly find their place in the digital world.

Considering its effectiveness, Blue Vest has assisted seniors in overcoming the digital divide by conducting numerous community initiatives, combining online and offline channels, and introducing new service models that allow these seniors to better fit into the modern information society. As part of its public welfare model, the organization establishes volunteer service stations with the title of "Digital Assistance for Seniors", trains different groups of elderly people, markets educational anti-fraud campaigns, and runs the Blue Vest mini-program, which aids older individuals in gaining smartphone familiarity and increasing their knowledge about telecommunications scams. Specifically in anti-fraud education where the national program uses "Silver Bell Mutual Aid" to empower senior volunteers to assist their fellow seniors in a more relatable and informal manner to help a majority of seniors to let go of their fear towards technology and eventually have a safe, reassuring experience online. All of these initiatives effectively increase seniors' comfort with digital devices and their perceived ease of use allowing the older generations to manage a number of digital solutions on their own.

Blue Vest also actively promotes corporate ESG strategy practices. Under the concept of "community," supportive networks for the elderly population are developed. Blue Vest's activities not only raise the digital competence of senior citizens but also do embed social responsibility as a priority within the corporate governance framework, which makes it quite an exemplary case under ESG governance. In addition, Blue Vest addresses and supports UN Global Goals focusing on "Reducing Inequality" (SDG 10), and "Decent Work and Economic Growth" (SDG 8). With the help of Blue Vest, more people can enjoy the benefits of technology and the elderly people can enhance their quality of life, which makes them feel more confident and valued. This inclusive program aimed at specific vulnerable groups provides a concrete practice for building a more inclusive and equitable digital society and serves as a valuable reference for the digital integration of more groups in the future.

8.3 "Cyber Mulan" Empowers Women's Philanthropic Outreach Globally

According to PwC's 2024 Global Women in Work Report, between 2021 and 2022, the gender pay gap widened in 20 out of 33 OECD member countries. It indicates that progress towards workplace gender equality remains slow, with achieving global gender pay parity still projected to take over 50 years. Over the past decade, the average score on the Global Women's Work Index rose from 56.3 in 2011 to 68 in 2022. The average score for OECD countries increased from 66 in 2021 to 68 in 2022. The majority of the progress in the Women's Work Index in OECD countries from 2021 to 2022 was driven by increases in female labor force participation rates (from 70.8% to 72.1%) and decreases in female unemployment rates (from 6.4% to 5.3%). However, the average gender pay gap widened from 13.2% to 13.5%, indicating that despite improved participation, women remain disadvantaged in terms of earnings in the labor market compared to men.

Employment is one of the key components found within the 'S' (Social) of the ESG metrics as it relates to encouraging talent growth towards sustainability. Ant Group has shown its willingness to foster employment programs targeting at-risk groups and women's technical skill development for better employment opportunities. Nevertheless,

ESG is not limited to the domestic ones. Companies that wish to extend operations to foreign markets must now also bear more social responsibilities by embracing institutional contexts and gaining legitimacy in foreign countries. Of the above initiatives, it is only Cyber Mulan that has so far taken its charitable activities abroad.

In 2022, Ant Group entered into a partnership with the UN Women to premiere the "Together Digital," which has been Ant Group's debut global charitable initiative. This partnership of five years duration will commence with a pilot phase lasting 27 months to be conducted in two countries China and Indonesia, and target women entrepreneurs in small enterprises operating in the digital age. The program provided training modules for women-led small businesses in Indonesia. At present, training camps have reached out to 2056 women entrepreneurs directly while 439 beneficiaries were enlisted in the mentorship program. Support for the camps was also obtained from the government of Jakarta and West Nusa Tenggara Province. UN Women carried out transnational studies on women-led business in China, Indonesia, Thailand, Rwanda, and Uganda, and launched the paper in June 2023 with more than 20 key actors taking part in the documents.

In all sectors, particularly in the formal sector, women's employment rates in Indonesia are significantly lower than those of men. Although the civil service promotes gender equality without discrimination, women in other jobs still face disparities in terms of income, benefits, opportunities, and career progression. Data shows that women earn less than men across all sectors. Recent public unemployment figures continue to show a higher number of unemployed women. Even when women are employed, they often work unpaid (71.5%). In 2005, male unemployment was 9.3%, while female unemployment was 14.7%. By 2007, male unemployment had dropped to 8.1%, and female unemployment to 10.8%. Despite women's active efforts to enter the labor market, registered job openings predominantly favor men (88.67% for men vs. 69.38% for women). Women's income accounted for 29.2% of total income in 1990, 28.3% in 2002, and 33% in 2009. Achieving the Millennium Development Goal of equal income share by 2015 seems distant, requiring concerted efforts to materialize.

Beyond supporting women's employment and demonstrating corporate social value, it presents an opportunity for Ant Group to break into the Indonesian market. Given the dominance of Shopee, Lazada, Tokopedia, and TikTok Shop in e-commerce, and J&T Express in logistics, starting a new major e-commerce platform or breaking into logistics is challenging. Initiating from a philanthropic standpoint, providing microloans to support entrepreneurship and employment is an effective way to enter the market.

8.4 International Experiences in Elder Care Welfares

Based on recent social trends and the focus of the United Nations on women's issues, Ant Group has extended its "Cyber Mulan" initiative globally. However, this is the only project that has been internationalized so far, a fact that shows how far Ant Group has remained in charity work so far. This section will examine the challenges that the elderly suffer in the digital context and try to gain help from international experiences in planning future international charity activities.

As the United Nations Chronicle states, the "silver economy" has a value of $15 trillion, while the global market value of eldercare technology is higher than $13 billion. Governments and private organizations agree on the fact that older people for example in developed countries should be viewed mostly as a market rather than a burden as it is particularly appealing to some industries, especially the technological and digital industries. The Organization for Economic Co-operation and Development (OECD) suggests that establishing a working-age population that includes the younger and older generations would help per capita GDP improve by an additional 19 percent over the next thirty years. Thus, it is imperative that every stakeholder takes full advantage and grabs the opportunities brought by the digital revolution in changing societies.

The main challenges faced by the elderly in the digital domain include:

1. Low percentage Internet and smart device penetration rate of elderly people among themselves. For instance, in Brazil, 85% of young people have smart mobile phones whereas only 32% of the old people own mobile phones. In comparison, in the Philippines, 74% of the youth have smart mobile phones, nevertheless only 27% of the elder generation possess such devices.
2. Minimal adoption and negative attitude towards the adoption of digital products and services amongst the elderly. As it is highlighted by the Index of the Tomorrow's EU Relies on Digital Economy &

Society Report, only 35 percent of 55–74 aged people in the EU have the required level of basic digital skills. Pew Research Center found that 32% of Americans who don't use the internet say it's because it's too difficult to grasp, with 8% citing that they're "too old to learn."

3. Vulnerability to online scams and difficulties accessing convenient government services.

What international experiences are worth learning from to solve these issues?

1. For the issue of low penetration rates, the U.S. Senate passed the Digital Equity Act, allocating annual funding totaling $250 million to expand broadband coverage and help states implement digital equity plans. In Chile, organizations like Fundación Conectividad launched a "Telecommunications Connection" campaign, distributing specially configured smartphones to 1,750 elderly individuals. In South Korea, the Seoul Metropolitan Government and telecommunications companies offer a monthly plan for KRW 20,000 for citizens aged 65 and above. Singapore implements the Silver Infocomm Access Program, allowing low-income seniors to purchase a smartphone for $20 with a monthly internet fee of $5 for two years.
2. To tackle usage rates, Singapore's Digital Office launched comprehensive policies targeting the elderly, such as setting up digital service counters, appointing digital ambassadors, and conducting digital supermarket tours. Australia's Konnekt offers a video calling app specifically designed for seniors, allowing login with just a name and providing subtitles. Japan focuses on developing and promoting intelligent products for lifting assistance, mobility aids, walking support, automatic excretion handling, health monitoring, and wandering monitoring. In South Korea, the Seoul Metropolitan Government and companies have co-developed user-friendly unmanned terminals for digitally disadvantaged groups.
3. To address digital literacy, Singapore's Digital Office provided training services for over 100,000 seniors on video calling, accessing government digital services, and online payment, with more than 250,000 sessions. In France, the city of Nantes distributed 2,000

> free training vouchers for teaching specific groups, including seniors, how to set up email accounts, install printers, and fill out online forms. Chile's National Service for Older Adults implemented the "Elderly Digital Integration Program," establishing 32 senior digital centers across 16 regions equipped with 200 computers to teach seniors how to browse the web, manage personal data, and edit documents. In the U.S., the Senior Planet Online Centers have established over 30 learning branches offering computer and internet education services for adults aged 50 and above. In South Korea, the Seoul Digital Foundation started a digital skills training course using teaching robots, offering free training in five districts for residents of 17 welfare facilities for the elderly.

The elderly population in China faces similar challenges to those around the world. Most digital support for the elderly is implemented by the government, with limited corporate philanthropy. Additionally, "Blue Vest" proposes an innovative "caravan promotion" approach based on China's reality, which has not yet appeared in international experience. The mobile promotion method could serve as a bridge for building international philanthropic efforts. Drawing on international experiences and practices could better align with international philanthropy standards.

8.5 Conclusion

The "Wind Rider" supports girls' football teams in remote areas, encouraging more girls to participate in football, aligning with Quality Education (SDG 4) and Gender Equality (SDG 5). The "Cyber Mulan" is dedicated to enhancing women's competitiveness and autonomy in the digital economy era. By providing training resources and entrepreneurial support, it helps women acquire digital skills, thereby narrowing the gender digital divide globally. This initiative contributes to Gender Equality (SDG 5) and Decent Work and Economic Growth (SDG 8). The "Blue Vest" is a social welfare project aimed at helping the elderly bridge the digital gap. Through organizing various activities, providing digital skill guidance, and establishing integrated online and offline platforms, it assists the elderly in adapting to a digital lifestyle. The program works towards Reduced Inequalities (SDG 10) and Good Health and Well-being (SDG 3).

The "Wind Rider," "Cyber Mulan," and "Blue Vest" are three remarkable programs started by the Ant Group to meet its social obligations. These initiatives address different social issues but collectively promote the principles of ESG (Environmental, Social, and Governance) and the United Nations Sustainable Development Goals (SDGs). Whether it's through supporting girls' participation in football to foster inclusivity and diversity, enhancing women's digital skills to narrow the gender gap, or aiding the elderly in adapting to the digital world to improve their quality of life, the philanthropic efforts aim to advance fairness and harmonious development in society. These programs not only show Ant Group's engagement as a technology firm in the advancement of society but also its appreciation of the responsibility and the mission to promote sustainable development through tangible measures.

CHAPTER 9

Sustainable Fintech for Data Privacy Governance in the Global South: Zoloz E-KYC Applications from China to over Ten Countries in Southeast and South Asia

In the age of digital transformation, the stewardship of data privacy has emerged as a universal challenge, transcending the realms of technology, economics, and social ethics. As the digital economy continues to flourish, the balance between protecting personal information and nurturing innovation and inclusive growth has become more critical than ever. This section takes a deep dive into the role of sustainable FinTech, with a particular focus on Ant Group's trailblazing endeavors in data privacy governance, particularly through the deployment of Zoloz e-KYC technology across Southeast and South Asia.

The story opens by acknowledging the growing importance of the Chief protection Officer in corporate governance, which reflects the commitment of companies such as Ant Group to making data protection a strategic priority. After that, it explores the technological innovations Ant Group has implemented to foster trust via technology, highlighting the interaction between technological and human endeavors to protect privacy.

As we progress through the section, we confront the myriad challenges that come with the global footprint of technology, from international regulatory demands to political sanctions and media scrutiny. Each challenge underscores the intricacy of data privacy in an interconnected

J. Sun, *Sustainable Financialization*, Sustainable Development Goals Series, https://doi.org/10.1007/978-981-96-3580-1_9

world, where technology must serve as a dual-edged sword—facilitating connectivity while ensuring security.

The crux of this section lies in Ant Group's philosophy towards technology globalization, where data governance is more than just a compliance requirement; it's a catalyst for innovation. The ZOLOZ Trusted Identity Authentication Platform and its e-KYC technology are highlighted as case studies that illustrate the delicate balance of keeping data invisible for privacy while making it visible for value creation.

The section also underscores Ant Group's role in fostering inclusive cooperation through technology's global reach, the collaborative development of international standards, and the open-source movement that propels industry growth. These initiatives are not just corporate pursuits but are pivotal in sculpting the global digital economy's narrative on privacy governance and sustainable development.

9.1 The Emergence of a New Domain in Corporate Governance: Chief Privacy Officers

In the fast-paced digital transformation landscape, safeguarding personal privacy has become a critical issue for businesses and regulatory bodies worldwide. For digital technology companies and industries, the ability to handle data responsibly and protect user privacy is tantamount to having a robust governance defense system, which is essential for sustainable development. In 2017, Ant Group took the lead in establishing the role of "Chief Privacy Officer" (CPO) and the formation of a dedicated "Privacy Protection Office" within the Chinese industry. This pioneering move not only reflects Ant Group's high regard for user privacy protection but also signifies that Chinese enterprises have begun to elevate privacy protection to a strategic level. The first step in creating a thorough and organized privacy protection management system is creating a privacy officer position in response to the management and use of large amounts of user data. Throughout the gathering, processing, storing, and sharing of user data, they make sure the company follows strict privacy protection guidelines and employs comprehensive data security measures.

The primary responsibilities of Ant Group's Privacy Officer encompass, but are not limited to: formulating and enforcing stringent privacy protection policies to ensure all business activities comply with international standards and local laws and regulations; organizing internal training to enhance the privacy protection awareness among all staff members;

conducting regular privacy impact assessments to identify and address potential privacy risks; maintaining close communication with domestic and international regulatory agencies and actively participating in the formulation of industry standards; additionally, they are responsible for establishing an efficient data breach response mechanism to ensure that in the event of a data breach, losses can be swiftly and effectively controlled.

9.1.1 Technology as the Guardian of Trust: The Driving Force of Technology and the Establishment of Behavioral Norms

"Technology as the Guardian of Trust" has always been the philosophy of personal information protection practiced by Ant Group. In 2022, Ant Group's Chief Privacy Officer, Nie Zhengjun, pointed out that technology-driven approaches are the only way forward for personal information protection. Therefore, this CPO places great emphasis on the value that technology brings to privacy protection. In 2017, after the establishment of the industry's first Privacy Protection Office, the first colleague recruited by law-background Zhengjun Nie was a technical personnel.[1] Today, 70% of the Privacy Protection Office members are technology professionals, ensuring that the team can better apply technology throughout the entire lifecycle of personal information protection. In 2021, Ant Group specifically established a Board Committee on Privacy Protection and Data Security to coordinate data security and privacy protection efforts at the corporate level.

To bolster technology research and foster collaboration, Ant Group has formed a Personal Information Protection Supervisory Committee. This committee is made up of legal and tech experts who act as an independent watchdog, keeping personal information protection practices in check and further strengthening privacy protection evaluation and oversight mechanisms. Furthermore, Ant Group is at the forefront of developing privacy computing technologies, championing standardization in this domain to achieve a higher caliber of data security protection.

Ant Group's "2023 Sustainable Development Report" indicates that over the past three years, Ant has consistently maintained a high level of investment in scientific research funding. In 2023, Ant Group's R&D

[1] 21jingji, 2022, https://m.21jingji.com/article/20221103/1bb779b2c0f5e87592c620c7a292cd3d.html.

expenditure reached 21.19 billion yuan, invested in areas such as artificial intelligence and data elements, and for the first time, linked ESG assessments to executive compensation.[2] This demonstrates Ant Group's strong emphasis on technology and sustainable development.

How can "Technology as the Guardian of Trust" be better implemented? Chief Privacy Officer Zhengjun Nie has emphasized the "Three Concurrents" principle. The "Three Concurrents" principle means that (1) privacy protection and data security management plans should be formulated concurrently with product design; (2) privacy protection and data security management measures should be implemented concurrently with product development; and (3) privacy protection and data security management functions should be launched concurrently when the product goes live. Privacy protection and data security accompany product functionality like a shadow, effectively avoiding product modifications or shutdowns due to compliance issues in the later stages, which not only reduces corporate costs, wins user trust, but also reduces legal risks, helping enterprises maintain compliance and security in the rapidly changing regulatory environment.

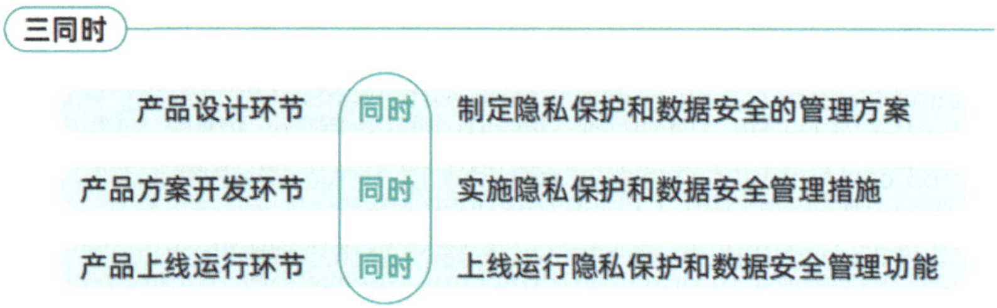

The "Three Concurrents" principle of Ant Group

[2] Ant Group, 2023, https://mdn.alipayobjects.com/huamei_f5zjmk/afts/file/A*_vcKR4qMeToAAAAAAAAAAAAADgx7AQ/%E8%9A%82%E8%9A%81%E9%9B%86%E5%9B%A22023%E5%B9%B4%E5%8F%AF%E6%8C%81%E7%BB%AD%E5%8F%91%E5%B1%95%E6%8A%A5%E5%91%8A.pdf.

9.2 Controversies and Challenges of Technology Going Global

9.2.1 International Challenges

In the global arena, conversations around data security and privacy protection are intensifying, presenting businesses with the urgent task of swiftly and adeptly recalibrating their strategies to keep pace with the tightening landscape of data protection. As data volumes swell, so too does the risk of data breaches. The IBM Cost of a Data Breach Report published in 2024 reveals that the average economic loss from global data breach incidents has climbed to $4.88 million, a figure that ups the ante for the professional capabilities of corporate cybersecurity teams. Compared to the previous year, the economic loss due to data breaches has seen a 10% increase, marking the steepest rise since 2020. Moreover, a staggering 70% of surveyed companies indicated that data breaches resulted in severe or very severe damage. The report also flags that data security vulnerabilities could heighten the risk of intellectual property theft. The associated cost of intellectual property theft has risen by nearly 11% year-on-year, with the cost per record now at $173. Especially with the spread of generative artificial intelligence technology and the infiltration of data and other highly proprietary information in hybrid environments, the protection of intellectual property faces greater challenges. As the application of critical data in various business environments increases, companies must re-evaluate the security protection and access control mechanisms for this data.[3]

9.2.2 Regulatory Compliance Pressure

The General Data Protection Regulation (GDPR), introduced by the European Union in May 2018, marked the beginning of a new era in data privacy. By limiting how businesses handle personal data, this historic law aims to protect personal information. The core idea of the GDPR is "data minimization," which forces companies to collect and use only the information necessary to achieve their particular goals, strengthening individuals' control over their personal data.

[3] IBM Newsroom, 2024, https://china.newsroom.ibm.com/2024-07-31-IBM-2024-,AI.

The GDPR empowers users with a range of rights, including the ability to access, amend, or erase their data, and to insist that companies halt the processing of their information. On top of that, the regulation has put in place a stringent protocol for reporting data breaches. In the event of a breach, businesses must notify regulatory agencies within 72 hours and, if necessary, notify the individuals whose data was compromised. In addition to encouraging businesses to strengthen their data security protocols in order to avert breaches, this required reporting guarantees that individuals impacted can promptly take remedial action.

When it comes to transferring data across borders, the GDPR raises the bar high. Companies that wish to move the data of European Union (EU) citizens outside the EU's borders must ensure that the destination country maintains a comparable level of data protection or implement adequate security measures, such as entering into standard contractual clauses or participating in frameworks like the Privacy Shield.

Companies like Ant Group risk severe administrative fines and harm to their reputation if they violate these strict legal regulations, which might have a disastrous effect on their financial performance and market ecological niche.

Along with technological improvements, the GDPR's compliance requirements also address organizational structure, internal management systems, staff training, and more. How to comply with international data and privacy standards while expanding internationally is an unavoidable and substantial barrier for Chinese businesses more accustomed to domestic data information security or privacy protection laws.

9.2.3 *Political Sanctions*

On January 5, 2021, former US President Trump signed an executive order prohibiting transactions through eight Chinese mobile applications, including Ant Group's Alipay and Tencent's WeChat Pay. The US government believes that Chinese technology companies pose a threat to US national security. "The relevant Chinese applications invade users' smartphones, tablets, and personal computers, collecting sensitive information involving personal identity and users' private life data." Entering the Biden administration era, this trend has continued and strengthened. President Biden signed the Executive Order on preventing the access of sensitive personal data of Americans and US government data by "countries of concern," which has explicitly listed China as one of them, prohibiting these countries from obtaining sensitive personal data of Americans or sensitive government data through investment, supplier relationships, and other means.[4] The executive order not only restricts the cross-border flow of data but also widely affects commercial activities involving potential cross-border data flow, posing stricter requirements for the operation of multinational technology companies in the United States.

9.2.4 Media Criticism

Ant Group, relying on its vast user base and advanced data technology, has built a complex ecosystem covering large-scale, diversified financial services such as payment, lending, and insurance. Taking payment services as an example, Alipay has cooperated with payment wallets in multiple countries overseas, such as Epay and Tinaba, while also actively exploring overseas business opportunities through the 2024 European Cup and the Olympic Games. However, it is this large-scale, data-intensive business model that has made Ant Group's data processing a focus of foreign concerns. Western media reports often amplify the potential issues of Ant Group in data security and privacy protection to question the sustainability and ethical compliance of its business model.

Foreign media and assessment institutions frequently report on the data exchange between Ant Group and its ecological partners, suggesting that it may infringe on user privacy and even imply that the data may be used for unintended purposes. For example, in a market research report provided by Oshum, data privacy protection, compliance with data regulations, and identity theft issues are considered weaknesses of Alipay.[5] Shazeda Ahmed and Adrian Fong question whether mobile payment systems can protect citizens' data, believing that their services may collect highly sensitive personal data.[6] In addition, an article published in The

[4] The White House, 2024, https://www.whitehouse.gov/briefing-room/statements-releases/2024/02/28/fact-sheet-president-biden-issues-sweeping-executive-order-to-protect-americans-sensitive-personal-data/#:~:text=The%20Executive%20Order%2C%20which%20marks%20the%20most%20significant,give%20those%20countries%20access%20to%20Americans%E2%80%99%20sensitive%20data.

[5] Osum, 2024, https://blog.osum.com/weaknesses-of-alipay/.

[6] Shazeda Ahmed andAdrian Fong, 2017, https://citizenlab.ca/2017/01/cashless-society-cached-data-mobile-payment-systems-protecting-chinese-citizens-data/.

Banker, Kimberly Long believes that unlike other Chinese companies facing stricter scrutiny (such as Huawei and TikTok), Alipay+ has not received the same level of significant regulation, and its international expansion should be given more attention by regulatory authorities.[7]

The perception in foreign media that Chinese companies have always had a close relationship with the government makes it difficult for Ant Group to shake off speculation about data flowing to the government. International media have repeatedly questioned whether the massive amount of personal and corporate data collected by Ant Group will be used to support government surveillance systems or policy formulation. For example, Jianguo Meng pointed out in The New York Times that China's largest internet companies often provide user information to Beijing, and government systems may at least partially rely on data collected by companies like Alibaba and Tencent.[8] Although Ant Group has officially stated many times that its data management strictly follows laws and regulations and will not disclose user data to any third party without permission, in the absence of full transparency, these guarantees are often difficult to resist external worries and questioning.

9.3 Embracing Challenges: Ant Group's Experience in Technology Globalization

This section takes a deep dive into the complex hurdles Ant Group has navigated and the strategic maneuvers it has employed to not only endure but flourish in our ever-globalizing world. It kicks off with an emphasis on fortifying data governance, a cornerstone of Ant Group's modus operandi. In an age where digital transactions are the norm, safeguarding personal data and managing user information securely have become mission-critical. Ant Group's strategy to tackle these challenges is epitomized by the ZOLOZ Trusted Identity Authentication Platform and its pioneering e-KYC Technology.

The ZOLOZ platform is revolutionizing identity verification by utilizing cutting-edge AI, biometric technology, and data analytics to create a seamless user experience globally. It represents Ant Group's

[7] Kimberley Long, 2024, https://www.thebanker.com/Alipay-international-expansion-raises-questions-over-security-risks-1718869786.

[8] The New York Times, 2018, https://cn.nytimes.com/business/20180105/china-alibaba-privacy/.

commitment to breaking down geographical barriers and promoting confidence in online services. The e-KYC technology showcases the adeptness at fusing proactive security with smart decision-making, ensuring that each transaction is secure, and user privacy is preserved. In addition to making previously complicated procedures easier, this technology has increased customer confidence in digital platforms and set the stage for improved global cooperation.

As we get close to analyze Ant Group's role in fostering inclusive partnerships and its contributions to shaping international standards, it is clear that this company endeavors align with the overarching objectives of global digital economy governance and sustainable growth.

9.3.1 Strengthening Data Governance

9.3.1.1 ZOLOZ Trusted Identity Authentication Platform and Its e-KYC Technology

Confronted with the security risks and challenges of the digital age, Ant Group's security technology has formed a product system with core technologies in identity security, data security, and intelligent decision-making, aimed at different risk control and security scenarios such as compliance, transactions, and industry.[9] ZOLOZ has created an intelligent, efficient, and secure identity verification system by integrating advanced artificial intelligence, biometric technology, and big data analysis, aiming to break geographical boundaries and provide a seamless and convenient digital identity authentication experience for users worldwide.

Among them, Ant Group's e-KYC/KYB (Know Your Customer/Business) technology is leading a revolution concerning data security and privacy protection. As the core technical solution of the ZOLOZ platform, the core value of e-KYC technology lies in its ingenious combination of proactive interactive security, multi-party security, intelligent decision-making and confrontation, full-graph security, and end-edge-cloud collaborative security, building an efficient and secure customer identity verification platform. With advanced technologies such as artificial intelligence and big data analysis, e-KYC can identify and fend off

[9] Ant Group, 2022, https://antdigital.com/community/articles/1578.

potential security threats in real time, ensuring the accuracy and security of each identity verification, even when processing massive user data, while also ensuring the privacy of each user is properly protected.

Along with cutting down on the paperwork and red tape that waste resources, the ZOLOZ Trusted Identity Authentication Platform and its e-KYC technology also increase customer trust in digital services. In turn, this strengthens the social environment and creates the foundation for improved global cooperation and communication. More importantly, Ant Group's security tech promotes transparency and adherence to regulations in data management, leading to a ripple effect on supporting its internal governance. Therefore, this practice of Ant Group can be viewed as a vivid demonstration of responsible business conduct, a key to maintaining a stellar corporate image, and a magnet for investors and partners alike.

9.3.1.2 *Invisible Information*

In terms of proactive interactive security, e-KYC, through intelligent algorithms of human-computer interaction, can dynamically adjust the verification process, achieve personalized security protection, reduce the false positive rate, and enhance user experience. Multi-party security is reflected in e-KYC's ability to support cross-institution and cross-regional data verification needs, ensuring the security of data transmission through encryption technology, while using technologies such as blockchain to ensure the immutability of data, enhancing the trust foundation of multi-party participation. Intelligent decision-making and confrontation mechanisms enable e-KYC to predict and prevent new types of attacks in complex and changing network environments through machine learning models, continuously optimizing its defense strategies. The full-graph security strategy allows e-KYC to use graph analysis technology to build a network of relationships between users and transactions, effectively identifying potential fraudulent activities and abnormal patterns. End-edge-cloud collaborative security ensures that every layer of protection from the user terminal to the cloud server is unbreakable, maintaining high-standard security performance even at the network edge nodes.

Diving into a new financial service, whether it is setting up an online bank account or diving into the world of cryptocurrency trading, often starts with an encounter with cutting-edge technology. You will begin by inputting some essential details—think full name, birthdate, home address, and how to reach you. Besides, to confirm the legitimacy of these details, you will need to upload a crisp image of an official ID, like a driver's license, passport, or ID card.

The next step is to prove that you are the real deal rather than some fraudster. The e-KYC system will lead you through a biometric verification process, which typically involves snapping a selfie. You might be asked to make a few facial expressions or blink to show you are not just a static image. The advanced software in the system will then verify that your biometric information and the ID photo you supplied match. To increase the accuracy and security of the verification, it may occasionally even access external databases, such as official government documents.

Throughout the e-KYC journey, managing risk and ensuring compliance are paramount. The system will automatically flag any connections to high-risk activities, such as being on a sanctions list or having a dubious financial past, all in line with global AML and CFT regulations. Once all checks are greenlit, your account springs to life, and you're free to revel in the convenience of financial services at your fingertips. But even after the initial verification, the e-KYC tech keeps a vigilant watch, quietly monitoring your account for any suspicious activity to safeguard your funds.

This entire procedure not only ramps up efficiency and eliminates the hassle of paperwork but also delivers a more secure digital experience for users.

9.3.1.3 Visible Value

ZOLOZ has recently placed its sights on international growth, creating a stir in the global scene with the goal of becoming the global leader in digital identity authentication. ZOLOZ is exporting not only its state-of-the-art technical expertise but also obtaining profound insights into the unique requirements of other markets by establishing strong relationships with financial institutions, governmental organizations, and industry partners worldwide. This understanding allows ZOLOZ to offer tailored solutions that empower local businesses to quicken their digital transformation journey, enhancing both service efficiency and security.

ZOLOZ has acted as a driver for change, especially in the area of financial inclusion, by using its e-KYC technology to close the digital gap for people in distant areas and developing nations. Globally, ZOLOZ is helping to create a more inclusive story of economic progress by granting equal access to financial services.

The development of ZOLOZ has always been accompanied by a high emphasis on security and privacy. ZOLOZ has not only obtained international certifications such as ISO27001 and ISO27701, proving its professional level in data management and privacy protection, but also effectively resisted new types of security threats such as "AI face swapping" through continuous technological innovation, such as the launch of the deepfake defense product Deeper, building a solid digital defense line for users. In addition, ZOLOZ actively fulfills corporate social responsibility and advocates the construction of an open and cooperative ecosystem. By setting up a "vulnerability reward" program, it encourages global "security geeks" to participate in vulnerability discovery and repair, enhancing the overall security of the platform while promoting knowledge sharing and technical exchange within and outside the industry.

With the use of e-KYC technology, Ant Group has not only given over 1 billion users a convenient and safe digital financial services environment since 2021, but it has also offered robust technical support to small and micro businesses as well as other institutions, assisting them in the digital transition. More importantly, Ant Group is sharing e-KYC technology and the security philosophy behind it with global partners through open cooperation. Besides, it is also jointly promoting industry data security and privacy protection standards nowadays. The ZOLOZ platform is not only a showcase of Ant Group's scientific and technological innovation but also an important promoter of the construction of a global digital trust ecosystem. So, how have they built a safer, more trustworthy, and inclusive digital future at home and abroad through continuous exploration and evolution?

9.3.2 *Promoting Inclusive Cooperation*

9.3.2.1 *Technology's Global Outreach in Practice*

According to China Network Technology, as one of Ant Group's core technology products going global, ZOLOZ has currently helped dozens of partners overseas to serve more than 100 million users. Its e-KYC technology solution has been adopted by dozens of overseas partners, including financial institutions such as Indonesia's largest bank Mandiri, Malaysia's largest bank Maybank, and the Philippines' largest digital bank CIMB, telecommunications operators like Globe in the Philippines, and electronic wallets such as Indonesia's Dana, Philippines' GCash, and Thailand's TrueMoney.

It is the cooperation with Ant Group that has enabled Bangladesh's "Alipay" bKash to contribute greater strength to the development of local digital inclusive finance. Six years ago, Ant Group invested in Bangladesh's electronic wallet startup bKash, which was their ninth local wallet partner found globally based on a shared inclusive finance philosophy, through sharing technology and experience. In Bangladesh, the cooperation between bKash and Ant Group is not limited to technology transfer and user growth but also deeply cares about the core issues of data security and privacy protection.

In a market like Bangladesh, which is full of potential but also challenges, establishing a data security and privacy protection system that meets international standards is crucial for winning the trust of regulatory authorities. bKash works closely with Ant Group to jointly commit to building an efficient and secure digital payment ecosystem. Through the introduction of Alipay's biometric technology and security procedures, bKash not only improves user experience and streamlines the user authentication process, but it also successfully guards against fraud and identity theft, preserving users' privacy and property protection. Every transaction is conducted inside a stringent regulatory framework by both parties adhering to local laws and regulations and prioritizing data security. In addition to increasing regulators' trust, bKash's dedication to compliance helps it establish a strong industry reputation, which opens the door for its recognition as a responsible fintech business.

bKash and traditional banks have moved from being rivals to becoming partners, with data security and privacy playing a big part in this change. bKash uses Ant Group's top-notch encryption to keep user data safe when it is being sent and stored, which cuts down on risks. This clear and trustworthy way of handling data has shown banks that working with bKash is a good idea. They are now teaming up on things like moving money and loans, which helps grow the kind of financial help available.

When COVID-19 struck hard in 2020, many Bangladeshi families found themselves in a difficult situation as their jobs dried up. The Bangladesh government wanted to help by giving money to families who lost their jobs because of the virus. But getting the money to each family quickly and safely was a big challenge. That is when Ant Group's ZOLOZ and bKash, Bangladesh's biggest mobile payment service, stepped in. They used e-KYC tech to let people open bank accounts from home. With e-KYC, users can prove who they are and set up electronic wallets on their phones without going to a bank. This was a big help during the pandemic when people needed to stay apart.

ZOLOZ's e-KYC solution let bKash check people's information, like their ID and face, to make sure they were real. This made opening accounts fast and safe, cutting down the time it took and stopping the risk of people having to meet in person.

During the pandemic, 5 million families in Bangladesh were able to register accounts remotely because of the use of this technology, which allowed them to get government assistance. The e-KYC technology under ZOLOZ has proven the enormous potential of financial technology in emergency situations in addition to guaranteeing data security and privacy protection. Not only has it enabled the government of Bangladesh to promptly address the demands of the populace, but it has also given other parts of the world invaluable expertise.

ZOLOZ was awarded the IDC FinTech Real Results Award for its technological innovation during the pandemic, which brought practical benefits to society.[10] This award not only highlights ZOLOZ's leading position in the field of data security and privacy protection but also reflects its contribution to ESG development and practice, especially at the social level. By supporting vulnerable groups with technology to overcome difficulties, it reflects the social responsibility of Chinese enterprises. In the future, as technology continues to innovate and the regulatory environment becomes increasingly mature, the cooperation between bKash and Ant Group is expected to continue to deepen, bringing more benefits to financial consumers in Bangladesh and the world, and also making a greater contribution to the achievement of sustainable development goals.

In the early spring of 2021, the Malaysian government worked with Touch 'n Go e-Wallet to issue economic subsidies to about 2 million young people aged 18 to 20 nationwide to reduce their living burden and boost the recovery of physical businesses. Users can receive the subsidy after completing identity verification and registration online through e-KYC. All this only requires users to touch their mobile phones and complete online identity verification, and they can receive care and support from the government in just a few minutes. This technology has greatly improved the progress of identity verification and registration, and the number of manual reviewers has decreased by nearly 80%. Ant

[10] Tech China, 2024, https://tech.china.com.cn/internet/20220302/385433.shtml.

Group ZOLOZ's e-KYC technology ensures that every subsidy can be accurately and correctly received by eligible beneficiaries, while also building a solid defense for users' data security and privacy protection through advanced encryption methods.

GCash in the Philippines, DANA in Indonesia, bKash in Bangladesh. With the help of Ant Group's e-KYC technology, these electronic wallet platforms have become effective assistants for the government and businesses to safely issue subsidies, wages, and vocational training allowances. In Southeast Asia, 70% of adults have not yet had access to bank accounts, but in the past two years, the rise of mobile payments has turned ordinary people's smartphones into a key to the financial world. Ant Group's e-KYC technology is like the fingerprint identifier on this key, ensuring that every user's identity is real and safe.

From Malaysia to Bangladesh, from the blue sea and blue sky of the Philippines to the island country of Indonesia, Ant Group's e-KYC technology has shown a global perspective on technology for good in Southeast Asia. It not only promotes financial inclusiveness and enhances economic vitality but also sets a high standard for security protection for the global financial technology industry.

9.4 Participating in the Development of International Standards

In the summer of 2024, the global technology community's attention was focused on Geneva, Switzerland, at the 27th United Nations Conference on Science and Technology (UNCSTD). This gathering brought together the brightest minds and decision-makers from around the globe to chat about the latest tech trends and how they are shaping our world. Two important worldwide standards, "Safety Testing Standards for Generative Artificial Intelligence Applications" and "Safety Testing Methods for Large Language Models," were introduced at this event by the World Digital Technology Academy (WDTA). The purpose of these standards is to provide a regular safety check for AI systems, which are becoming increasingly complicated, and the goal is to make sure that AI technology can contribute to society in a healthy and secure way.

Ant Group was a key player in crafting these standards. As a leading fintech company, Ant Group didn't just take the lead on putting together the "Safety Testing Methods for Large Language Models"; they were also heavily involved in shaping the "Safety Testing Standards for Generative Artificial Intelligence Applications." By collaborating with WDTA and other international partners, Ant Group is actively working to establish a series of standards that meet global norms and keep up with the rapid advancement of technology.

The UNCSTD guidelines set the norm for future AI applications, not only a safety guide for the worldwide AI sector. Ant Group's contribution extends beyond the technological domain. It vividly demonstrates how Chinese businesses can contribute positively to the global tech scene by addressing global issues and pursuing a more secure, dependable, and inclusive digital environment. With these guidelines in place, we can anticipate a time where technology will continue to benefit people while prioritizing data security and user privacy.

9.5 Open-Source and Industry Development

In July 2022, Ant Group officially announced the open sourcing of its independently developed trusted privacy computing framework "Secret Flow," which took six years to create. The Secret Flow privacy computing framework, crafted by Ant Group based on years of practical experience and technological accumulation, covers nearly all mainstream privacy computing technologies, including but not limited to homomorphic encryption, multi-party secure computation, and federated learning. This framework is designed with security and openness at its core, aiming to address data security and privacy protection issues in data circulation from a technical perspective, enabling the maximization of data value in a secure and controllable environment.

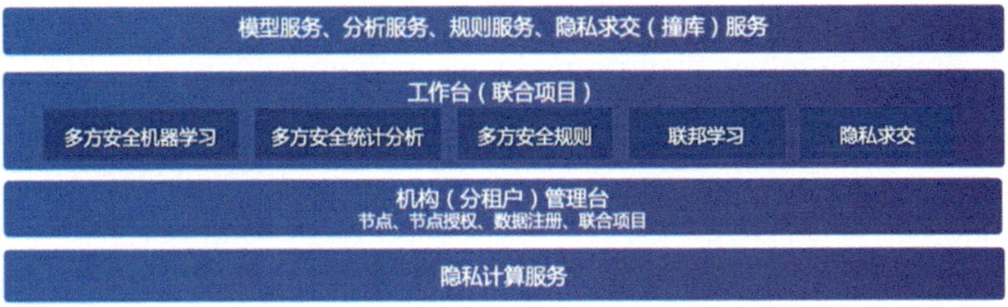

Product Structure of Secret Flow

At the World Artificial Intelligence Conference (WAIC) Trusted Privacy Computing Summit held in September 2022, Ant Group launched the Trusted Privacy Computing "Secret Flow Open Platform" (referred to as the Secret Flow Open Platform), open to global users. It allows direct use of product functions without the need to call or develop code, helping users explore privacy computing application scenarios at a low cost.

After the Secret Flow privacy computing framework was made publicly available in July, Ant Group has made another important advancement in the open privacy computing technology space with this step. Following the release of the first industrial-grade full-data AI detection platform in the industry and the open sourcing of the high-performance graph database TuGraph, this is yet another significant event at the World Artificial Intelligence Conference.

At the launch, Ant Group's Vice President and Chief Technology Security Officer, Tao Wei stated: Faced with the challenge of data security and privacy protection, trusted privacy computing technology is a core technical solution. Only by enhancing the technical level of the entire industry and jointly building a positive technical ecosystem can we promote the large-scale implementation of technology in the entire industry and promote the healthy development of the industry as a whole.

In the plenary session of the conference's industry forum, Wei Tao proposed that the large-scale application of artificial intelligence in the industry urgently needs to address issues such as security and trust, collaborative consensus, complex association analysis, storage and computing scale explosion, and energy consumption reduction. There is a need to deeply explore the "root technologies" of large-scale AI applications, such as privacy computing, blockchain, graph computing, distributed databases and computing infrastructure, and green computing.

By open sourcing "Secret Flow," Ant Group not only provides a powerful toolkit for developers worldwide, allowing them to use the code of "Secret Flow" for free, but also takes a substantial step in promoting technology benefits and the co-construction of the industry ecosystem. Because developers can more easily achieve secure data sharing and joint analysis without worrying about the risk of data leakage or misuse, thus greatly reducing the entry threshold of privacy computing technology and promoting the widespread application and innovation of technology. The role of the open-source "Secret Flow" privacy computing framework in promoting privacy protection and data security is reflected in many aspects. First, it strengthens the security of data circulation, ensuring that personal privacy and corporate secrets are effectively protected during the data usage process. Second, through open-source, developers and researchers worldwide can participate in the optimization and iteration of the framework, forming an open technical community and accelerating the development and maturity of privacy computing technology. In addition, open-source also helps to establish a more fair and transparent industry environment, promote cooperation between different enterprises, avoid the phenomenon of data islands, and enable the entire industry to develop in a healthier and more sustainable direction.

The maturity of the Secret Flow Open Platform has been verified, and it has achieved successful application in the fields of medical health, insurance claims, joint credit risk control, and smart city governance.[11]

In the complex field of medical health data, Secret Flow has built a data fusion platform for the First People's Hospital of Jiande City, Zhejiang, China, to enhance the value of medical data security. This case was selected as one of the excellent cases of the 2021 "Galaxy" big data privacy computing of the China Academy of Information and Communications Technology. Secret Flow, through Alibaba Cloud's medical big data management platform, serves the hospital to increase the number of samples and expand the scale of data while protecting patient privacy. At the same time, it can also optimize clinical diagnostic pathways, accurately

[11] Secret Flow, https://secret-flow.antgroup.com/.

predict patient disease grouping and reasonably allocate medical insurance quotas, and reduce the workload of feedback analysis, promoting the sustainable development of the medical and health industry.

In the field of insurance claims, the application of Secret Flow in Ant Insurance, as a representative case, won the 2022 IDC China Financial Industry Technology Application Scenario Innovation Award. Based on this exploration, a health insurance customization multi-party data joint analysis solution was built, covering medical data in more than 10 provinces and cities nationwide. The highlight of this application is that it can intelligently assist insurance companies in reducing the risk of wrong compensation, improving the efficiency of compensation, and promoting the concentration of resources to inclusive medical care.

In the field of joint credit risk control, Secret Flow and Shanghai Pudong Development Bank have developed a complete set of risk assessment solutions for retail loan business, using Secure Multi-Party Computation technology, by building a series of cryptographic tools and decentralized deployment architectures to strictly protect data privacy. Secret Flow uses the data of Pudong Development Bank and its partners to improve the effectiveness of the model with partners while ensuring the security and privacy of the data. This technology can obtain more accurate and efficient models and decisions through the secure collaborative joint modeling of multi-party data, enhance the bank's independent risk control ability, and achieve ultra-fast loan approval. It can identify and control risks in the pre-loan, loan, and post-loan stages of financial credit, preventing the issuance of billions of high-risk loans, increasing the identification of more than 300,000 low-risk customers, and expanding the scope of inclusive credit services.

In smart city governance, Secret Flow cooperated with a certain number of exchanges to help the city achieve dual effects of convenience and safety. Government data usually involves social security, housing fund, taxation, life, transportation, etc., and these data usually belong to different departments, and there are serious data islands. There are real obstacles such as coordination difficulties and complicated approval procedures in sharing these data. In this process, the paradigm provided by Secret Flow, while protecting personal privacy and ensuring data security, orderly promotes the circulation and application of data in some fields, thus solving the problem of government data islands and improving government governance capabilities.

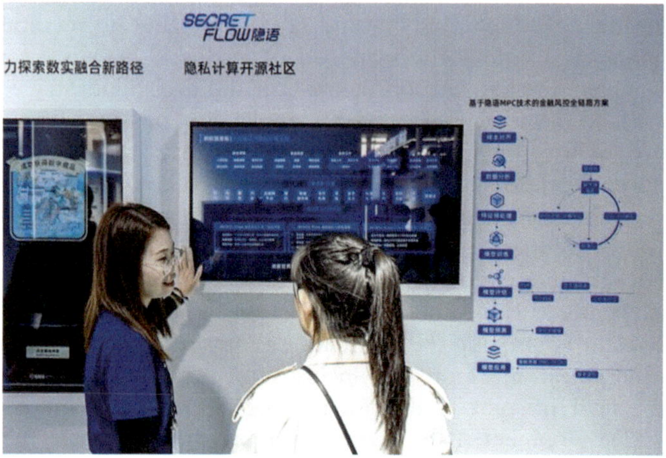

Secret Flow in the 2023 China International Financial Exhibition (*Source* Geekpark)

Two years after the open-source (as of July 2024), "Secret Flow" has achieved a community scale of more than 20,000, with more than 3,000+ Stars, and more than 100+ community contributors. At the same time, it has cooperated with more than 50 institutions such as Tsinghua University, Zhejiang University, and the National University of Singapore in scientific research, teaching, and open-source dimensions, and has jointly promoted the development of data industry technology, standard promotion, and benchmark case construction with more than 30 industry ecological partners. It has contributed a total of more than 30 industry awards and honors, and more than 60 domestic and foreign standards, helping more than 2,000 learners with privacy computing courses. This move by Ant Group also reflects its corporate image of being responsible to society. The application of privacy computing technology can enhance the public's confidence in data usage, promote social fairness and inclusive growth; the open-source strategy can also encourage positive competition and cooperation within the industry, help build a more stable and healthy digital economic ecosystem, and provide more Chinese experience for global data governance and the sustainable development of the digital economy.

9.6 Industry-Academia-Research Integration in Privacy Governance: CCF-Ant Financial Privacy Computing Special Research Fund

In July 2022, at the open-source release conference of Ant's trusted privacy computing framework "Secret Flow," the China Computer Federation (CCF) and Ant Group jointly announced the launch of the "CCF-Ant Financial Privacy Computing Special Research Fund" under the framework of the "CCF-Ant Research Fund."[12] The fund will invest millions of dollars in scientific research funds over the next month and open 27 research topics for global scholars with innovative thinking and research capabilities. The "CCF-Ant Research Fund" was jointly initiated by CCF and Ant Group in 2020. The fund, targeting global scholars, has established multiple research directions, including artificial intelligence, privacy computing, green computing, blockchain, basic systems, and databases.

The privacy computing special issue is the second phase of the release, with this fund planning to invest more than 3 million yuan in scientific research funds, establishing a total of 17 research topics. The research topics focus on key technologies for the circulation of data elements, with 12 cutting-edge exploration topics centered around privacy computing + large models, privacy security enhancement, and hardware acceleration of privacy computing. The integration of privacy computing technology with AI algorithms and computing power is emphasized to construct "Privacy-Enhanced AI." At the same time, 5 practical verification topics are carried out, using the "Hidden Language" open-source framework as the research vehicle, conducting research around the application scenarios of privacy computing industry, and forming technical capabilities and methodologies that are feasible for landing, using technical means to help data elements enter the market.

Since 2022, from the perspective of regulations, policy requirements, or technical maturity, the entire field of data circulation is about to bid farewell to the era of plaintext data and embark on a new journey of data ciphertext era. Privacy computing is considered a key technology that balances the circulation of data elements and protects the security of personal privacy. It covers multiple disciplines, including cryptography,

[12] Alibaba Cloud, 2023, https://developer.aliyun.com/article/1374994.

machine learning, and trusted hardware, and includes various technical routes such as multi-party secure computation (MPC), federated learning (FL), and trusted execution environment (TEE).

Following the announcement of this plan, in 2022 and 2023, more than 30 privacy computing special topics from 26 universities, including Zhejiang University, Shandong University, Wuhan University, Peking University, East China Normal University, Sichuan University, Shanghai Jiao Tong University, Southern University of Science and Technology, and Hong Kong Baptist University, have successfully obtained funding. These topics cover various aspects of privacy computing, including but not limited to secure multi-party computation, federated learning, homomorphic encryption, and privacy-preserving recommendation algorithms. The pathways for translating the research results of privacy computing into practical applications are extensive, involving multiple industries and fields, including healthcare, finance, supply chain, and micro-data security. The research of these topics aims to promote the development and practical application of privacy computing technology, thereby further advancing the level of privacy computing research and digital governance in China.

9.7 Conclusion

Against the background of data security and privacy protection increasingly becoming a challenge of the times, Ant Group's practice demonstrates the importance of taking technological innovation as a core capability, using industry-leading technologies to prevent, identify, and respond to various risks, and effectively ensuring data security and user privacy.[13]

In business practice, Ant Group leverages its data security and privacy computing capabilities to achieve open and development between highly sensitive and high-value public and social data that is usable but not visible, serving the construction of Digital China. For instance, it has helped the Hangzhou municipal government build the "Zhijiang Chain," creating a city-level data trusted collaboration network. In 2022, the

[13] Ant Group, https://mdn.alipayobjects.com/huamei_f5zjmk/afts/file/A*_vcKR4qMeToAAAAAAAAAAAAADgx7AQ/%E8%9A%82%E8%9A%81%E9%9B%86%E5%9B%A2023%E5%B9%B4%E5%8F%AF%E6%8C%81%E7%BB%AD%E5%8F%91%E5%B1%95%E6%8A%A5%E5%91%8A.pdf.

"Zhijiang Chain" platform has provided evidence services for 568 digital transactions worth 1.091 billion yuan at the Hangzhou International Digital Exchange. Globally, Ant's trusted authentication history has served over 500 million internet finance users, ensuring the security of billions of transactions. This rich experience in security confrontation stems from Ant's years of practical experience in risk confrontation.[14]

The e-KYC technology solution has been adopted by dozens of overseas partners, including financial institutions such as Indonesia's largest bank Mandiri, Malaysia's largest bank Maybank, and the Philippines' largest digital bank CIMB, telecommunications operators like Globe in the Philippines, and electronic wallets such as Indonesia's Dana, Philippines' GCash, and Thailand's TrueMoney.[15] In Bangladesh, more than half of adults do not have bank accounts. During the pandemic, Zoloz's e-KYC technology also helped 5 million families open accounts remotely, enabling the government to distribute financial aid to struggling families who lost their sources of income through the mobile payment platform BKash.[16] Zoloz's e-KYC technology has helped dozens of partners overseas serve more than 100 million users, reducing the average identity verification time from several days to 3 minutes and increasing the verification success rate from 65% to over 90%.

Public information shows that ZOLOZ has served more than 2,000 customers in dozens of industries such as finance, bulk trade, travel, and e-commerce in China, and has served hundreds of partners including banks[17] and electronic wallets along the Belt and Road in overseas countries, serving more than 100 million users.[18] In the international authoritative research institution Forrester's report on anti-fraud technology service providers in the Asia-Pacific region, "The Forrester Wave™: Enterprise Fraud Management In Asia-Pacific, Q3 2023," Ant Group and its

[14] Ant Group, 2024, https://antdigital.com/products/trustid.

[15] Sina, 2022, https://k.sina.com.cn/article_2118746300_7e4980bc0200194nw.html.

[16] Chinaz, https://www.chinaz.com/2021/0914/1303837.shtml.

[17] Ant Group, 2023, https://www.antgroup.com/news-media/press-releases/1697076678000.

[18] https://www.jiqizhixin.com/articles/2023-03-24-7.

commercial brand ZOLOZ achieved first place in the Asia-Pacific region and were rated as the "leaders" in data security in the region.[19]

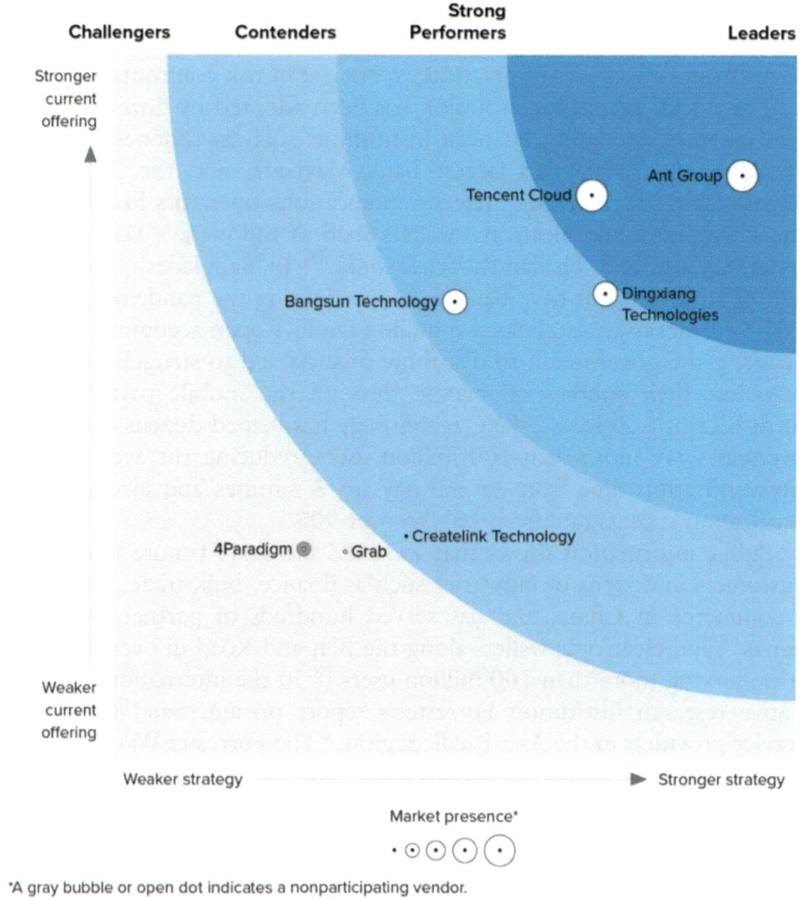

Ratings of seven Asia-Pacific data security companies in the Forrester reports (*Source* Forrester)

[19] Forrester, 2023, https://www.forrester.com/report/the-forrester-wave-tm-enterprise-fraud-management-in-asia-pacific-q3-2023/RES179818?ref_search=0_1728533003223.

In industry promotion, Ant Group uses its technology and experience to help partners improve management levels together, by actively participating in the formulation of standards related to data security and privacy protection and the open-source of products, it promotes the improvement of the protection level of the entire industry. The e-KYC has passed the global strictest data security SOC 2 Type 2 audit, which means that ZOLOZ has reached the international top-level in aspects of data security, service security, and service availability.[20] In August 2023, the International Telecommunication Union released the international standard in the field of privacy computing technology for the first time, and Ant Group was one of the formulators. At the United Nations Conference on Science and Technology in 2024, Ant Group took the lead in formulating the "Security Testing Methods for Large Language Models" and participated in the formulation of the "Safety Testing Standards for Generative Artificial Intelligence Applications." Ant Group has obtained the international ISO 27001 certification and ISO 27701 certification, covering third-party payment, mobile payment, online financial platforms, credit services, etc., covering more than 90% of Ant Group's business.[21] In addition, in July 2022, Ant Group announced the open-source of the trusted privacy computing framework "Secret Flow," aiming to make it easier for developers worldwide to access privacy computing technology and accelerate the application of privacy computing in various scenarios, providing greater help for the implementation of privacy protection in the industry. Since its use, "Secret Flow" has integrated a comprehensive set of privacy protection technologies such as multi-party secure computing, federated learning, homomorphic encryption, differential privacy, zero-knowledge proof, and trusted execution environment, and has been applied in various scenarios such as medical, insurance, and risk management.[22] According to the 2022 "Global Privacy Computing Technology Invention Patent Ranking (TOP100)" jointly published by IPRdaily, a

[20] https://www.sohu.com/a/540362264_121104381.

[21] Ant Group, 2024, https://www.antgroup.com/esg/open-ecology-folding-panel-details/007,2024-7.

[22] Ant Group, 2023, https://www.businesswire.com/news/home/20230725059163/en/.

third-party global intellectual property institution, and the incoPat Innovation Index Research Institute, Ant Group's privacy computing patent count reached 1,152, ranking first in the list for two consecutive years.[23]

This chapter examines Ant Group's impact on the global IT scene, including how it is influencing international standards and disseminating its open-source innovations, such as the Secret Flow framework. Ant Group is undoubtedly a major force in the global digital economy, and its success goes beyond simply being tech-savvy. Their efforts highlight privacy, security, and sustainability in our digital environment and are directly related to the United Nations' sustainable development goals.

The aforementioned cases and practices also demonstrate that a comprehensive strategy that combines ethics and technological advancement is essential, and it is in line with the worldwide conversation surrounding digital economic privacy and sustainable growth. These efforts serve as a reminder that via a combination of creativity, regulations, and collaboration, we can build a flourishing digital future that upholds our rights to privacy and data.

Navigating the complex landscape of data privacy in today's digital economy can be challenging, but the above story serves as a reference on leveraging technology to create a world that is not just safer and more equitable, but also sustainable.

[23] IPRdaily, 2022, http://www.iprdaily.cn/news_30714.html.

CHAPTER 10

Sustainable Fintech for Inclusive Partnership and State Capability Building in the Global South: Paytm in India and Easypaisa in Pakistan

10.1 Introduction

In the context of Sustainable FinTech, the dimensions of an inclusive and effective partnership as well as the building of appropriate state capability are among the most important factors that would determine the extent to which the benefits of financial technologies would be spread across a wider and more diverse section of the population. Inclusive partnerships bring together various stakeholders—governments, the business sector, the civil society, and the people at the grassroots level—to find solutions that are both efficient and just. Such partnerships further enable better delivery of FinTech solutions to distinct populations most of whom are disadvantaged and underserved by the regular financial institutions.

On the contrary, state capability building is concerned with assisting the governments to regulate, manage, and promote FinTech. For instance, a capable state will have adequate regulatory frameworks, infrastructure and policies that support the growth of FinTech. This includes measures for data protection, consumer protection rights, and digital skills for the citizens. By doing so, governments can instill confidence among their people over FinTech products and promote their uptake enhancing the levels of financial inclusion and economic empowerment.

J. Sun, *Sustainable Financialization*, Sustainable Development Goals Series, https://doi.org/10.1007/978-981-96-3580-1_10

While these elements are not the only factors for the success of FinTech, they certainly rank high. There is a clear two-way interaction between the factors: inclusive partnerships and state capability building which minimizes the level of excesses in the distribution of the benefits that accrue from the application of the financial technology and also that the existing systems are not only appropriate but also sustainable.

To demonstrate the relevance of these principles we will examine two case studies: Paytm in India and Easypaisa in Pakistan. These examples demonstrate the complementary nature of building capabilities of the state and forging partnerships in inclusive initiatives to the success and the impact of FinTech initiatives in these countries.

Paytm, one of the most prominent Indian digital payment companies, has developed unique offerings that changed the way millions of Indians handle their financial dealings. Through strategic partnerships with the government, banks, and small businesses, Paytm has created a comprehensive ecosystem that caters to a wide range of users. The Indian government's support, particularly through initiatives like the Digital India campaign, has been instrumental in promoting the adoption of digital payments and creating an enabling environment for FinTech innovation.

Likewise, Easypaisa in Pakistan, owned by Ant Financial, has performed exceedingly well towards increasing the scope of financial services and uplifting the lives of millions of Pakistanis. With local partners like the Sindh Education Reform Program (SERP), Easypaisa has not only rendered easy and safe financial operations but has also assisted in social welfare projects. The Pakistani government's attempts to improve the regulatory structures and facilitate digital finance have also boosted Easypaisa's effectiveness making it one of the essential contributors in the country's transition toward a comprehensive and sustainable economy.

These case studies will shed light on how state capability building and inclusive partnerships can be appropriately utilized in ensuring the success of FinTech initiatives and enhancing the quest for financially inclusive and sustainable systems.

10.2 Paytm in India

10.2.1 Severe Financial Repression

Despite being home to 1.2 billion people, India's telecommunications and financial services lag far behind its massive population. For instance, the peak number of fixed telephone lines in India was only around 50 million despite having so large potential market. This figure in recent years has gone down, though, as mobile technology becomes more popular and makes landlines practically obsolete. Also, both personal and business broadband subscription rates show average figures. While internet access is expanding globally, many regions in India still face significant challenges in obtaining information and staying connected.

The Indian government has also come up with policies like "One Person, One Bank Account" to promote financial literacy, but they have been relatively ineffective, mostly due to a combination of high poverty levels, excessive concentration of bank branches, and insufficient numbers of payment use cases. Nationwide, only about 1.2 million merchants have POS machines, while the number of credit cards and debit cards stands at roughly 20 million and 300 million, respectively. The number of actual users who make such transactions is however in the tens of millions at most.

Furthermore, as the world's second-largest developing country, India's issues go beyond just infrastructure. There are also more fundamental structural constraints within the socioeconomic system. In the financial domain, the country has suffered from what may be called as "financial repression" for a considerable time, characterized by limited outreach of financial institutions, a narrow product range, and stagnation in product innovations. Even international payment giants such as Visa and Mastercard had issued less than 18 million cards in total by the year 2015, even though they have operated in India for decades. This not only illustrates the practical constrains that multinational companies encounter in penetrating into the emerging markets but also points to the lack of supply of modern financial services in India.

Despite the fact that India has performed exceptionally well in some advanced technology areas, like IT Outsourcing, there is still a great deal of scope for developing the basic telecommunication networks and the various forms of inclusive financial services.

10.2.2 The Founding and Challenges of Paytm

Paytm, short for "Pay Through Mobile," is a digital payment and financial services company based in Noida, India. Established in 2010 by its parent company, One97 Communications Limited, Paytm has developed to become one of the largest mobile payment finances in the nation and can sometimes be referred as India's "Alipay." Its founder and CEO Vijay Shekhar Sharma commenced with a simple mobile prepaid recharge website and has now grown to start managing a proper financial service. At its core, Paytm offers mobile wallets, bank transfers, bill payments, and a wide range of online and offline shopping experiences. Paytm Wallet allows people to deposit money that they can use for various transactions. Money is transferred in a real-time from one bank to another by UPI which is the most effective financial system in India.

Merchants find Paytm useful as they provide the customers with various services to help facilitate customer payments such as the QR code collection, POS terminal solution, and e-commerce integration. This makes it easier for small businesses to take advantage of digitalization hence fostering financial inclusion. Beyond basic payment functions, Paytm has expanded into a broader spectrum of financial services. For example, it works jointly with numerous other financial establishments to provide micro and personal loans, which belong to a class of short-term loans, directed at real consumers' and merchants' needs. It has also rolled out "buy now, pay later" facilities, which enabled customers to buy goods on credit and pay for them in portions, which lowered their financial pressure. In addition, Paytm began offering retail broking which opened up the stock market to Paytm users of the app, and also online games and ancillary entertainment services were introduced to enhance customer interest and activity on the platform.

Nevertheless, it would be fair to conclude that in the beginning stages of the development of Paytm, the company operated in a tough environment. These were caused mainly by rather poor development of the mobile internet environment in India, the low level of banking population, and high price sensitivity of users for the use of smartphone and data.

To begin with, the mobile internet structure in India has always been viewed as pathetic. In April 2015, around 100 million people were said to have been offered broadband services in India, which is almost negligible considering the entire population. In addition, 4G networks were still in their early phases of implementation, making the internet connections in many regions weak and intermittent. This created a significant obstacle for digital payment applications that need a consistent and quick data connection to implement. Even through some urban parts, where most of the population is located, the network seems inconsistent or weak at times allowing no smooth functionality in transactions.

Secondly, looking at the statistics provided by the World Bank in 2016, possibly 47% of India's population including women of vast use did not have a formal bank account. This was more so in second and third-tier cities where banking services may not even be available. People having no formal banking services, had no exposure to and knowledge of e-payment system, thus, less acceptance of non-cash payments. In this situation, cash on delivery was the only option most people would accept for paying for online purchases. For example, around 72% of Flipkart's transactions in major cities and up to 90% in smaller cities were made via cash on delivery. Clearly, changing consumer habits to adopt new payment tools was a difficult task.

Additionally, the smartphones commonly used in India were often mid- to low-range devices with limited memory, storage, and processing power. Given that users were highly conscious of their mobile data usage, they would typically open apps only when absolutely necessary. This implied that to cope with the prevailing situation, Paytm had to make sure that its software was optimized and as simple as possible for the end user. Studies done by the Paytm team reveal that for every additional step taken in the transaction process, there was approximately 10% reduction in the total volume of transactions. Thus, it became a key technical question how to achieve a full functionality, while streamlining the users' experience as much as possible.

Finally, educating the market was another crucial aspect. A lot of possible users were frightened to use the new technology especially regarding the safety of their cash transfer. Therefore, besides technical efforts, huge amount of funds needed to be spent on managing the publics thoughts and confidence towards electronically invested assets. But in view of these people having overall low level of financial literacy in most areas of the country, it took this particular exercise a very long time to deliver any encouraging results.

In summary, throughout the initial years, Paytm went through a lot of pain, including establishing the required technology, changing user habits, and further enhancing the market understanding. However, with resolve, the management was able to overcome these hurdles and carve a niche in the fintech market, both within India and outside the country.

In India, the process of opening a bank account and completing the Know Your Customer (KYC) verification is exceptionally complex and time-consuming, posing significant challenges for digital financial service providers like Paytm. In difference to China, where the entire promotional process for opening an account and completing KYC is straightforward and uncomplicated, in India it is quite the opposite. For instance, customers have to sometimes provide fixed asset documents to banks as a prerequisite of opening an account. This rigorous procedure not only raises the bar when it comes to entry into the financial systems, but also makes it difficult for new users to go the formal banking way.

For Paytm, even though it has attempted to simplify the onboarding process for new users, it still faces the issue of low bank card penetration. This implies that those who are ready to make use of electronic payment services are themselves unable to go through the KYC process because of the absence of requisite bank cards. In order to remedy the situation, Paytm had to implement other solutions, including establishing offline kiosks or working with third parties, to allow the users to authenticate themselves through biometrics like fingerprints. Unfortunately, these steps raise the operating costs to a considerable extent.

According to Gao Hang, a senior director at Ant Group's International Business Unit, whether setting up service points independently or finding partners to assist with face-to-face identity verification, the entire process, which includes information checks, signatures, and photo-taking, costs approximately $5 per person on average. In remote areas with rugged terrain and poor transport systems, the actual cost is certainly higher, for instance, travel and logistics costs tend to increase. Also, in residents who stay in two types of infrastructure limited space(s), it would be almost impossible to have any form of a financial institution, let alone sophisticated technology for registration of personal data, let alone fund management. Thus, regardless of the significant efforts made by Paytm to expand its target to the unbanked communities, its growth has been severely impacted by these impediments.

In this situation, to which the company itself added successfully, to its early stages, promotional efforts of Paytm were not fruitful. As for 2014, Uber's collaboration increased the visibility of Paytm, though the audience remained small. By 2015 Paytm was able to boast only approximately 25 million registered users.

The defining moment came after Prime Minister Narendra Modi, during his speech, took the opportunity to promote the initiative called "Digital India." Citizens were urged by Modi to "learn how the digital economy works, learn different ways to use e-banking and bank accounts, learn the effective use of various banking apps, and learn how to run businesses without cash." He further proclaimed that the country was in a shift and is moving towards a "cashless society."

When in 2016, the government led by Modi announced the demonetization of 500 and 1000 rupee notes largely used in the economy, it aimed to reduce corruption, choke the funding of terrorist organizations, and eliminate fake currency. Following the demonetization, Paytm saw a record-breaking surge in daily transactions, with app downloads increasing over a thousandfold and the average number of transactions per user skyrocketing. Within four months, the number of registered users grew to 122 million. This explosive growth was not only in the user base but also in transaction frequency and volume. As more people turned to digital payments to cope with the cash shortage, Paytm became an essential part of daily life, widely used for everything from everyday purchases to bill payments and even small loans.

However, the rapid increase in business volume also meant that Paytm had to contend with increasing backend pressures. Achieving system stability and maximizing the response time became an enterprise on their part. Paytm had to increase server capacity and improve the system's architecture to handle data traffic growth and improve processing efficiency. In addition, the large number of new accounts created, especially determining how to prevent fraud, needed more resources and technical solutions. Lastly, as the firm's platform gained prominence, so did the issues of regulatory compliance, and ensuring appropriate laws and regulations are adhered to in the face of increasing growth pressures also became a major area of focus for management.

10.2.3 The Entry of Ant Group and Technological Empowerment

The connection between Paytm and Alipay goes back a long way. Vijay Shekhar Sharma, the founder of Paytm's parent company, has shared the story of his early inspiration from Jack Ma. As of 2009, when Vijay began operations at Paytm, it was a mobile recharge company. After learning in

2011 that Alipay was integrated into Taobao, Vijay sought encouragement from Jack Mas' speech and decided to launch a similar platform in India where mobile shopping and paying went hand in hand.

In October 2014, about a month after the IPO of Alibaba, Vijay made a trip to Hangzhou and interacted with Jack Ma, Daniel Zhang, Lucy Peng, and Eric Jing and affected their ideas on financial inclusion. Back then, plans were already in place for Alibaba to venture into India's financial market when Paytm having 20 million customers was already the largest mobile payment service provider in India.

It was followed up in March 2015, when Ratan Tata, an Indian businessman, put money into Paytm, while that same month the Alibaba Group outlaid $575 million to invest. Later on, Ant Financial acquired a 25% stake in One97 Communications and Paytm also acquired operational capital financing from ICICI Bank.

Alibaba and Paytm on September 29, 2015 issued a phrase on how the Alibaba Group and its financial institution Ant Financial will together invest in Paytm. Alibaba termed it as a strategic investment.

Thereafter, Alibaba took the opportunity to actively prepare Paytm for its practical application. They sent employees of Paytm for training and paid mentoring to Hangzhou, and many people from the teams based in Hangzhou frequent the teams in Delhi and vice versa. For the most part, Paytm assimilated and modified the systems, structure and operating procedures of Alipay.

Moreover, Ant Financial became elaborately integrated into the processes of Paytm—developing products, marketing, managing risks and providing security, all while distributing and receiving growth. Yang Xinyun, a specialist from Ant Financial's International Business Unit in India, said, "We shared all the experience we've gained over more than a decade with Alipay. Since many challenges arise during product development, we passed on our past experiences to help them shorten their development cycle."

For Paytm, the partnership with Ant Financial had at least two significant implications:

1. Technological Architecture Upgrade: Without delving into complex technical details, the massive data processing capabilities of Alipay's backend were far beyond what Paytm could achieve on its own. With Alipay's current architecture, it set a record of 120,000 transactions

per second during the 2016 Singles' Day event. Without Ant Financial, Paytm would have been limited to designing its infrastructure for handling only a few million transactions daily, and scaling up would not have been as simple as just adding more servers.

2. Payment Scenario Adoption: The ability to create mobile payment scenarios does not spring forth in a moment, it comes about after a long period of practical experimentation & exposure. For instance, movie ticket purchasing, which is a common one in India, was a scenario Paytm repeatedly sought Ant Financial's assistance in. Nowadays, purchasing of movie tickets has become one of Paytm's leading categories of Merchants Revenue generators, particularly in an area where going to the movies is a cultural trend.

Vijay described the collaboration with Ant Financial as akin to taking an MBA course. Before partnering with Ant, Vijay expressed his desire to learn from Ant's proven successful models, spanning payments, wealth management, banking, insurance, and credit. This underscores the importance of inclusive finance practices for developing countries.

For instance, in risk management, starting from August 2015, Ant Financial's international risk management team frequently traveled to Delhi to work closely with Paytm. Ant Financial brought the latest concepts and technological frameworks to Paytm. It was discovered that Paytm's development bottleneck around 2015 was due to its weak technology foundation. The application architecture, scalability, and stability were fragile, and without a strong technical base, product and feature upgrades often hit roadblocks.

Once, Paytm saw the potential of offering incentives for user-to-user transfers, however, such practices encouraged fraudulent activities, such as creating multiple user accounts to game the system and gain in reward, something that always proved detrimental to business, and thus caused Paytm to place a ban on the transfer feature. Moreover, owing to a sudden spike in transactions it was also unable to take part in mass promotional activities such as Singles' Day and expand its business in other areas.

Initially, Paytm's risk management team consisted of fewer than ten people, responsible for both risk and policy analysis. They operated manually, receiving daily emails from the data mining team listing suspicious activities, and then conducting investigations via phone calls. By 2015, with over 20 million users, it was surprising that Paytm still relied on such traditional methods.

Wu Hui, a senior expert from Ant Financial's International Business Unit, recalled an instance where Paytm lacked project management experience. When a project was about to start, Wu suggested appointing a project manager, and the CEO pointed to an employee, saying, "You're the project manager now." Paytm also did not have a habit of documenting, and product requirements were often communicated verbally or via email, leading to a lack of systematic clarity. The Ant team had to start from the basics, teaching Paytm how to document and organize product requirements.

Ant Financial provided consultancy and shared product and technical knowledge with Paytm. For instance, when Paytm noted high levels of fraud, Ant Financial advised changing their algorithms and rules which helped a lot.

Over the years, the two companies have seamlessly integrated their operations as they have worked together. As the user base and the number of transactions increased dramatically from millions to tens and hundreds, the complexity of retaining system and security integrity along with effective risk management and feature development became a significant concern. Nevertheless, due to the assistance from Ant Financial, Paytm was able to grow its business at a fast pace as well as run a number of system upgrades and transformations.

Vijay believes that the most important lesson Paytm learned from Alipay is the philosophy. "In wealth management products, we see how Ant Financial serves a broader audience. Now, Paytm focuses mainly on offline services, and so does Alipay. We can learn from each other in this area," he said.

10.2.4 Financial Inclusion and Governance Enhancement

Paytm has effectively contributed to India's digital evolution, being the first to introduce QR codes as a mode of payment, enabling small business owners, and enhancing financial literacy among millions of people and bringing them into the mainstream economy. A recent tweet highlighted the extent of Paytm's market penetration, showing that people from all walks of life are now able to use digital payments.[1] One user

[1] NDTV Profit Brand Studio, 2023, https://www.ndtvprofit.com/bq-brand-studio/deeply-rooted-in-india-social-media-users-hail-paytms-qr-strength-that-powers-indias-small-merchants.

shared a picture of an elderly woman selling cow dung, with a prominent Paytm QR code displayed. The caption read, “Digital India at its best.” Another Twitter user, Vaibhav, shared a photo of a woman in Bangalore selling corn and peanuts, with a Paytm All-in-One QR code on her cart. Vaibhav wrote, “Paytm is now deeply rooted in India. It’s the first rain in Bangalore, and this lady doesn’t have to worry as much anymore.”[2]

Paytm empowers grassroots vendors with seamless, cashless digital payment solutions (*Source* X)

With design innovations such as the Paytm All-in-One QR and Soundbox, smaller merchants and street vendors, in turn, can now receive payments digitally thus avoiding the challenges of managing cash.[3] With the assistance of the All-in-One QR code, these small merchants can

[2] Ibid.

[3] Ibid.

effortlessly receive digital payments as the payment system is simplified. Cash transaction is unnecessary as cash customers can scan the QR code for any transactions, and this reduces the headache of counterfeits as well as the problem of change. This not only enables these small players to enhance the efficiency of transactions but also cuts down on operational costs which assist in their faster adoption to the digitalized economy. Since an increasing number of small merchants are embracing Paytm, the process of digitalization of the entire market has been boosted to a great extent, propelling the economic development of India.

Paytm has yet again pushed the envelope in the integration of technology and financial governance with its latest offering, Paytm UPI Lite. This feature enables users to carry out small-value transactions for up to 200 rupees at a lightning-fast speed and with hardly any chances of the transaction not succeeding. This change is incredibly useful to users, especially low-income earners and those residing in rural areas, who regularly perform many small-value transactions. With the introduction of Paytm UPI Lite, an individual is now able to pay for everyday small purchases like groceries or transport easily and instantly. Such a user-friendly payment process not only improves user experiences but also expects higher uptake of digital payments by the masses and hence increase the coverage of financial services. Furthermore, the advancement of the Paytm UPI Lite will facilitate the decrease of cash exchanges leading to overall lesser opportunities for money laundering, black market engagements as well as improved financial management of the nation. A government that relies heavily on such cashless transactions will be in a better position to monitor the movement of money, hinder criminal actions, and with a large percentage of the public embracing digital payments, will understand the economic trends better and thus enhance its policies for better governance.

Due in large part to Paytm's QR codes and a host of other products geared towards Indian consumers and merchants alike, India's payment landscape has transitioned from being cash intensive to being digitally focused. Moreover, Paytm Payments Bank, a subsidiary of Paytm, is a leader in the UPI P2M (Person-to-Merchant) payment space, serving as the largest acquirer and collector bank.

It can be said that Paytm's revolutionary solutions have transformed India's payment and business landscape, making digital transactions accessible across the country. The Paytm wallet was among the first few mobile devices widely used for complete transactions when it was first released in 2014. The year 2015 saw the introduction of the QR code payment system by Paytm in the Indian market, a game-changer for vendors particularly petty shop owners who could now conveniently receive digital payments without an internet connection, a POS machine, and no initial investment.

In 2020, Paytm launched the Soundbox, an advanced payment acceptance device that uses IoT technology to provide real-time audio confirmation of completed payments. The device enabled seamless transactions between payers and customers, ensuring both sides are satisfied. Many merchant stores today would have the Soundbox embarrassment, in fact, it has become the norm.

10.2.5 Conclusion

The penetration of Paytm has enhanced financial inclusion of the low economic communities, small merchants, and street vendors by implementing smart technology such as the All-in-One QR code, Sound box, and UPI lite. There has been a surge in payment possibilities as these instruments and services have enabled users who have been economically marginalized to be able to receive and make use of digital payments. In smaller towns, small traders and even some farmers have started receiving payments electronically through the use of Paytms QR codes. Digital transformation like this not only alleviates the income risk that small traders face, but it also aids rural residents in cashless transactions which subsequently boosts the local economy. On top of that, other spheres of partnerships that Paytm had, such as with Uber, have increased areas of influence and how drivers and passengers could pay. These kinds of collaboration not only help in furthering the scope of digital payments but also improve the coherence of the whole ecosystem. With an inclusive approach to partnerships, Paytm has helped many people integrate in the major economy with positive implications on society as a whole.

These goals are consistent with those of the United Nations Sustainable Development Goals (SDGs), in particular "No Poverty" (1), "Decent Work and Economic Growth" (8), as well as "Reduced Inequalities" (10). Such activities have a social dimension as framed by the ESG initiatives which seek to improve the economic engagement and conditions of life for the disadvantaged segments through technology and social cooperation.

The innovative payments of Paytm have not only transformed payment methods especially for micro merchants' but also empowered the governance framework of the nation. By encouraging cashless transactions in the country, Paytm has managed to reduce the cash intensive economy making the need for money laundering, counterfeit currencies, black markets, and criminal activities unnecessary thus contributing towards "Peace, Justice and Strong institutions" (16). For instance, digital payment systems will be beneficial in tracing the movement of money thereby enhancing the soundness of the countries' financial regulations policies. For example, Paytm UPI Lite, which allows quick and almost risk- free small-value payments, gives the Government a better picture where the volume and frequency of such payments enabling precise formulating. Furthermore, Paytm's analytical capabilities have made the government more intelligent on the things that matter. Given the overwhelming amount of transaction data, the government can get clearer signals of the economy's direction and seek undermined policy actions in a timely manner meeting the goal of Industry, Innovation, and Infrastructure (9). These initiatives not only enhance individual financial education, but also improve the health of society's finance, which meets the aspects of soundness of governance in ESG framework, which seek to enhance national governance through more reliable and effective control and the knowledge of the population through the provision of education.

Paytm, who aggressively supports Indian Financial Inclusion, has provided smarter solutions to businesses in regard to offline payment devices, credit, etc., in order to bring 500 million underbanked or unbanked Indians into the economy. This is in sync with the vision of the government to make India a digitized and knowledge-based economy. As a result, Paytm has been able to create strong digital infrastructure which is using technology in the advancement of financial services and products in India.

10.3 EasyPaisa in Pakistan

10.3.1 The Current State of Fintech Development in Pakistan

Pakistan is currently undergoing a process of economic liberalization. Pakistan is in fact a top tier economy among some emerging top ten countries. They have quite a significant growth potential, and that's because of their youth population, developing internet infrastructure, dismally low banking density, expanding e-commerce, and relevant policies issued by regulators. All these factors create an environment with markets opportunities in abundance.

Pakistan boasts a large and diverse banking system, comprising state-owned banks, domestic private banks, multinational banks, and specialized banks. However, the penetration of banking services remains low, with 77% of the areas not covered by banking services, and only 10 branches per 100,000 adults, which is below the Asian average of 16.38. The availability of bank loans to the commercial private sector and individuals is also very small, as extensive government borrowing stifles private capital. As a result, the small and medium-sized enterprises (SMEs) in the region, which are approximately 3.3 million and produce a considerable one-third of national GDP, cannot obtain formal lines of credit. According to figures for credit issued over the month of June 2021, released by the State Bank of Pakistan, the investment/loans issued by banks in the country was around 20.22 trillion rupees. By borrower classification, government sector investment and loans were 12.39 trillion, which is roughly 61.3%, private sector (commercial) loans were 5.17 trillion which is around 25.6%, and the personal loans stood only at 0.67 trillion which is approximately 3.3% of the total investment.

The demographic is huge, because there are about 200 million people, out of which around 100 million are under the age of 35 providing a constant supply of labor. Notably, though, the internet and smartphone penetration in Pakistan is quite low and is also quite backward in payment channels and logistics. Cash is used to complete purchases by over 90% of the population of Pakistan. As such, the majority of transactions made in Pakistan are made in cash. Pakistan is also relatively new to the e-commerce age and the industry has not yet exhausted its full potential.

Since the economic liberalization, the Pakistan Telecommunication Authority (PTA) has gradually relaxed regulatory policies for the telecom sector, attracting significant domestic and foreign investment and promoting the construction of advanced infrastructure. Such policy shifts have contributed to the swift growth in the number of operators and users of the internet, thus causing the further penetration of digital technologies. As per the figures released by the PTA, by 2023 the number of mobile phone prepaid users in the country went over 180 million and there were almost 100 million internet users, the majority of whom were using smartphones to connect to the web. There is, therefore, a huge untapped market for fintech companies looking to invest in the country considering this growing population of netizens. On the other hand, the incidence of banking patronage remains bleak—this comes in at just twenty percent of adults who own a bank account meaning a huge segment of the population still operates in cash. There is a readily available opportunity for Fintech companies to bridge this gap by providing easy mobile payment options, transfer or e-bill to a wider audience and integrate them into the formal economy. In addition, such a swift growth of e-commerce in Pakistan has also opened up new avenues for the expansion of the fintech industry.

Due to an increase in the adoption of online shopping, the popularity of faster and secure modes of payments has remarkably gone up. E-commerce and fintech companies working together not only improve the payment's process of consumers but also help in the growth of the entire digital economy ecosystem in all levels.

Banks and other regulators are also on the frontline advocating for fintech advancement. The State Bank of Pakistan (SBP) has recently developed some practices that support innovation without neglecting financial stability. For example, SBP executed the "National Financial Inclusion Strategy" on January 23, 2020, whose objective is to have 50% of the population have access to financial services by October 2025. In addition, SBP has also initiated a number of regulatory sandbox models under which fintech firms can trial their products and services in a contained manner. These approaches not only enable fintech firms to operate under a more permissive regime but enhance the supply of more foreign capital into the economy as well.

10.3.2 The Belt and Road Initiative Accelerates Pakistan's Move Towards the Mobile Payment Era

In July 2018, during a visit to China, Tariq Bajwa, then Governor of the State Bank of Pakistan (SBP), observed the widespread use of mobile payments in the daily lives of Beijing residents. Impressed by the convenience that mobile payments brought to the public, he expressed an urgent desire to "bring the Chinese experience home" so that the people of Pakistan could also benefit from the tangible results of inclusive finance.

Easypaisa is Pakistan's largest internet finance application, with approximately 20 million users. The most recent version of the product was developed and endorsed by the Chinese ant financial company. Ant Financial, for instance, said independently back in March 2018 that Easypaisa had already teamed up with Alipay with the intention of turning Easypaisa into Alipay of Pakistan with the help of the Chinese-based company. The newly launched version of Easypaisa is the new era of Easypaisa as it also features advanced technologies such as QR code scanning and microloans that enable online-to-offline transactions, which will in turn widen the scope and breadth of financial services for the displaced individuals as well as the small and medium-sized enterprises in Pakistan.

"The new blockchain-based remittance service is a bold attempt to transform the landscape of financial services in Pakistan and bring about a breakthrough, which is a top priority on the agenda of the State Bank of Pakistan," said Tariq Bajwa. "This collaboration demonstrates that the cooperation between China and Pakistan has expanded beyond traditional infrastructure projects to include the tech-financial sector."

According to Pakistani media, the release of the new version of Easypaisa not only enhances the efficiency of overseas remittances to Pakistan and improves the consumer environment but also aids in combating illegal financing and financial crime. It is the properties of the Blockchain which make remittances secure and transparent; within the confines of protecting the user's privacy, all remittance details are encoded on the blockchain system and this wipes out the possibilities of channels for illegal remittances and worldwide financial abuse.

There have been positive developments in e-commerce in Pakistan over the past years as conditions conducive for inclusion of inclusive digital technologies are improving with time. Internet usage within Pakistan has been on the rise, with an estimated number of 30 million users active and this number expected to rise to 56 million by the year 2019. The availability of more internet connections, the lower price of smartphones, and the fast-paced growth of mobile 3G and 4G networks have all combined to make it easier to reach large numbers of mobile internet users. Interestingly, there is a large segment of the Pakistani population that uses and prefers the cheaper Chinese mobile brands, increasing mobile phone usage among the lower and middle classes as more people can use mobile broadband services.

At the policy level, the Pakistani government has launched supportive strategies such as the National Financial Inclusion Strategy of 2015 and the SBP Vision 2020. It is projected that the e-commerce market in Pakistan will experience explosive growth. Statistics show that the total value of the e-commerce industry in Pakistan was $100 million in 2016, and this figure is expected to continue rising, reaching over $1 billion by 2020.

In 2018, the Chinese-inspired shopping festival "Singles' Day" also made its debut in Pakistan. According to statistics, during the trading "Singles' Day" event, the average daily money transfer through Easypaisa on the most popular site, Daraz, was about three times the amount that it is in practice. Within the context of the "Singles' Day" celebration, which happened on November 11, the sales volume was reached a staggering 5.6 times the daily average transaction amount.

Alongside such positive trends, there is also an increasing appreciation of the vast market potential. The remittances from Pakistanis living abroad are worth a decent amount with estimates moderating at approximately $20 billion per year, contributing to roughly 7% of Pakistan's GDP. Carrying out such transfers is not only an important means of sustaining Pakistan's economy but also allows millions of Pakistani families to sustain themselves. Unfortunately, both slow remittance transactions and delays in remittance across countries have been lingering challenges in Pakistan. World Bank statistics indicate that there are 100 million people in Pakistan who do not have the most basic access to financial services. Many such market gaps do exist in Pakistan, and many such potentials and demands provide opportunities. This in future opens pathways for cooperation between the government and the business of China and Pakistan.

10.3.3 Mobile Payment Technology Brings a Sense of Achievement to Various Sectors in Pakistan

With the growth of the Belt and Road Initiative, more and more Chinese Internet entrepreneurs are going out not only bringing modern and convenient ways of life to local residents, but also development, training, and job creation to the countries that host them. In the case of Pakistan, this is very visible as the mobile payment technology and internet-based finance solutions are changing the economy and the social structure of the country for good.

The application of mobile payment technology in Pakistan has greatly facilitated people's lives, enhanced work efficiency, and significantly reduced the high costs associated with time. In the past, common Pakistanis had to go to the banks or related organizations for basic transactions like transfers, paying utility bills, or recharging mobile phones which was costly in terms of time and energy and as a result, long queues were a common sight. Now, with the help of mobile payment applications like Easypaisa on a mobile phone, these processes can be accomplished within several finger touches. This shift does not only enhance the payment experience of individuals but also offers companies better and more effective tools for dealing with finances. For example, smaller vendors can now accept payments from customers through mobile payment systems without worrying about cash and improve the control over their income and outgo. At the same time, the popularity of mobile payments has also led to an increase in e-commerce activities with a lot of small businesses and people starting to offer goods and services for sale on the Internet, further boosting the growth of the digital economy.

Pakistan has also been helped by microfinance service providers who rely on the internet for their services which has been greatly beneficial to the needs of entrepreneurs and business owners in all industries. In the past, micro and small enterprises (MSEs) and individual entrepreneurs faced several challenges with borrowing loans including long and tedious application processes, time-consuming approval cycles, and high barriers to entry. However, these issues have been resolved thanks to the support of internet finance service providers. Consumers are able to take a whole loan application in a span of a single phone call and will be able to access the required funds without any hassles at all. For instance, an Ant Financial-owned company Easypaisa has a microloan portfolio which enables its users to apply for loans and get them within a single call. These instant loans do not only assist startups and individual merchants in meeting their short-term cash flow requirements but they also stimulate the desire to start a business. For a lot of people living in rural areas who cannot be able to reach local banks providing microfinance helps them fulfill their life goals.

Moreover, the spread of mobile payments and microloans contributes to greater financial inclusion. According to World Bank data, as of 2017, only about 25% of adults in Pakistan had bank accounts, with women accounting for just 7%. Mobile payments as well as Internet finance platforms have helped extend financial access to more people, especially those who have been out of the formal system for a long period. Through basic phone activities even those without bank accounts are able to perform payments, send and receive money and borrow which improves their lifestyles and believe in the financial system.

Farmers, for example, preferred to buy seeds, pesticides, and fertilizers with cash because the cost of sending a package in rural areas of Pakistan was twice as high as in cities due to high shipping and logistics costs. The gradual development of online payment technologies, accompanied by end-to-end control of logistics processes, substantially decreased transaction expenses. At the current rate of development, it won't be long before farmers in remote regions of Pakistan can conduct low-cost online transactions and enjoy fast offline services via their mobile phones.

By combining "local partnerships + technological empowerment," integrating local expertise with Chinese technology, local companies can drastically reduce research and development time and costs. With such a model, Alipay was able to outgrow localized editions of "Alipay" in the countries and regions along the belt and road initiative which include: India, Pakistan, Bangladesh, Thailand, the Philippines, Malaysia, Indonesia, and South Korea. Although these localized versions could be different when it comes to the names, it is the case that all of them receive the technological empowerment from Alipay thus enabling the local people to enjoy the privileges of the digital inclusive age. In the longer term, Ant Financial, after replicating the models of Southeast Asia, hopes to establish acquirement programs for human resources in Pakistan targeted at internet innovation technologies. Such training programs are likely to cover e-commerce, fintech, and cloud services and are primarily focused on the development of novel ideas within the region.

10.3.4 *Blockchain "Breaks the Ice" to Facilitate Remittances for Overseas Workers*

In July 2018, during his visit to Beijing, Tariq Bajwa, then Governor of the State Bank of Pakistan (SBP), made a special trip to Sanyuanli Market to experience China's mobile payment system firsthand. Upon entering the market, he noticed that over 90% of the stalls had Alipay QR codes displayed. Whether it was for a few cents' worth of vegetables or high-end seafood costing hundreds or even thousands of yuan, consumers could easily scan and pay. The market was filled with the clear sound of female voices announcing, "Alipay received * * yuan." At stall number 27, Tariq Bajwa stopped and, using the phone of Ant Financial's Chairman and CEO, Eric Jing, tried scanning a QR code to purchase a bag of raisins for 33 yuan. The vendor, Mr. Huang, shared how beneficial Alipay was: "The

Alipay QR code is very useful; it not only makes our business more convenient, but I also use it to reimburse medical expenses, and my boss uses it for loans." According to Eric Jing, millions of small merchants across China have become the smallest yet integral part of the digital economy. A simple QR code has become their ticket to the digital inclusive era.

Inspired by this visit to China, just 168 days later, on January 8th, the Governor of the SBP inaugurated the country's first blockchain project in Islamabad. This project which is a digital inclusive solution from Alipay, marked the introduction of the blockchain technology-based cross-border remittance system in South Asia.

Pakistan is among the top 10 recipients of remittances in the world, as it gets around US$20 billion per year through its over 10 million nationals working abroad. For most of the time, these wage earners faced lots of problems and difficulties in transferring money back home. Everyday remittance services forced them to line up in banks and agencies that operated only within hours that were inconvenient to them, which was very expensive. But now everything is changing gradually with the emergence of fintech and developments under the Belt and Road Initiative.

Alipay and Valyou, a Malaysian remittance service provider, are working with each other to make this possible. This makes it possible for Pakistani overseas WEPs to send remittances at ease, faster and in a secured manner. People send money across borders whenever they want and from wherever they want through a mobile app regardless of time and place. This enables Pakistani workers based in Malaysia to remit money directly to Pakistan to their Easypaisa accounts, the mobile payment platform introduced by Ant Financial in Pakistan. Such a smoothened experience of remittance makes it more convenient to use and simplifies and speeds up cross-border transfer processes.

Alongside convenience, this new remittance scheme has a high cost-benefit ratio. Typically, remittance channels charge high fees and offer less favorable exchange rates, however, this is not the case with the Valyou and Easypaisa partnership. User data available indicates that people can save as much as 50% of the remittance costs. Such is the great advantage to many of the Pakistani families who depend on remittances for survival. Moreover, the transparent transaction process and real-time fund transfer feature enhance user security, allowing them to instantly check the status of their remittances and confirm receipt, reducing the risks associated with intermediaries.

The effective execution of this new service is made possible thanks to the support of the relevant authorities. Both the SBP and the Central Bank of Malaysia (BNM) have been supportive and rendered some policy assistance that was required. Such moves help advance the challenges of financial interdependence of the two countries and provide a good example for other countries as well. In fact, Alipay has already started marketing the same models of a digital inclusive finance services in a couple of countries along the Belt and Road so as to allow more people to take advantage of the features of fintech.

Pakistan also views this change in the dynamics of cross-border remittances as more than a transformation in ways of sending money but also as a major development on the socioeconomic fronts. To begin with, it significantly enhances the standard of living of foreign workers and their families and eases them in managing and spending their earnings. Secondly, to the extent that remittance costs come down, the volume of funds that would be remitted to Pakistan would be able to boost the overall level of consumption and investment in the economy. Lastly, this digital form of financial service also assists in furthering the cause of financial inclusion whereby more people are brought into the mainstream formal economy and are offered a wider array of financial facilities.

Easypaisa's introduction of the mobile savings service, Khushaal Munafa, and related insurance product, Khushaal Beema, has not only significantly enhanced financial inclusion in Pakistan but also played a crucial role in improving national governance. Easypaisa provides access to an attractive low-risk investment option, which pays a 9% return on deposits and provides life and accidental death insurance for those who have traditionally been unable to access formal financial services. Especially for residents in rural areas, this innovative service model not only allows them to get acquainted with modern forms of financial management but also restores their faith in the financial system in general. Numerous people, prior to the usage of Easypaisa, had no knowledge about insurance and its advantages, now are able to get basic insurance policies through simple operations with their accounts, which increases their capacity to tolerate perils. This not only enhances the living standards of the people but also helps maintain social order.

Furthermore, Easypaisa's collaboration with the Sindh Education Reform Program (SERP) further demonstrates its potential in enhancing national governance. With the help of Easypaisa's services, the government of Sindh will be able to target educational subsidies to families living in the remote, rural areas while focusing on the education of girls, as they aim to target around 400,000 beneficiaries. This reduces the risk of corruption and waste as funds are channeled to the correct intended beneficiaries and thus enhances the effectiveness of government programs. More significantly, the use of the digital payment mechanism assists the government in monitoring the movement of funds more effectively, assessing program performance, and making timely interventions when the need arises. In this manner, Easypaisa not only enables the government to provide a better-targeted distribution of social assistance but also allows for increased transparency and accountability of public finances and overall governance.

From a broader perspective, the success of Easypaisa reflects the important role of fintech in promoting financial inclusion and national modernization. By the end of the year 2020, Easypaisa registered over 8.1 million active mobile wallets and transacted 1.5 trillion Pakistani rupees. These numbers suggest that there is certainly an increasing and promising trend of utilization of digital payment solutions in Pakistan as a core infrastructure for both the economic and social enhancement of the country. By offering customers affordable and effective financial solutions, Easypaisa not only stretches the boundaries of financial accessibility but also fuel consumption, investments, and greater economic activity. At the same time, the government can use these systems to gather essential information that can be used to develop better and more reasonable policies and to enhance the quality and delivery of public services. To summarize, Easypaisa's operations in Pakistan are a good illustration of how much fintech can improve financial inclusion and country's governance and serves as a useful case study for other regions in the world.

10.3.5 Conclusion

The establishment and widespread adoption of Easypaisa have undeniably brought about significant changes for Pakistan and its people. Also, under the Belt and Road initiative, with Ant Financial's support in both finance and technology, Easypaisa will continue expanding and bringing more comfort and change to the country.

Easypaisa has significantly enhanced financial inclusion in Pakistan through the introduction of mobile savings services like Khushaal Munafa and insurance products such as Khushaal Beema. These services provide secure and attractive savings options and basic insurance coverage to those who were traditionally excluded from formal financial services. Particularly in rural areas, many residents, who previously had little understanding of insurance and its benefits, can now access these services through simple account operations. This innovative service model helps them manage their finances better, strengthens their trust in the financial system, and improves their ability to withstand risks. Additionally, Easypaisa's collaboration with the Sindh Education Reform Program (SERP) further demonstrates its potential in fostering inclusive partnerships. Through Easypaisa, the Sindh government can efficiently and transparently distribute educational allowances to families in rural areas, especially supporting the education of 400,000 girls. This initiative ensures that funds reach beneficiaries directly, reducing the potential for corruption and waste, and enhancing the efficiency of government programs.

These endeavors correspond squarely with the targets set out in the United Nations' Sustainable Development Goals (SDGs), in particular "No Poverty" (1), "Quality Education" (4), and "Reduced Inequalities" (10). Such efforts reflect the social dimension of Environmental, Social and Governance (ESG) framework as well, which focuses on enhancing economic inclusion and life improvements of marginalized communities through technology revolution and social collaboration.

Easypaisa's role in enhancing national governance is equally noteworthy. By providing convenient and low-cost financial services, Easypaisa has expanded the reach of financial services, stimulating consumption, investment, and economic growth. Such revenues not only improve the living standard of the citizens but also bring in more tax revenues to the government thus enabling more funding of public services and social welfare programs to the citizens. In addition, the funds which are distributed using the digital payment platform of Easypaisa are managed more efficiently by the government, making it easier to monitor the fund's flow, project effectiveness, and changing policies in tackling fundamental issues. For instance, it can be shown that by providing educational allowances within Easypaisa, the government ensures that funds are utilized for the intended purposes and thus are able to track the funds throughout the process. This visibility and traceability, in turn, significantly enhance the regulatory powers and the decision-making processes of the government and can greatly aid in curbing corruption while at the same time enhancing the level of public service delivery. Furthermore, Easypaisa's data analytics capabilities provide the government with valuable insights. By analyzing large volumes of transaction data, the government can gain a more accurate understanding of economic trends and make timely policy adjustments. This data-driven decision support system is crucial for improving overall governance.

These efforts align with the SDGs, with particular reference to "Decent Work and Economic Growth" (8) and "Peace, Justice, and Strong Institutions" (16), and they also reflect the governance aspect of the ESG framework, emphasizing transparency, efficient regulation, and data analysis to enhance the overall governance of the nation.

Conclusively, the interventions instituted by Easypaisa in Pakistan demonstrate the possibilities of fintech in advancing financial inclusion and good governance in achieving sustainable development. The further expansion of Easypaisa which is backed up by the Belt and Road Initiative and Ant Financial, tend to bring even more advantageous outcome to the citizens of Pakistan towards building an inclusive, just and prospering society.

PART V

Alipay+ for SDGs—Digital Infrastructure for Sustainable Tourism, Decent Work and Inclusive Development

CHAPTER 11

Sustainable Fintech for State Capability and Inclusive Tourism: Cross-Border Payment Infrastructure for 28 Million International Travelers Via Alipay+ Ecosystem in Thailand

11.1 The Fragmentation War of Digital Payments in Southeast Asia

Due to the booming economy in Thailand, the use of smartphones has become commonplace within the population. The Internet in the Twenty-First Century absolutely is part of the daily lives of people around the world and in Thailand it is even more accessible. It has shaped a culture of making mobile payments especially among the youth. Therefore, there is a partnership between retailers and international companies to enable mobile payments within retail stores because such opportunities exist. As the trend of making mobile payments rises among the youth population in Thailand, worldwide firms are trying to tap into the market and have started offering services in collaboration with retail chains.

The revenue of the mobile payment sector in Thailand stood at USD 32.75 Billion in the year 2023 and it is expected to register a CAGR of 14%, reaching USD 72.49 Billion in the next five years. The development of mobile users resulting from more regular activities seeking smartphone ownerships is adding to business expansion in other industries including financial services using mobile wallets which are cheaper and easier to

J. Sun, *Sustainable Financialization*, Sustainable Development Goals Series, https://doi.org/10.1007/978-981-96-3580-1_11

use. TrueMoney's customer base in Thailand and LINE Pay in Japan will increase upto more than 200% from 2020 to 2025. The occurrence of the COVID-19 pandemic made a number of changes to how payments were done such as reducing the number of cash payments, changing people's perspectives from the option of paying in a store to paying online, and the adoption of swift payment options. Such transformations created new windows of opportunity for mobile payment service provider segments in Thailand.

Essentially, the tourism sector, which is one of the cornerstones of the economy of Thailand, has opened avenues for rapid growth of mobile payments in that country. According to estimations, as of 2023, Thailand's overall population is about 66.1 million people. The tourism segment of the economy of the country in question has for many years been one of its key revenue earning sectors. Specifically, in the whole year of 2023 (from January 1 to December 31), the total amount of foreign guests visiting Thailand came to 28,042,131 people with revenues totaling an impressive 1.2 trillion Thai baht. Of these, the top five countries that sent the highest number of tourists to Thailand were: Malaysia (4,563,020), People's Republic of China (3,519,735), South Korea (1,658,688), India (1,626,720), and Russia (1,481,878). It is clear that the main market target for Thailand's tourism is the Asia-Pacific region. This also emphasizes the reason why the Thai government is actively seeking to expand the acceptance of the digital wallet in the Asia-Pacific region.

Almost 10 million Chinese who are used to QR codes and digital payments visit Thailand each year which creates strong market demand for strengthening electronic payment capabilities. Not only does this improve the economic modernization level of the country and the region from an ESG (Environmental, Social, and Governance) development perspective but it also empowers people. It creates new supporting growth points for convenience and for comprehensive development of the society.

11.1.1 Inconvenience Arising from the Fragmented Payment Methods

Unlike the duopoly of WeChat Pay and Alipay in mainland China, the mobile payment market in Southeast Asia, led by Thailand, was highly fragmented before 2022. Zeng Jia, a 32-year-old frequent business traveler from China, often encounters this complexity. Before each trip, she exchanges cash for the local currency. After her expenses, she's left with small change that isn't worth converting back, gradually turning her into a collector of Southeast Asian coins and notes.

She explains, "Many places in Southeast Asia accept digital payments through local apps, which may not be interoperable. I don't want to download a new app, go through verification, and bind my card every time I visit a different place—it's just too much hassle. Cash is universally accepted, but sometimes you end up with a few coins or small bills that are hard to spend, so they become keepsakes or wait to be used on the next trip."

A significant part of its population is youth, and for this region with over 650 million, the potential for the growth of the digital economic aspect is enormous, which has seen many players coming to the region. There is no denying the fact that the competition among the digital payment providers in the Southeast Asia region is very high, considering the presence of local banks, tech companies, and telecommunications. Due to this competition, a good number of the players have already exited this contest.

In Thailand, for example, several mainstream payment methods include TrueMoney, Thai QR, and Rabbit LINE Pay.

TrueMoney, a digital wallet launched in November 2016, was co-founded by Ant Financial and Ascend Money, a Thai payment company. It enables more than 50 million TrueMoney users to make money transfers or cashier payments at convenient locations, such as the 7-Eleven, Family Mart, CP Fresh Mart, and True Shops. Additionally, users can shop via the internet, recharge games, and pay more than seventy types of bills like utility bills, transportation fares, taxes, and insurance premiums. For this reason, TrueMoney is also referred to as "the Alipay of Thailand." TrueMoney, which currently has presents in Thailand, Vietnam, Cambodia, Myanmar, and Philippines operates in the e-money continuum covering various countries in South East Asia. With over 65,000 agents, TrueMoney has the largest regional agent network and ranks as Thailand's top e-wallet. TrueMoney allows payments at 7-Eleven, payment of bills, payments in stores, payment of credit card bills, utility payment, purchase of insurance, payment for subscriptions to Netflix and shopping either Google Play or Apple Store major cineplex tickets, vending machine payment, game recharge and mobile data top-up, or by attaching a debit or credit card.

In Thailand, most people have a mobile application to make payments through a QR code. Many people use the Thai QR code to make payments, with 75% of consumers having used or are using it. Other Thai QR supporting banks include Bangkok Bank, Kasikornbank, Siam Commercial Bank, UOB, and Krung Thai Bank. Thai QR is used throughout the country not limited to small stores, street vendors and even large department stores, and e-commerce platforms.

Rabbit LINE Pay is a popular digital wallet in Thailand. In March 2016, LINE, the Japanese social app giant and the most popular messaging app in Thailand, partnered with Rabbit, a smart card provider for public transportation and retail, to launch Rabbit LINE Pay. This integration expanded offline payment scenarios. Rabbit LINE Pay offers QR code payments, online shopping, bill payments, and money transfers, allowing users to pay with their account balance or linked debit and credit cards. As of May 2021, Rabbit LINE Pay had over 8 million registered users and 300,000 merchants.

The Rabbit card is the only transport card accepted on the BTS Sky Train in Bangkok. When integrated with LINE Pay, Rabbit cardholders can link their card to the LINE app. Real-name registration is available at BTS station counters. Once completed, Rabbit LINE Pay users can view and use the QR code for payments within the app, as well as for other online payment functions. Additionally, they can use the physical card for offline payments, with the card's balance shared with the Rabbit LINE Pay e-wallet. However, while it provides convenient payment options, some users have questioned its limited use cases (Fig. 11.1).

In the Malaysian market, the ride-hailing app Grab launched GrabPay at the end of 2016. By 2018, it had become one of the largest mobile payment platforms in Southeast Asia. Nowadays, GrabPay is available in six countries. Gojek, similarly, which also started as a ride-hailing service, entered the payment space and launched GoPay in 2015.

Furthermore, even the state-owned businesses are penetrating into the digital payment sphere. In July 2019, a group of Indonesian state-owned companies came together and launched the payment platform LinkAja for the objectives of becoming the most advanced digital payment service provider in the nation.

Nonetheless, the fierce rivalry for space within the digital payment's sector in Southeast Asia's region hasn't come without bodies. Vcash, one of the first e-wallets which was part of Malaysia's mobile provider Digi, was shut last November 2019 within two years of its launch.

Fig. 11.1 Digital Payment Methods Supported by FamilyMart in Southeast Asia (*Source* Internet)

The battlefield in the Southeast Asian payment market is nothing short of cutthroat, and in the modern age of capitalism it seems a few players are aggressively trying to expand their working domain by merging with or buying out competitors. In March 2018, mPay, one of the three big mobile network operators in Thailand owned by AIS, had launched a partnership with Rabbit LINE Pay. They planned AIS's Rabbit LINE Pay service to become the number 1 mobile payment in Thailand within 3 years timeframe.

11.1.2 The Dilemma of Central Bank Capacity Building Behind Digital Payments

The digital payments landscape in Southeast Asia is characterized by a mismatch between the demand and the availability of supply. Compared to Western countries, credit card penetration in the region is relatively low, with people generally relying on cash and bank transfers. This has sparked the emergence of online banking and E-wallets as alternative payment methods. In fact, according to PwC's 2019 Global Consumer Insights Survey, six out of the ten regions with the highest adoption of mobile payments are found in Southeast Asia. It is evident that the trend for mobile payments is growing especially in Thailand making foreign investors to scramble to enter the market. These firms are now collaborating with retailers to provide services in physical stores, matching the trends of younger customers who prefer more advanced, efficient, and electronic payment systems.

According to Google, Temasek, and Bain & Company, the overall transaction volume of the Southeast Asian digital payment industry is anticipated to increase at a rate more than that of five times between the span of 2019 till 2025 where the target would be $114 billion. As of 2023, the mobile payments industry in Thailand is worth about $32.75 billion US dollars with a CAGR of 14% which is expected to reach $72.49 billion US dollars in five years. Also, the volume of transactions carried out on the internet is rapidly growing. The trend increases the market needs of consumers for the digital payment service providers (Fig. 11.2).

The Thai government initiated in 2016 the introduction of the growth and transformation of the Digital Economy and Society Development Plan in Thailand. Until recently, cash payments were the most popular method of conducting financial transactions in Thailand. Statistics, however, show a gradual decrease in the use of cash. In 2022, it was estimated that cash payments comprised 50 percent of all point of sale transactions. This figure dropped by 7% when compared to the previous year. Even so, Thailand is one of the regions with the heaviest use of cash in Asia. Networked cash is preferred due to the presence of the underbanked, and unbanked populations who constitute 45% and 18% respectively of the population. The underbanked segment is the segment with one savings account while the unbanked have no access to any financial service or product. The government however has initiated strategic activities that would already see Thailand position itself in the twenty-first century on a number of irreconcilable elements.

As an effort to encourage the use of digital payments the Thai government introduced the National e-Payment Master Plan which gives the infrastructure and connectivity framework necessary for advancing Thailand's digital payment ecosystem. In January 2017, the government and private sector launched a national project Promoted to be PromptPay. Every lawful holder of a Thai bank account, be it a Thai national or

Fig. 11.2 Forecast of the Market Size of Thailand Mobile Payments (*Source* Mordor Intelligence)

Foreign nationals, can register for this service since they possess an accounts which the banks have already authenticated. A national ID number and/or a mobile phone number are sufficient for us to be able to identify him when using PromptPay for making interbank transfers. Four months after its launch, over 4 billion Thai baht was transacted by PromptPay over 7.5 million transactions. By June of 2017, the figure was at least 28.3 million Thais registered in the system. As it stands, it is the most preferred payment system in Thailand, with PromptPay registering an estimate of 67 million users by 2023.

Also in 2017, the companies Mastercard, Visa, and UnionPay International worked together to launch the first standard QR Code supported by the Bank of Thailand. The Thai market is moving towards a cashless economy as the government of Thailand in conjunction with the Bank of Thailand promotes the use of digital payments. The Bank of Thailand has proclaimed as follows, "QR code technology allows to broaden the possibilities for consumers in terms of financial services, increases efficiency in the performance of business operations and lowers the cost of business." The standardization of QR code system of Thailand began with domestic payments and was eventually extended to 5 commercial banks including Siam Commercial Bank, Kasikornbank, Bangkok Bank, Krung Thai Bank, and TMB Bank. Recently, connections to Indonesia, Malaysia, Vietnam, Cambodia, Singapore, and Japan through cross-border interbank functionality have been established within the QR code payment system.

Thus, this particular integration of payment systems into the application is expected to provide in more ways than just a few, an incentive for users to conduct bank transfers via PromptPay. Likewise to that of the Central Bank's National QR Code Standard of the Philippines (QRPh) as well as that of Indonesia's Quick Response Code Indonesian Standard (QRIS), this low-cost and convenient mechanism allows businesses to necessitate their clients to simply need to use one QR code that they can scan and make payment. This system recorded about 13.39 million transactions in 2020, marking an increase of 49.14% from 2019 figures. There were about 7.4 million QR code payment acceptance points across the terrain of Thailand as of August 2020.

Out of these positive statistics, it should be understood that PromptPay is more of an integration of interbank transfer with the intention of encouraging people to use less cash and make lower bank transfer fees. In particular, the target consumers of PromptPay are Thai banks account's Legal person who can be identified by mobile phone number, ID number, company registration number, or by a number of a digital wallet. By the guidance of the user's manual provided by the Bank of China, in the process of using PromptPay, users are expected to first have registered their proxy accounts which can be a Thai ID number of registering phone number only for payment purposes. Also, registered users are provided the facility to transfer funds to one another just by mobile phone numbers and the interbank transaction charges are less. The Government also has provisions to use PromptPay for the purpose of making tax refunds or distribution of social benefits.

In as much as PromptPay is useful for both day to day local payments and for small businesses and customers, the Bank of Thailand's main aim is to meet payment needs for tourists, especially from Southeast Asia, who need respect and convenience while on the go. PromptPay is a solution that demonstrates the desire and will of Thai government and the Bank of Thailand to capacity build but its core focus remains to assist the citizens of Thailand. Without a doubt, as a strategic project, prompt Pay has inched the country towards the achievement of a cashless society. At this rate, however, Prompt Pay might remain to be viewed as a public service for quite some time with the backing of banks and other financial service institutions and not meeting the urgent call of mobile payment systems.

Furthermore, another barrier that contributes to the low success of PromptPay is its low acceptance by Thai merchants. Being accustomed to digital payment methods with a fixed price structure is hard for the historically price negotiating Thai market. This is also applicable to the Thai market where local merchants prefer cash transactions thus leaving very few cases of digital payments such as PromptPay among street vendors. In an objective sense, such a situation creates an opportunity for international providers of payment systems to enter the market. Innovations by social commerce platforms like LINE are adapting by integrating bargaining features into digital payments, offering flexible pricing through payment links.

11.2 Bridging the E-Wallets: Unleashed Power of Digital Payments

As early as 2014, Alibaba, along with its subsidiary Alipay, signed a cooperation agreement with Kasikornbank in Thailand for the purpose of expanding business between the two nations. The focus was on assisting Thai small enterprises to market their products to Chinese customers through various Alibaba platforms. At the same time, the consumers in China would be able to buy Thai products like jasmine rice, tropical fruits, and handicrafts from Alibaba's online platforms that would source them directly from Thailand suppliers.

Under this arrangement, local Thai retailers could be connected to the Alibaba through Kasikornbank and reach targeted customers through promotional activities tailored for them. The partnership also made it possible to connect Alipay with Kasikornbank's payment gateway. This made it possible for customers to pay directly with Alipay when they made purchases from Thai vendors. After establishing a strong presence in Europe and Japan, it seems that the web payment system Alipay has found its next destination—Thailand—and this latest partnership promises to be equally advantageous.

The Tourism Authority of Thailand (TAT) in October 2022 signed a letter of intent with the Chinese payment platform in anticipation of its integration into Thailand increasing the influx of Chinese visitors to the country. In 2019, nearly 10.5 mn Chinese tourist traveled to Thailand, many of whom may have had access to Alipay, which is already operational in the region. Yet it was just the beginning of the Alipay buildout in Thailand and the rest of Southeast Asia. First, Ant Financial only engaged with local merchants to secure customers among traveling Chinese. At the same time, it invested in local partnerships to deploy local branded e-wallets, with Ant Financial's group giving technology and expertise for Alipay's success in other countries. Since 2015, Ant Financial started looking for foreign partners to develop local copies of Alipay, with India, Thailand, and others under its ambit. This underwent a gradual metamorphosis to its globalization strategy model of "1+9," where "1" signified China and "9" signified nine other countries. Out of these nine countries, four are based in Southeast Asia.

This strategic plan enabled the Ant Financial to utilize its technological experience and support local e-wallet providers in order to enhance the digital payment system infrastructure of each country. Through these associations, Ant Financial not only increased their worldwide presence but also adjusted its offerings to the specific requirements and regulations of each market.

11.2.1 Alipay+'s Implementation of E-wallet Practices in Thailand and Malaysia

Alipay+ practices are well-represented in Thailand and Malaysia as well. Information was revealed in Kasikorn's press release regarding Alipay's partnership with Kasikornbank in 2017 and aims at promoting QR code technology to Thai citizens so as to allow them to be able to receive payments from the Chinese tourists.

Alipay's partnership with Kasikornbank in Thailand is a mutual benefit arrangement as it allows Kasikornbang to increase its QR code system's attractiveness to retailers by giving them more potential customers.

The Nation, a Thai publication, disclosed this arrangement, which is strategically designed to broaden Alipay's influence in Thailand and enhance the attractiveness of Kasikornbank's QR code system to retailers by expanding their potential clientele.

While launching a press conference to publicize this alliance, Kasikornbank declared that it was the first Thai bank to sign up for QR codes through the application. Now the bank plans to push this service to a 200,000 store network across Thailand as a way of utilizing the newly launched service.

Blow are four case studies of Kasikornbank's implementations below, featuring Central Department Store, Chatuchak Weekend Market, Suvarnabhumi International Airport, and 7-Eleven convenience stores.

Central Department Store: Enhancing Shopping Experience for Chinese Tourists with Convenient Payments

For many Chinese tourists to Bangkok, the Central Department Store is an attraction in itself. Central began collaborating with Kasikornbank in 2017 in order to implement QR code payments via Alipay. To various merchants positioned in the mall, the store distributed Alipay's QR code payment terminals together with the corresponding training and technical implementation. To support the new payment means, they offered several sales promotions such as a discount or a freebie if a customer used Alipay for payments at the store.

Since the beginning of the partnership, according to the Central statistics, the volume of expenditures of Chinese tourists increased approximately by 30%. The exposure contributed to income for the mall as well, making it even more profitable in peak tourist times. By working with Kasikornbank and Alipay, Central not only increased its own competitiveness but also participated in the digitization of the retail industry in Thailand. This partnership brought in the strength of digital payments and allowed more consumers to access effective and safe payment services and merchants to have more opportunities. In the first place, Central owes a lot of its brand images to the penetration of Alipay in China which in turn brings more traffic in potential customers.

Chatuchak Weekend Market: Traditional Bazaar Meets Digital Payment Trends

The Chatuchak Weekend Market which is among the largest outdoor markets in the country having around 15 thousand stalls selling clothes, handicrafts, and other items has introduced cashless payments owing to the surge in the number of the Chinese visitors. In partnership with

Kasikornbank, the market administration put up several Alipay QR code payment terminals at many stalls. Many small vendors were trained in the method of receiving payments via QR codes. The market administration also engaged in activities aimed at increasing the number of participating merchants such as tourism—Alipay Payment Discount Days to encourage tourists to use Alipay.

As more and more Chinese tourists take to Alipay, the number and value of purchases made by them keeps increasing. For small businesses, it not only alleviates the challenges of changemaking but also increases the security and efficiency of the transaction. Even more crucial, through Kasikornbank and Alipay, Chatuchak Weekend Market harmoniously coupled a conventional marketplace with advanced payment technologies, providing a boost towards the digital transformation of the market. This partnership energized the potential of digital payments enabling many small vendors to partake in the digital economy enabling the market to thrive. The ease of transacting through Alipay also motivated the younger consumers to the market urging them to spend more.

Suvarnabhumi Airport: Seamless Payments Enhance International Traveler Satisfaction

Suvarnabhumi Airport, Thailand's main international aviation hub, welcomes millions of international travelers each year, with a significant portion being Chinese tourists. To enhance the travel experience for these visitors, duty-free shops, restaurants, and retail stores within the airport started accepting Alipay. Kasikornbank also set up dedicated service desks to provide payment consultation and support for Chinese tourists. Additionally, the airport management launched several promotional activities, such as extra discounts or loyalty points for using Alipay, to attract more Chinese tourists to use this payment method.

Through the collaboration with Kasikornbank and Alipay, Suvarnabhumi Airport not only improved its service levels but also enhanced its international competitiveness. This partnership leveraged the power of digital payments, allowing more Chinese tourists to enjoy convenient and efficient payment experiences, thus promoting the economic development of the airport.

7-Eleven: Efficient Payment Services Anytime and Anywhere

7-Eleven is Thailand's most prominent convenience store chain and has numerous outlets throughout the country selling everyday products and services. To further meet the needs of their customers, including Chinese tourists, 7-Eleven has joined forces with Kasikornbank, which enables Alipay payments in all 7-Eleven outlets throughout the country. People performing payments can do so by simply scanning the QR code. As for the implementation, 7-eleven conducted the required training and technical assistance and put up Alipay payment notices around the stores for visibility.

For local residents, Alipay payments have made it much easier for the local population to process transactions anytime without breaking a sweat. Working with Kasikornbank and Alipay helped 7-Eleven grow its competitiveness, but more importantly helped digitalize Thailand's retail industry which aided the industry's growth.

Four case studies illustrate that, whether in large shopping centers, traditional markets, international airports, or convenience stores, the application of e-wallets has significantly unleashed the power of digital payments. These partnerships have improved the payment methods and overall experience of shopping for Chinese visitors and the other users. More gratifying is the fact that this advancement in technology has catalyzed the advancement of the retail and service industries in Thailand, thus raising its competitiveness. The collaboration between Kasikornbank with Alipay in extending payment facilitations enables merchants to have better business prospects thereby advancing digital economy growth.

As part of its ambitions to expand in Malaysia, Maybank has teamed up with the Ant Financial Services Group in 2017. This relationship was intended to give Alipay users in Malaysia the ability to make cashless transactions with selected local vendors. For the time being, Maybank has secured 37% of the Malaysian market's merchant sales volume, making them the market leader in the region's merchant sales volume.

DuitNow QR, operated by PayNet, which is Malaysia's national payment network and central financial infrastructure, serves as the National QR Standard. It allows participating merchants to receive instant payments from the customers of the participating banks and e-wallets in Malaysia and other linked countries through a unified QR code. In payments apart from the current transactions; the two firms Ant group and PayNet also planned to assist in each other's digital marketing campaigns in 2023 to promote business directly to users' wallets, turning them into a more effective marketing tool.

In September 2018, Ant Financial was authorized by Bank Negara Malaysia to set up a joint venture with Touch 'n Go Sdn Bhd, which will provide mobile e-wallet services. With the Touch 'n Go e-wallet, users can now scan QR codes to enter the subway making Malaysia the second country in the world to enable QR code entry mobile payment for public transport. According to Alipay+, more payment partners are expected to join the bandwagon. As of 2023, tourists from eight Asian markets using Alipay+ partner e-wallets can pay over 1.8 million merchants in Malaysia has locations using QR codes.

Alipay has also been placing focus on other regions in Southeast Asia with the rollout of digital payment products. In July 2016, Ant Financial participated in the Series D funding round of Singapore-based fintech company M-Daq. In April 2017, ANT entered its first collaboration with Indonesian media group Emtek, which launched the co-venture DANA in March 2018 to provide financial services to BBM, the second-largest messaging app in Indonesia, together with the latter. That same year, Ant Financial also made a strategic investment in Mynt, the fintech subsidiary of the Philippines telecommunication provider Globe, to enhance the GCash service.

As of 2019, citing Reuters, among other transactions Ant Financial had acquired up to 50 percent of shares in Vietnam e-wallet eMonkey which had been granted all the operational licenses by the State Bank of Vietnam (SBV).

In October Ant Group added another country to the list of its partners after the launch of the campaign initiative SingapoReimagine with Alipay+, which further timezone brings the company closer to Singapore. This program allows six e-wallets from five Asian countries to use QR codes for payments in Singapore which aims at promoting tourism and shopping in Singapore for foreign visitors.

11.2.2 Integrating E-wallets to Facilitate Travel across Asia with a Mobile Device

Currently, Alipay+ has made significant strides in building digital payment capabilities across Southeast Asia and beyond. It has integrated 22 e-wallets,[1] including: Alipay in Mainland China, AlipayHK, Macau's Alipay and Mpay, the Philippines' BillEase, HelloMoney, GCash, and Maya, Malaysia's Boost and Touch 'n Go e-Wallet, Singapore's Changi Pay, EZ-Link Wallet, and GrabPay, Indonesia's DANA and Doku, Mongolia's Hipay, South Korea's Kakao Pay, Toss Pay, and NAVER Pay, Thailand's LINE Pay and TrueMoney wallet, and Italy's Tinaba. Additionally, Alipay+ has partnered with several banks in Southeast Asia to build a more convenient digital ecosystem. These partners include: BPI (Bank of the Philippine Islands), K PLUS from Kasikornbank in Thailand, MyPB from Public Bank in Malaysia, and OCBC Digital in Singapore. In the "buy now, pay later" field, Alipay+ collaborates with multiple financial platforms such as Akulaku PayLater, Klarna, and Kredivo to offer flexible buy-now-pay-later and installment payment solutions.

While Alipay has enabled Chinese consumers to travel globally with just their smartphones, Alipay+ extends this capability to Asian consumers, allowing them to use their preferred e-wallets across borders. Since the collaboration between the Singapore Tourism Board (STB) and Ant Group in October 2022, many small blue signs with the Alipay+ logo have appeared in shopping malls, hawker centers, and taxis in Singapore. These signs not only feature the familiar Alipay and AlipayHK for Chinese tourists but also e-wallets from various Asian countries. This means Chinese tourists in Singapore can continue using Alipay while those from the Philippines, Malaysia, Thailand, and South Korea can transact using their respective e-wallets. The emergence of Alipay+ makes it easy for consumers who want to use their preferred e-wallet as well as merchants who are willing to accept payments from a variety of e-wallets.

Alipay+'s projects are much more complex than the monotony of being called an app for payments. It positions itself to provide an efficient answer to cross-border payments which have been around for a long time. This is how Zhao Ying, President Ant Group International Business, portrayed the company in Singapore: "Alipay+ is by no means just another payment app; it is a next generation innovative cross-border

[1] Alipay+, 2024, https://www.alipayplus.com/mobile-payment-provider-list.

payment and marketing solution which is a culmination of years of Alipay technology and expertise."

This "one-stop" service connects global merchants to a unified QR code, indicating that Alipay+ is not a fragmented, consumer-facing mobile wallet but a comprehensive digital solution that brings more overseas customers to local businesses. As a B2B technology application, it addresses the post-COVID-19 market pain points: "more," meaning connecting more overseas consumers, and "more convenient," meaning upgrading digital marketing for global merchants in the digital age.

From a fundamental perspective, Alipay+ connects over 2.5 million global merchants on one end and more than 15 global digital payment tools on the other. Leveraging its robust technical capabilities and resources, Alipay+ integrates with e-wallets worldwide, providing standardized access and payment methods, creating a mutually recognized, interconnected, and highly efficient system. Thus, Alipay+ plays a central role as a technology provider and a clearing center within the e-wallet alliance.

The technological advantages of Alipay+ provide interoperability of e-wallets now, especially useful to travelers during offline transactions. Local payments are possible for independent e-commerce platforms thanks to Alipay+ which consolidates multiple e-wallets into one point of sale. It is a B2B service product which does not serve directly consumers but business and thus works under a B2B2C structure. Alipay+ traders who are used to such a payment method don't even have to change their shopping habits; the logo means they can always take out their mobile phones and pay using a suitable method regardless of international borders. For the merchants, this kind of solution is very efficient logistically as well as economically and improves the service's efficacy levels. In the global marketplace, this will enhance consumer satisfaction in the digital era and enlarge the chances for all merchants around the world.

In late March 2022, San Marino, among the oldest microstates recognized within Europe, became the first European country to accept the offerings of Alipay+'s cross-border payment. Envisioning the future, Fabio Righi, Minister of Industry of San Marino emphasized, Digital technology will make us stronger! "San Marino has always been perennial open economy. In order to enable the foreign tourists visit us more; we need our merchants to offer services which are more up to date from a customer's stand point. For Asian tourists, the shopping and payment process is bland, and they can pay using the payment methods they already use, and we can try to innovate from there." In the first week of April of the same year, over 70,000 overseas merchants out of which more than 60,000 retailers and e-commerce operators from Malaysia, 15,000 GS25 stores in Korea, and one of the largest drugstore chains operating in Germany and Austria—Müller joined Alipay+.

According to the vision of Ant Group who provide cross-border payments services and global travel, "Travel the world with one wallet." In particular, Alipay+ allows cross-border digital payment and marketing services for merchants around the world, linking them with consumers in the mobile age. As of 2023, more than 25 international mobile payment services providers are cooperating with Alipay+, serving over 500 million overseas users and having partnerships with over 8 million global merchants and 800,000 in China alone across 57 nations and territories, benefiting both consumers and merchants.

11.3 A Broader Vision: Capacity Building and Digital Governance in Southeast Asia

The Asia-Pacific region was already placed high with areas capable of generating a lot of income from mobile payments in 2020, but this has now grown more so due to the COVID-19 havoc. Being able to adapt the use of mobile payments was only influenced by the COVID pandemic, especially in countries such as Thailand.

11.3.1 Unified Payments Facilitate Favorable Conditions for Tourism Consumption in Thailand

Apart from the impact of the COVID-19 pandemic, Thailand's economy mainly relies on tourism with no less than 18% of the GDP. Thai statistics data indicates that in 2023, Thailand had more than twenty-eight million overseas tourists that translated to total revenue from tourism succumbing to up to two trillion Thailand Baht approximately four hundred eighteen billion RMB, where international tourists accounted for around 1.2 trillion Thai Baht. Malaysia when rankied against other international tourist source countries emerged the best contributing over 4.4 million tourists while China came second with 3.4 million followed by Korea, India, and eventually Russia in that ranking.

Within the first ten months of 2023, according to Ms. Sutawan, Minister of Tourism and Sports, preliminary assessments by the ministry found that the number of Chinese tourists increased by 44,744, a 72.49% increase, following the implementation of visa-free policies. Starting from the end of 2023, the "Singapore-Malaysia-Thailand" golden tourism route countries successively introduced visa-free benefits for Chinese tourists. Comparatively, during the 2024 spring festival, these three destinations recorded a growth of 80% in tourist spending when compared to that of the potential May Day Golden Week when the visa-free policy had not yet been enforced. From Alibaba's Alipay Southeast Asia travel data, 580% more was spent on altering that spring festival than in 2023. Spending was highest in Thailand, but Malaysia significantly increased spending, emerging as the three countries' greatest spending increase.

Wang Zhuo, Vice President of the Marketing Division of the China Tourism Association, mentioned in a media interview that the number of Chinese tourists visiting Thailand is influenced by factors such as flight availability, destination image, and specific consumption patterns. At the moment, there are also new shifts in the spending and travel patterns of Chinese tourists. A shift seems to be occurring in the preferences of younger travelers from China. These travelers desire out-of-the-box experiences, deep cultural immersions, and local life experiences while the other group of usual native Chinese tourists travel for social sceneries at cheaper prices. The Tourism Authority of Thailand (TAT) observed a gradual decrease in large tour groups from China, with independent travelers and small groups of 6 to 8 people becoming the norm. Gary Bowerman, founder of the travel marketing company Check-in Asia, believes that the Chinese outbound travel industry is going through a "transition period," with changing trends and sentiments, as "more Chinese consumers are looking for different experiences."

In May 2024, the Tourism Authority of Thailand (TAT) and a subsidiary of Ant Financial Alipay, made an announcement at the "TAT x Alipay+, Ant International & TAT, The Only One Amazing Thailand to the World 2024" project launch on how they were going to enhance their partnership further to support the d-Hub initiatives by the Alipay+.

With this initiative, it is further desired that the Chinese users of Alipay be provided with promotional offers for newly launched products and travel services as a means to enhance digital innovation and marketing efforts with a view to boosting Chinese tourists into Thailand.

Chattan Kunjara Na Ayudhya, Deputy Governor for Asia and the South Pacific of TAT, stated that according to TAT statistics, in the first four months of 2024, the number of Chinese tourists visiting Thailand reached 2,351,909, indicating that Thailand remains a top choice for Chinese tourists. Alipay+ consumption figures during the Labor day period indicated that spending through alipay increased by 161% as regards the same period in 2023. Also, TAT is focused on improving creativity and modernizing the tourism industry. For this reason, it is TAT that has initiated the Alipay+ d-hub project as well as the Amazing Thai e-Card which has so far been offered to the Chinese tourists while it intends to extend the services to Alipay users from other countries in future.

The purchasing power of Chinese Tourists is on the rise continuously with the quality of experiences also being sought after. Techo Bo, the manager of a family-run hotel in Thailand, said, "Chinese tourists coming to Thailand now are more open to diverse experiences and spend a lot on food and activities. Our Chinese clients even rent out entire islands for birthdays or weddings, or just for a proposal." Wolfgang Arlt, CEO of the China Outbound Tourism Research Institute, believes that today, outbound tourists from China are looking for new experiences at reasonable prices, "not just for leisure, but also to incorporate business, health, education, and family time into their travels."

Alipay+ was formally introduced in the Thai market to accommodate the growing number of Asian tourists. Numerous well-known shops and brands operated by Central Retail Group are among Thailand's most popular shopping destinations for foreign visitors.

Piyawan Leelasompop, Head of Corporate Marketing, Central Retail Corporation PCL, commented, "This partnership between Central Retail and Ant Group to launch Alipay+ in Thailand will increase digital payment channels for Central Retail customers, delivering comprehensive services to serve consumers who prefer cashless payments. The company aims to target Asian tourists travelling to Thailand, providing convenience for those shopping at brands under Central Retail via Alipay+ with their local mobile wallet. Additionally, users can also receive promotions and special privileges provided by Central Retail and international mobile wallet providers, delivering next-level shopping experiences. There are also plans in the pipeline to expand payment services, accommodating tourists from other countries, including in Asia and Europe."[2]

[2] Central Retail, 2023, https://www.centralretail.com/en/newsroom/news-and-activities/641/central-retail-joins-forces-with-ant-group-to-launch-alipay-delivering-seamless-digital-payment-services-for-asian-travellers-in-thailand.

"The launch of digital payment service Alipay+ will elevate the CRC Ecosystem by bringing the best technologies and systems from around the world to drive inclusive growth for our customers, brands, and partners as well as deliver next-level experiences for consumers in every dimension. This is in line with the CRC Retailligence strategy, in which we aim to drive leaping growth within 5 years and become Asia's Next-Gen Omni Retailer," concluded Piyawan.[3]

Sittipong Kittiprapapong, General Manager for Global Merchant Partnership in Thailand at Ant Group, also emphasized the importance of Alipay+'s collaboration with Central Retail in providing convenient payment services. Central Retail's vendors can now accept digital payments from Asian tourists, offering international shoppers a seamless platform that makes their spending experience in Thailand as easy as it is in their home countries.[4]

11.3.2 The Governance Prospects Imparted to the Southeast Asian Financial Market

In the nutshell, after invading the Southeast Asian region, Alipay+ was active in uplifting the local digital payment markets and the tourism sector by connecting with dozens of e-wallets, partnering with banks in a variety of industries, and investing in and taking ownership in a number of other financial payment domains.

As well as improving the convenience and overall security of payment systems, the initiatives have also addressed the capacity building of the demand side of the financial market in a much broader sense.

Firstly, in terms of B2B (business-to-business) capacity building, Alipay+ has greatly expanded users' payment options by connecting with local e-wallets, making cross-border payments more accessible. Users can use their local e-wallets for cross-border transactions without the hassle of currency exchange and complex payment processes. From the perspective of security, Alipay+'s router banking partnerships help to ensure that payments processes are secure and compliant therefore allowing for user trust to be enhanced. Also the technological extension from Alipay+ has also stimulated the technological enhancement of the local payment

[3] Ibid.

[4] Ibid.

platforms to the overarching market reach and through market-based stimulation raised the overall tech level in the payment market.

Though many B2C applications of Alipay's services are apparent, company representatives frequently point out that travelers can easily make payments with tools they are familiar with, as well as making it more convenient for people without knowledge of a different language. This ease of payment brings about higher consumerism tendencies, which in turn positively impacts the local tourism industry, alongside other supporting service branches. With the help of big data analysis, it becomes easier for travel companies to fulfill tourist needs and provide those services more effectively. The success of Alipay+ in reaching Asian market A has had the effect of encouraging users to use as well as accept digital payments. This in turn increases the size of the total market. What's more, Alipay+'s penetration to other markets also heightened the standards of security and service, which led to a more civilized expansion of the local markets and thus more trust in the whole market.

As a business serving B-end customers, but indirectly C-end users, Alipay+ does not only run its own business but also has to meet with the needs of all other involved entities including merchants and partners as well as consumers themselves. According to Alipay, such strategic focus and alertness on target markets together with creativity has been the secret behind avoiding too much exploration cost and attaining positive externalities fast. These efforts not only enhance the refinement of Alipay+'s own business, but also offer great insight for other enterprises on how to transform and develop.

Ant Group's Alipay+ has gradually shaped and improved consumer payment trends and even the regional financial systems of southeast Asian countries. It has exported to other countries within Southeast Asia its expertise on mobile payments to boost the efficiency of central banks within the region. This extends past just the use of mobile payment systems, it also includes helping develop the fintech ecosystem and subsequently the level of provision of financial services in the whole region.

Firstly, the success of Alipay+serves as a valuable reference for Southeast Asian countries. With a rapidly growing user base and transaction volume—covering hundreds of millions of users by 2024, with both monthly active users and transaction frequencies on the rise—more local financial institutions are recognizing the importance of digital transformation and actively seeking collaboration opportunities. By adopting advanced payment technologies and operational models, these institutions can better cater to the increasing and diverse needs, while also enhancing their competitiveness and service efficiency.

Secondly, Alipay+'s investment and control strategies have directly contributed to the development of financial infrastructure in Southeast Asia. For instance, in countries such as Singapore and Malaysia, Alipay+ has established strategic partnerships with several local banks in order to create products and services that fulfill the needs of the local market. Such strategies are externally oriented and not only cut the time of implementation of new technologies but also save time and money for each of the participants.

As early as 2019, Siam Commercial Bank (SCB) signed a partnership agreement with Alipay+, making it possible for users of SCB's mobile app, "SCB Easy," to make payments anywhere in the world. In addition, SCB started catering to Chinese tourists who could now pay Thai merchants through Alipay, without requiring any cash or exchanging currency. Meanwhile, to improve user experience, SCB together with Alipay+ developed a range of QR-based offline retail and e-commerce payment solutions. This joint efforts not only enlarged the scope of the services of SCB but also contributed to the upgrading of the payment system in Thailand, setting an example for other financial institutions and merchants to pursue digitization. More people, particularly the youth and micro businesses, would benefit from affordable and efficient financial services, rejuvenating the economy as a whole.

Krungsri Bank reached an agreement with Alipay+ in 2018, initially deploying a large number of Alipay acceptance terminals in major tourist cities and popular attractions in Thailand, and training merchants on how to use these devices to accept Alipay payments. In addition to this, Krungsri Bank partnered with a number of hotels, shopping malls, and restaurants to enable them accept Alipay payments. For the Chinese tourists, it meant that the majority of the shopping places in Thailand accepted a payment method that most of them are used to. At the heart of the agreement was to enable Chinese tourists to have a better payment experience and at the same time, improve the payment options available to Thai merchants. This improved payment method led to more tourists from China opting to travel to Thailand which was a good thing for Thailand's tourism sector in terms of growth.

Thirdly, the expansion of Alipay+ in the local market has spurred the emergence and growth of a host of emerging fintech companies.

These startups often focus on addressing specific pain points, such as financing difficulties for micro and small enterprises, inaccurate personal credit assessments, and more. They leverage cutting-edge technologies like big data analytics and AI algorithms to offer innovative solutions.

Ant International launched Bettr, a platform providing microloan services for small businesses. Bettr announced its first embedded finance collaboration with Asset World Corporation (AWC), a comprehensive lifestyle and real estate group in Thailand, to offer digital lending services to SMEs on AWC's platforms. Yang Peng, CEO of Ant International, stated, "Bettr's mission is to make financing simpler and easier for micro and small businesses, as they will be the new drivers of regional and cross-border trade." According to estimates by the International Finance Corporation (IFC), there is a $5.2 trillion annual financing gap for approximately 65 million businesses (about 40% of formal MSMEs) in developing countries... The thriving SME economy in Thailand made it a strategic market for Ant International from day one. With Bettr, we can now extend our financial services by providing accessible financing, helping small businesses to grow and develop.[5]

Bettr's first collaboration is with Asset World Corporation (AWC), a comprehensive lifestyle and real estate group in Thailand, which is also a member of the TCC Group, one of Thailand's largest property conglomerates.[6] The initial phase of the partnership will involve AWC's food wholesale center, Phenix, which is expected to open on June 26th. Following this, the collaboration between Bettr and AWC will expand to include merchants across AWC's hotels, retail, wholesale, and other business sectors. Wallapa Traisorat, CEO and President of AWC, stated, "This collaboration is expected to benefit sellers, buyers, and AWC suppliers who conduct both online and offline businesses at Phenix, and ultimately, it will extend to the merchants associated with AWC's hotel, retail, and wholesale as well as commercial operations." Unlike direct digital lending, Bettr, launched by Ant Group, primarily focuses on embedded financial services, which means integrating financial services or products into non-financial platforms. Through the establishment of an embedded finance partnership, Bettr will enable small and micro enterprises under AWC

[5] Business Wire, 2024, https://www.businesswire.com/news/home/20240618515438/en/Ant%EF%BF%BEInternational-Unveils-Global-SME-Digital-Lending-Service-Through%EF%BF%BEEmbedded-Finance-to-Accelerate-Financial-Inclusion.

[6] Ibid.

to be the first to enjoy digital loan services. This approach allows for a seamless integration of financial solutions within the existing business ecosystems, providing a more tailored and convenient experience for the end users. It is thanks to such a vibrant ecosystem that the Southeast Asian financial market continues to show resilience in the post-pandemic era.

Last but not least, Alipay+ brings more than just capital and technology; it also focuses on talent development. To ensure long-term sustainable growth, the company actively engages in educational initiatives. It hosts various seminars and training programs to impart the latest industry knowledge to professionals. It is also noted that the company has established target funds to finance the most promising research projects or new business ideas and intends to breed a greater number of specialists with international outlook.

The Electronic World Trade Platform (eWTP) Initiative which is still a murky construct was developed to ease the entry of small and medium-sized enterprises (SME's) in the global market. Since its inception in 2016, this concept has been put into practice in countries and regions such as Malaysia, Thailand, and Ethiopia. In Southeast Asia, Ant Group which is one of the key players in the eWTP program has been in touch with various government agencies and education institutions in Malaysia to implement several programs for training of entrepreneurs and technical personnel in the region. For example, Ant Group partnered with the Malaysian Digital Economy Corporation (MDEC) to implement e-commerce and fintech workshops and training on several occasions. These courses targeted students who desired to learn about basic e-commerce operations such as starting an e-commerce business and more advanced topics such as data analysis market positioning and mobile payment technology as a catalyst in promoting business effectiveness. By such means, eWTP contributed to the promotion of trends in the economy that would sensitize the training of people in Malaysia to be relevant and competent in the use of modern business methods and technologies. In another dimension, eWTP further enhances the entrepreneurial activities of young business people in the country by providing them with mentorship, funds to execute their ideas, and international market access, triggering the further wave of innovations in the country and the wider Southeast Asia region.

In view of particularly addressing the challenges startups and emerging tech projects are facing in Southeast Asia, Ant Group has rolled out the Southeast Asia Digital Economy Accelerator Program in collaboration with several partners. The aim of this program is to locate and assist highly-fragile businesses and individuals with great potential but limited resources to help them succeed in a turbulent marketplace. According to the program, the participative companies are expected to receive support in terms of capital and a wide range of services from the network of Ant Group including market coverage and access to China and other international markets, product development, mentorship, marketing strategies, and financial investments. A distinguishing feature of the program is targeted support for women and lower-income minorities as well in an effort to eliminate gender and racial barriers in the startup environment. Ever since its establishment, several tens of startups—seed, series-A or series-B from agri-tech, and other sectors to fintech—have gone through the accelerator. In a relatively short period of time, they managed to renew their products, broaden their audience, and grow quite rapidly. For instance, an Indonesian fintech startup worked to enhance access to financial services in rural locales and has grown its outreach substantially after joining the accelerator and is now able to offer its mobile payment to tens of thousands of customers situated in undeveloped regions.

Alipay+'s practices reflect the ESG (Environmental, Social, and Governance) principles, particularly the "S" (Social) aspect, in the context of modern digital financial markets. It focuses on building partnerships between the company and its market stakeholders, and undertakes capacity building to optimize outcomes. From a sustainability perspective, Alipay+ emphasizes cooperation with local businesses and institutions, achieving win-win results. This exploration of industrial innovation and consumption-production models also aligns with the United Nations Sustainable Development Goals (SDGs), driving long-term economic and social development in local communities and providing an innovative reference for modern digital financial governance.

CHAPTER 12

Sustainable Fintech for Inclusive Tourism and Economic Development: Boost, Mypb, Touch 'n Go E-Wallet and Alipay+ Ecosystem in Malaysia

12.1 Under the Pandemic: The Surge of Digital Payments in Malaysia

12.1.1 Malaysia Embraces the Digital Payment Era

12.1.1.1 Use of Digital Payments Spurred by the Pandemic

The transition into the digital payment era is an unintended consequence of the COVID-19 pandemic for Malaysia. Previously, payment transactions in Malaysia were heavily reliant on cash. However, under the Movement Control Order (MCO) and various other restrictions from March 2020 to December 2021, retail activities shifted to online spaces—leaving Malaysian consumers with no choice but to embrace a more digital lifestyle.[1] Data from the Central Bank of Malaysia shows that the annual growth rate of electronic payments between 2019 and 2020 (during the outbreak of COVID-19) was 25%, double the average annual growth rate of the previous three years.[2] In 2021, the astonishing increase in cashless

[1] Statista, 2024, https://www.statista.com/topics/10601/digital-payments-in-malaysia/#topicOverview.

[2] Bank Negara Malaysia, 2022, https://www.vixio.com/insights/pc-malaysias-digital-payments-growth-soars-payment-preferences-shift.

J. Sun, *Sustainable Financialization*, Sustainable Development Goals Series, https://doi.org/10.1007/978-981-96-3580-1_12

transactions made Malaysia one of the most developed payment markets in Southeast Asia, becoming the second-highest country in the region in terms of cashless payment adoption, just behind Singapore. Data from the same year indicates that there were over 3 million new digital consumers in Malaysia, with 98% of Malaysian businesses already accepting digital payments.[3]

The pandemic has given Malaysia's march towards a cashless society a serious boost, shaving off about three years from consumer expectations.[4] This change has really been accelerated by COVID-19 by a good three to five years. In this sense, Malaysia is at the forefront, followed by Singapore and the Philippines. More than half of Southeast Asia's consumers—52% to be exact—report having fewer chances to use cash than before the pandemic hit. This trend is especially noticeable in countries that had to implement strict movement controls and nationwide lockdowns, like the Philippines with 68%, Indonesia at 64%, and Malaysia at 59%.[5] Furthermore, 43% of Southeast Asian consumers now choose cashless payments due to health and safety concerns, indicating a substantial shift in regional behavior.

As the e-commerce market in Malaysia continues to expand, digital payment systems such as credit cards, online banking, and electronic wallets have gained immense popularity among consumers. Cashless payment methods now hold a substantial share of the e-commerce payment landscape in Malaysia, with bank transfers and credit cards leading the way.[6]

The accompanying graph highlights the most favored payment methods in the country. In addition to bank transfers and credit cards, electronic wallets have become increasingly popular among a wide range of Malaysian consumers, followed by cash payments and the Buy Now, Pay Later (BNPL) option. According to a Statista survey on digital payments, the percentage of respondents using digital wallets jumped from 15% in the first quarter of 2019 to an impressive 68% by the first quarter of 2022. This trend is expected to continue, with the number of digital wallet users projected to reach 25.5 million by 2025.[7]

[3] Google, 2021, e-Conomy SEA, e-Conomy SEA 2021 (google.com).

[4] Visa, 2022, Consumer Payment Attitudes Study 2022 (visa.com.my).

[5] Visa, 2021, CPA 2021 Report R8 (visa.com.my).

[6] Statista, 2024, https://www.statista.com/topics/10601/digital-payments-in-malaysia/#topicOverview.

[7] 2C2P, https://2c2p.com/blog/malaysia-payment-methods.

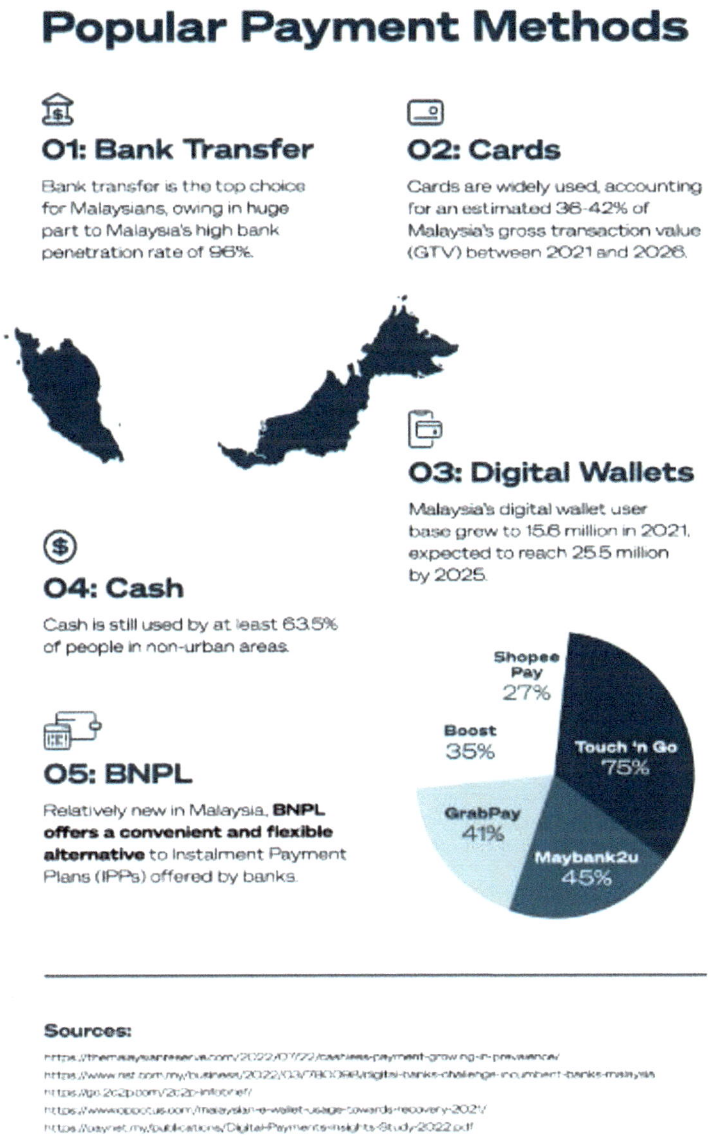

The most popular payment methods in Malaysia in 2022 (*Source* Internet)

Malaysia's fintech industry has grown quickly as a result of the consumer market for electronic payments. In fact, even after the pandemic, the widespread tendency toward cashless payment systems like electronic wallets has not changed. Along with the appeal to the young population, high internet penetration rate, and regulatory backing, this development is ascribed to the emergence and competition of multiple digital payment platforms in Malaysia.

Among the myriads of electronic wallet options, Touch 'n Go, GrabPay, and Boost have emerged as the most favored choices among Malaysians. In addition to these, Malaysians also prefer using e-commerce payment platforms such as ShopeePay and LazadaWallet for their transactions (54%), making them the top-performing platforms in this domain within the country.[8] Notably, Touch'n Go e-Wallet (hereinafter referred to as "TNG e-Wallet") is the largest electronic wallet in Malaysia and is also one of Ant Financial's key customers in the Malaysian market. The company behind it is TNG Digital (hereinafter referred to as "TNGD"). Starting as a commuter convenience, TNG e-Wallet has evolved into a fintech solution that has impacted the lifestyle of more than half of the nation's population.[9] Currently, Touch 'n Go users can make payments directly at over 300,000 merchants.[10]

[8] Uob, 2022, https://www.uobgroup.com/techecosystem/news-insights-consumer-research-malaysia-cn.html.

[9] 21jingji, 2024, https://www.21jingji.com/article/20240612/herald/a501b2b8b9230061004228e1740c2e07.html.

[10] Commission factory, 2023, https://blog.commissionfactory.com/ecommerce-marketing/payment-methods-in-malaysia#Most-Popular-eCommerce-Payment-Methods-in-Malaysia.

TNG, GrabPay, Boost are popular payment methods among Malaysian merchants (*Source* Internet)

12.1.1.2 Governmental Support for the Popularization of Digital Payments

Embarking on a transformative journey, the Malaysian government has laid out an ambitious roadmap in its "Vision 2030 for Shared Prosperity," delineating the nation's strategic economic objectives and the policies that will steer its future growth. At the heart of this vision are digital transformation initiatives, which are set to propel the digital economy forward through a trio of pivotal actions: fostering the pervasive integration of digital technology, nurturing the growth of indigenous tech firms, and drawing in high-caliber investments.

To spearhead these digital endeavors, the Ministry of Communications and Multimedia of Malaysia has established the "Malaysia Digital Coordination Council," tasked with the critical role of orchestrating and executing the nation's digital strategy.[11] In 2020, amidst the economic upheaval caused by the COVID-19 pandemic, Malaysian Prime Minister Muhyiddin Yassin unveiled a comprehensive $58 billion stimulus package. This initiative included a significant digital component, with the Ministry of Communications and Multimedia (KKMM) and the Malaysia Digital Economy Corporation (MDEC) collaborating to offer over $138 million in complimentary internet services to the Malaysian populace.

Furthermore, to bolster connectivity and enhance digital infrastructure, an additional $92 million was earmarked to expand internet reach and bolster network performance, ensuring robust and reliable telecommunications networks. These strategic investments are aimed at catalyzing the growth of digital economy sectors, such as e-commerce and digital payments, and are pivotal in laying the groundwork for a resilient economic rebound in the face of the pandemic.[12]

In the realm of digital transactions, the Malaysian government has repeatedly rolled out incentives to boost the uptake and spread of electronic wallets as a preferred mode of payment. Since 2020, with the aim of hastening the growth of digital wallet transactions and broadening financial inclusion, the government has initiated the "e-Tunai Rakyat" (e-TR) program, allocating a substantial fund of 450 million Ringgit. This program is designed to incentivize a wider adoption of electronic payments, especially among those who are new to the digital payment landscape, by offering financial assistance for topping up electronic wallets.

Under the e-TR program, every eligible adult Malaysian citizen, aged 18 and above with an annual income capped at 100,000 Ringgit, is entitled to a 30 Ringgit subsidy for topping up their electronic wallets. This strategic move not only reduces the initial hurdle for users to dip their toes into electronic payments but also spurs a greater number of individuals to experience the ease and convenience of digital wallets. To ensure that the subsidy is widely disbursed and targets those most in need,

[11] Chinadaily, 2022, https://cnsubsites.chinadaily.com.cn/wic/2022-09/26/c_174247.htm.

[12] Adyen, 2024, https://www.adyen.com/zh_CN/payment-methods-guides/asia-pacific/malaysia.

the government has implemented stringent eligibility criteria, prioritizing lower-income households and individuals.

Building on its efforts to digitize payments, the Malaysian government has launched the "Less Cash" campaign, nudging the populace towards more user-friendly and secure digital payment options over traditional cash transactions. Through multifaceted outreach and partnerships with digital wallet providers, the government is actively spreading the word about the benefits of electronic payments.

The "e-Penanja" program, which debuted in 2020, has been an ongoing initiative to further this cause. By offering newcomers a modest subsidy for electronic wallet top-ups, the program aims to entice more individuals to give digital payments a try. These subsidies, ranging typically from RM30 to RM50, are designed to ease the entry for first-time users and encourage them to discover the ease that electronic wallets can bring to their daily transactions.

Moreover, the Malaysian government extends financial incentives to low-income individuals to encourage digital wallet registration, a critical step in reaching the 19% of Malaysians who are unbanked. This inclusive approach is pivotal in bridging the financial divide and integrating more citizens into the digital economy.

12.1.2 Shortcomings of the Digital Ecosystem and Development Dilemmas

12.1.2.1 Security Concerns: Electronic Wallets Have Not Established Trust with the Public

The proliferation of electronic wallets in Malaysia is still in its infancy. According to the Nielsen Payment Landscape Report, although 67% of Malaysians use some form of cashless payment (debit cards and online banking being the primary methods at 63% and 57% respectively), only 8% of the population uses mobile wallets as a means of payment.[13]

Security appears to be the paramount concern. The report reveals that a significant 46% of non-users consider security issues as the primary barrier to embracing electronic wallets. In the realm of security concerns, an alarming 59% of individuals fear that electronic wallets could facilitate

[13] Nielsen, 2019, https://www.digitalnewsasia.com/digital-economy/nielsen-sees-security-concerns-main-barrier-e-wallet-adoption.

credit and debit card fraud, while 38% are anxious about transaction omissions. The risk of counterfeit mobile wallet websites is a concern for 36% of respondents, and 39% are apprehensive about the leakage of banking information. Besides, 30% of individuals are worried about the theft of their phones and the subsequent misuse for shopping. Other apprehensions include malware and software issues, as well as the specter of data breaches.

How can we mitigate these security-related apprehensions? Enhanced security measures are the key. Nielsen's report highlights that 62% of respondents have greater faith in electronic wallets that incorporate biometric features, while 56% are in favor of employing one-time passwords (OTP) for credit and debit card transactions. In addition, 38% of individuals advocate for two-factor authentication to be integrated into the security protocols of electronic wallets. Meanwhile, 27% prefer that payment details be stored locally on their devices rather than on company servers, indicating a preference for control over their sensitive financial data.

12.1.2.2 Insufficient Integration: Dispersion of Electronic Wallets in Number and Function

The electronic payment industry in Malaysia comprises eight major electronic payment institutions, leading to a relatively fragmented market, which contrasts sharply with the highly consolidated market in China. While industry fragmentation to some extent reflects the rapid expansion and development of Malaysia's electronic payment industry, preventing industry monopolization, it also results in intensified competition among institutions. This is detrimental to resource integration, establishment of authority, and standardization of the industry, leading to a broad but shallow development of the electronic payment sector.[14]

Analysis suggests that financial technology startups in Southeast Asia can be simply categorized as: (1) To-Consumer: Payment companies (covering all Consumer users), digital banks (primarily serving users without debit cards), and lending companies (mainly targeting users with debit cards and bank accounts); and (2) To-Business: Lending companies

[14] Thepaper, 2020, https://m.thepaper.cn/baijiahao_8208891.

(focused on small and medium-sized enterprises), and SaaS companies (aimed at large and medium-sized enterprises).[15]

This analysis posits that the application of To-Consumer financial technology is primarily in payment and lending. After several years of vigorous development, the payment segment in the To-Consumer has become relatively crowded, especially in the electronic wallet domain. Each Southeast Asian country has its own leading players, and a trend of traffic consolidation around these top players has begun to emerge. The wallet wars continue, but it is no longer an opportunity for new entrepreneurs. The lending segment remains a key investment theme at this stage, with startups emerging continuously, focusing on cash loans, consumer loans, student loans, Buy Now Pay Later, and other models. On this basis, digital banks have also become a hot track, with the head effect beginning to show, and new players continue to enter. In addition to payment and lending, intelligent advisory and Insurtech have also begun to emerge but are still in the exploratory phase. In terms of To-Business, the payment segment is mainly reflected in B2B payments and cross-border payments. The Southeast Asian market in this field is still dominated by overseas players, but local startups are also growing and becoming major participants. In the lending field, the main business model is to use services such as fee control, supply chain, and digital transformation as entry points to obtain the data flow and fund flow of Business users as a basis for risk control, thereby providing unsecured loan services to small and medium-sized enterprises.[16]

Post-pandemic data suggests a significant shift towards digital payments in Malaysia, with 78% of the population now utilizing electronic wallets. According to a PayNet survey, the most popular e-wallets among Malaysians are Touch 'n Go, leading with 83%, followed by ShopeePay at 58%, Boost at 49%, GrabPay at 45%, and MAE at 36%. The fierce competition among these providers indicates a saturation in the local digital payment ecosystem, which has resulted in a fragmented landscape for the digital payment industry in Malaysia. This fragmentation is further highlighted by the fact that 78% of Malaysians possess more than one e-wallet, with users favoring a variety of options to cater to different needs.

[15] Huxiu, 2022, https://m.huxiu.com/article/708947.html.

[16] Ibid.

12.1.2.3 Dilemma of Inclusiveness: The Convenience of Digital Payments Has Not Been Universally Beneficial

Peter Schiesser, Group CEO of PayNet, stated that Malaysia's payment industry appears to be divided along regional, demographic, and class lines, with younger and wealthier individuals in urban areas showing greater acceptance of digital payments. A similar divide exists in the attitudes of businesses towards digital payments. Businesses in the northern, central, and eastern parts of Malaysia clearly prefer the use of digital payments, while on the east coast and in the southern regions, the majority of businesses still lean towards cash.[17] The report indicates that businesses adopting digital payments are often influenced by customer and supply chain preferences, geographic location, as well as the type and size of the business. "This reflects a larger systemic design issue, where redundancy and alternatives have not been created, and the impact of the system on vulnerable groups and the disenfranchised is not understood. There is a big difference between digital-first and digital-only," said Dr. Rachel Gong, a senior researcher at Khazanah Research Institute.[18]

While the move towards a completely cashless approach has its benefits—chief among them being traceability, transparency, and security—it cannot be denied that the rise of cashless payment systems has made life more difficult for those on the margins and for those who are not adept at adapting to technology.

"Through cashless transfers, consumers can transfer money, pay bills, and purchase goods and services from home, markets, or shops while maintaining social distancing. It also enables many governments to quickly, efficiently, and securely distribute social assistance [financial aid] to the accounts of recipients through cashless channels," said Rekha Reddy, senior financial sector specialist at World Bank Group. She also added that "we need to recognize that certain groups may face barriers when using certain types of contactless or cashless methods and ensure that these individuals are not excluded."

"We need to act swiftly to avoid exacerbating the challenges brought about by the pandemic. The latest World Bank's Malaysia Financial Inclusion Index (Findex) data shows that 76% of individuals aged 15 and above

[17] Vixio, 2022, https://www.vixio.com/insights/pc-malaysias-digital-payments-growth-soars-payment-preferences-shift.

[18] The Edge Malaysia 2022, https://theedgemalaysia.com/article/cover-story-cashless-quandary.

indicated that they made or received digital payments last year. However, the number drops to 61% for the B40 group; 60% for rural areas; 53% for women; and only 39% for those with only primary education," also said Rekha. That is to say, the risk of exacerbating a digital divide loom, and a sudden cessation of cash payments might pose significant challenges for certain demographic segments.[19]

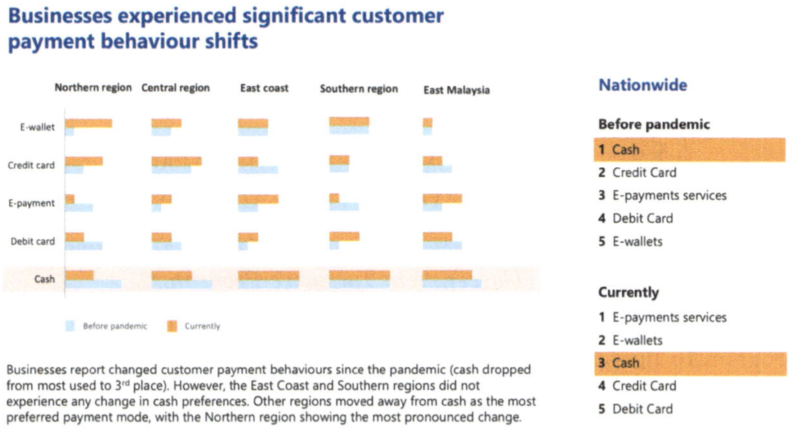

Regional differences in payment preferences in Malaysia before and after the COVID-19 (*Source* PayNet)

Rashaad Ali, CEO of the Social and Economic Research Initiative (SERI) stated that bridges of acceptance must be built, which often involve accessibility, education, and information. "If the necessary infrastructure is not in place, it is counterintuitive to expect digital payments to expand their reach. This includes the availability, accessibility, and affordability of the internet (especially mobile internet). Compatible smartphones will also be a requirement," said Rashaad. She also added that "the barrier to entry for cashless payments is much higher compared to cash, so if we want the benefits of cashless payments to reach everyone, we must ensure that the barriers are lowered."

This challenge is not unique to Malaysia—the country has 8% of the population without bank accounts, approximately 2.6 million women,

[19] Ibid.

youth, and low-income individuals—but it is particularly pressing considering the rapid shift towards electronic payments.

Debit cards and online banking were the most popular payment methods prior to the pandemic outbreak, with 63% and 57% of transactions occurring through these channels, respectively, according to the Nielsen 2019 Malaysia Payment Landscape report. These were mainly used for regular expenses, such as telephone and internet bills. Nevertheless, credit cards were utilized in just 27% of transactions, and mobile wallets were used in even fewer transactions (8%).[20] The majority of users were between the ages of 35 and 64.

12.2 Alipay+ Empowers the Capacity Building of Malaysia's Digital Ecosystem

12.2.1 Convenience + Security: Touch 'n Go E-Wallet Supports Government Inclusive Programs

In 2019, the Ministry of Finance of Malaysia launched the "ePemula" program to distribute relief funds specifically to young people, with an estimated 2 million Malaysians aged 18 to 20 expected to receive 150 Ringgit. For this, the Ministry needed to collaborate with four local electronic wallets, including Touch 'n Go e-Wallet. During the collection process, to prevent the misuse of funds and other dangerous situations, the government required verification of the identity information of these two million young people. However, according to Foo Yong Jin, Chief Risk Officer of TNG Digital, there was no leading e-KYC (online identity verification technology) in the Malaysian market at the time. Their needs focused on simplifying the process, high accuracy, and cost control. At this juncture, ZOLOZ, the security technology brand of Ant Group, played a role. According to Ant Financial, ZOLOZ relies on Ant's 20 years of risk control experience and has a multi-layer security system, including facial algorithm confrontation, facial guard device confrontation, link transmission confrontation, image anti-counterfeiting confrontation, and security strategy confrontation, which can effectively intercept attacks by images, videos, and deepfake technology.

Teh Huey Tzi, Chief Risk Officer of Touch 'n Go e-wallet, stated that in the first step of user application, the system needs to determine whether

[20] Ibid.

the user has completed the online identity verification to prove "you are you" and "eligible to receive" to prevent the misappropriation of subsidy funds. "This is where the e-KYC technology of ZOLOZ under Ant Group comes into play, determining whether the user has passed the online identity verification at the first critical step. If so, our platform will immediately link this user with the government's data system to determine the user's eligibility to receive; if not, it will guide the user into the e-KYC process," said Teh Huey Tzi. Throughout the process, the applicant's identity information, age, and student status data are stored in the government system and are not shared with the platform.[21]

Teh Huey Tzi revealed that during the communication process with the Malaysian government, they were very concerned about how to smoothly distribute funds to the correct user groups. As a platform, they also wish to avoid any fraudulent activities against government funds. "This is why the e-KYC platform is very important in this situation. We must comply with the central bank's requirements, and the central bank also has the right to conduct audits when necessary," she said.[22] With the help of e-KYC technology, Touch 'n Go also gained an opportunity to overtake during the "ePemula program." According to its CRO Foo Yeong Jin, Touch 'n Go accounted for 70% of the market share in this project, ultimately acquiring 5 million customers.[23]

Foo Yong Jin also introduced a risk control scenario of account theft: in 2021, we faced a significant challenge where the dark web used APKs to steal SMS verification codes for login and transactions. After integrating Ant Financial's risk control verification technology, we quickly converted SMS verification to facial recognition verification. Within a few months, we successfully reduced the proportion of account theft by 98%, achieving this with just this one step.[24] "ZOLOZ's technology played a key role for us," said Ni Xiang, CEO of Touch 'n Go e-Digital. "At that time, there were four electronic wallets working on this project, but Touch 'n Go e-Wallet was the preferred choice because of its high accuracy and best experience, ultimately accounting for 70% of the market share. After this

[21] Tech. China, 2022, https://tech.china.com.cn/roll/20220414/386737.shtml.

[22] Ibid.

[23] Sina, 2024, https://finance.sina.cn/2024-07-11/detail-incctkpc8691048.d.html.

[24] 21jingji, 2024, https://www.21jingji.com/article/20240612/herald/a501b2b8b9230061004228e1740c2e07.html.

campaign, we became the first electronic wallet approved by the central bank to use online identity authentication, and we successfully acquired 5 million customers in this entire project."[25]

According to relevant statistical data, after the two parties reached a cooperation, 75% of TNG e-Wallet users completed registration through ZOLOZ's e-KYC online identity verification technology, which greatly improved the registration progress and reduced the number of manual reviews by nearly 80%. In fact, since the outbreak of the pandemic, ZOLOZ has helped electronic wallets in multiple countries, including Malaysia's Touch 'n Go e-wallet (Bangladesh's bKash, the Philippines' GCash, Indonesia's DADA, etc.), to distribute funds and subsidies. It has not only ensured the security of fund flows but also helped reduce the spread of the pandemic and improve the lives of the public. As Chen Jidong, Chief Product Officer of Ant Group's security technology, said, "Since the pandemic, the acceleration of digital transformation and upgrading in Southeast Asia has been rapid, and we are very pleased to be a witness and participant in this trend. When we see our partners using Ant's e-KYC technology to help local governments distribute subsidies to small and micro groups, and help companies issue salaries online, bringing convenience and security to many people, we feel a great sense of achievement, and we also hope that the technology can help more ordinary people."

Touch 'n Go adheres to the purpose of improving financial inclusiveness in Malaysia. Over the past two decades, it has led the digital transformation of Malaysia towards a cashless society through the Touch 'n Go e-wallet, introducing digital solutions and redefining the role of financial technology in the lives of Malaysians nationwide.[26] With the support of e-KYC, TNG e-Wallet helps small and micro enterprises obtain financial support more quickly, supports the growth and development of local enterprises, thereby creating more employment opportunities and promoting sustainable economic growth. At the same time, it enables more people to conveniently and securely receive digital inclusive services,

[25] Ifeng, 2024, https://news.ifeng.com/c/8aCDr7kleMN.

[26] The Star, 2023, https://www.thestar.com.my/metro/metro-news/2023/03/30/safe-and-secure-ewallet-for-malaysians.

and marginalized groups and low-income families can also enjoy higher-quality financial services, which helps reduce socioeconomic inequality and enhance Malaysia's financial inclusiveness.

12.2.2 *Boosting Local Tourism: MyPB Goes International from Hong Kong*

Public Bank Berhad, a frontrunner in Malaysia's banking sector, has launched MyPB, a cutting-edge digital payment and banking application crafted to offer seamless online banking services at customers' fingertips. MyPB brings the convenience of mobile banking to life, empowering users to execute financial transactions on the go, thereby amplifying the reach and ease of accessing banking services. The application has bolstered its security measures to safeguard users' transactions and personal data with robust features.

MyPB embraces a multi-layered authentication process, allowing users to transact with peace of mind through methods such as fingerprint recognition, Touch ID, Face ID, or a 6-digit SecurePIN. The app not only fortifies transaction security but also safeguards users' personal information and funds. MyPB also nudges users towards adopting the latest app features, including PB SecureSign, taking an extra step towards securing transactions.

Beyond facilitating a shift from conventional to digital banking, MyPB is Public Bank's stride towards environmental sustainability by minimizing the reliance on paper and physical materials. This move aligns with the bank's broader commitment to corporate social responsibility (CSR), which encompasses market development, workplace enhancement, community upliftment, and environmental sustainability. Through MyPB, Public Bank is not just enhancing its service delivery but also deepening its community engagement, offering educational resources, and actively participating in community service initiatives.

In 2024, Alipay+ added 14 electronic wallets, including MyPB, allowing users to make payments in Hong Kong using their local applications.[27] As part of the Alipay+ network, MyPB allows its users to make seamless QR payments in over 90% of stores in Hong Kong. At restaurants and snack bars in Hong Kong, MyPB users can conveniently pay for

[27] Takungpao, 2024, https://www.takungpao.com/finance/236136/2024/0427/965941.html.

meals, enjoying fast checkout services whether at high-end restaurants or street food stalls. For transportation, MyPB users can pay for transportation costs in Hong Kong using MyPB, including taxis, subways, busses, and other means of transport.

Alipay+ enables MyPB users to make payments in multiple countries and regions while enjoying transparent and competitive exchange rates. Through this service collaboration mechanism, Alipay+ works with e-wallets like MyPB to improve cross-border payment experiences and boost operational effectiveness.

Cooperation between MyPB and Alipay+ (*Source* Alipay+)

This integration is a milestone, representing the most significant rollout of international payment applications in Hong Kong's offline commercial network through cross-border collaboration. Venetia Lee, CEO of AlipayHK and General Manager of Ant International's Greater China region, eloquently captured the essence of this initiative, stating, "When we pay like locals, we can travel and experience culture like locals." MyPB stands out as a key player in the new wave of electronic wallets from Malaysia that have joined the Alipay+ payment network, signifying a major step forward in digital payment diversity. Since 2023, the influx of tourists to Hong Kong has been on the rise, propelled by the dynamic global promotional campaigns of the Hong Kong Tourism Board. The swift expansion of the electronic payment market is in lockstep with this surging tourism demand and trend, providing a seamless payment experience for visitors and bolstering the local economy.

The Integration of MyPB into the Alipay+ Payment Network Boosts Hong Kong's Tourism Industry (*Source* SoyaCincau)

Through the Alipay+ payment network, MyPB provides merchants with more cross-border payment solutions, promoting the digital transformation of multiple industries, including tourism, retail, and catering. With the ability to accept more international customer payment methods, merchants have the potential to increase their sales, especially from Asian tourists who frequently use digital wallets. Merchants do not need to open separate accounts for each payment method; they can accept multiple electronic wallet payments through Alipay+, simplifying the cross-border payment process and improving payment efficiency. Merchants capable of accepting international payment methods have a competitive advantage in the market, attracting more foreign tourists and enhancing customer satisfaction. Alipay+ also provides marketing tools, such as Alipay+ Rewards, through which merchants can offer discounts and promotions to attract new customers and increase the proportion of repeat patrons.

The collaboration between Alipay+ and MyPB is a win-win venture that resonates amidst the burgeoning resurgence of Hong Kong's tourism sector. MyPB's strategic expansion within the region not only enriches the payment options for its users across a variety of consumer scenarios but also propels the global outreach of Malaysia's digital payment infrastructure. This alliance brings into play a substantial potential user base and transaction volume for electronic wallets like MyPB, which is a significant stride in the digital payment landscape.

At the same time, this inclusive foray into the digital economy aligns seamlessly with Malaysia's ambitious vision of fostering a digital economy and a cashless society. It is a step that bolsters Malaysia's competitive edge in the regional and global digital economy, echoing the country's digital economy blueprint that aims to transform Malaysia into a digital economy by 2030. By embracing such partnerships, Malaysia is not only enhancing its digital payment ecosystem but also reinforcing its commitment to digital inclusivity and economic progress, which are crucial to its long-term economic strategy.

12.2.3 *From Digital Payments to Super Apps: The Evolution of Electronic Wallets Through Integration*

12.2.3.1 *To-Business: D-store Creates Partnership Opportunities for Merchants*

At the 2022 Singapore Fintech Festival, Ant Group made a significant move by launching Alipay+ D-store, a digital storefront solution that equips merchants with an array of tools to digitize their payment, operational, service, and marketing endeavors. This initiative is designed to assist merchants in establishing a comprehensive "one-stop shop" across various digital platforms and social media, thereby unlocking new revenue streams and enhancing service delivery. Moreover, it empowers businesses, particularly those in the service industry with a physical presence, to operate more efficiently and effectively.[28]

Businesses can set up digital storefronts in the form of mini-programs on digital platforms. Digital stores can also be created as H5 pages, which can be quickly and easily integrated online and on social media to attract customers. Several digital platforms, like as Alipay, Touch 'n Go e-Wallet, and AlipayHK, have already cooperated with Alipay+ D-store. Touch 'n Go e-Wallet, a leading electronic wallet in Malaysia, initially a payment method, is now transforming into one of the super apps in Southeast Asia. The figure below illustrates how Ant Group's D-store assists business merchants in creating convenience within this largest electronic wallet in Malaysia.[29]

[28] Ant Group, 2024, https://www.antgroup.com/en/business-development/digital-globalization-tab-details/alipay-detail.

[29] Alipay+ D-store, https://www.antom.com/dstore.

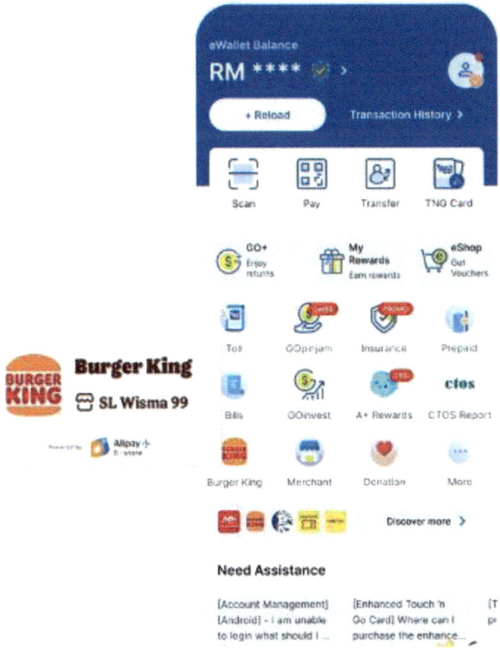

In Touch 'n Go e-Wallet, "one-stop merchants" such as Burger King (*Source* Alipay+ D-store)

Dr. Huang Xiaoling, Global Head of Alipay+ D-store, believes that leveraging digital technology is crucial for the service industry to improve operational efficiency and respond to rapidly changing market conditions. She added, "The development goal of the Alipay+ D-store solution is to make it easier for physical businesses to achieve large-scale digitalization, thereby better serving and attracting digital-savvy consumers by providing a seamless omnichannel experience." "We welcome more digital platforms to cooperate and jointly accelerate the digitalization of the global service industry."[30]

[30] Techwireasia, 2022, https://techwireasia.com/2022/11/ant-groups-alipay-d-store-solution-to-accelerate-digital-stores/.

The technology of Alipay+ D-store not only creates partnership opportunities for merchants across various industries but also brings more value to financial and other products. Alan Ni, CEO of TNG Digital Sdn Bhd, stated that Alipay+ D-store's reliable products and team are helping us grow from the leading electronic wallet in the country to the next super app in the region, offering more services to meet the needs of different users in various scenarios.[31]

In Touch 'n Go e-Wallet, "one-stop merchants" such as Burger King, accessed through Alipay+ D-store, are examples of how integration is transforming the electronic wallet into a comprehensive platform for both consumers and businesses.

12.2.3.2 *To-Consumer: DuitNow QR Provides Cross-Border Convenience for Users*

PayNet Group Sdn Bhd, a payment network company in Malaysia, and Ant Group have recently signed a memorandum of understanding that will soon enable cross-border payment acceptance between DuitNow QR and the Alipay+ network. This means that Asian travelers with electronic wallets supporting Alipay+ will be able to enjoy seamless payments at DuitNow QR merchants, while Malaysian electronic wallet users will be able to make overseas purchases at Alipay+ merchants.[32]

Gary Yeoh, Chief Commercial Officer of PayNet, remarked on the pivotal timing of the PayNet and Alipay+ collaboration, noting, "This partnership between PayNet and Alipay+ is forged at a vital time when overseas travel and tourism have fully resumed, and rapidly increasing. Consumers and merchants in both countries who are already accustomed to digital payments expect the same seamless and hassle-free experience when traveling overseas. They also expect this to be conducted at competitive exchange rates. This collaboration will address both these needs. Since Malaysia has long been a favorite and popular destination with Asian tourists, and vice versa, this collaboration will likely result in higher trade growth between the two nations. We do see the proliferation of seamless international payments at scale sooner than we think. It is just a matter of time."

[31] Ibid.

[32] Soyacincau, 2023, https://soyacincau.com/2023/08/29/all-malaysian-ewallets-with-duitnow-qr-will-soon-support-cashless-payments-in-more-countries-via-alipay-network/.

This synergy not only enhances the payment convenience for MyPB users across various consumption scenarios but also propels the internationalization of Malaysia's digital payment services, tapping into a substantial potential user base and transaction volume for electronic wallets like MyPB. This inclusive digital economy cooperation is in lockstep with Malaysia's long-term aspirations to foster a digital economy and a cashless society, thereby bolstering Malaysia's competitiveness in the regional and global digital economy landscape.

As Malaysia's tourism sector bounces back, this collaboration could not have come at a better time. The first quarter of 2023 saw a surge of 4.5 million visitors, predominantly from ASEAN nations, signaling a promising start. With this momentum, Malaysia is on track to meet its ambitious goal of welcoming 16.1 million tourists this year, which is projected to translate into a whopping 49 billion Ringgit in tourism earnings.

In the midst of this tourism resurgence, businesses are tasked with embracing the new digital landscape, particularly the widespread adoption of localized digital payment solutions like electronic wallets, which have become the go-to for everyday transactions. The trend towards QR code payments, driven by Southeast Asian users, is also set to skyrocket, with an anticipated growth of 590% by 2028.[33]

Datuk Seri Haji Mohd Shukri Abdul Rahim, Group Chief Executive Officer of PayNet, said that the collaboration is timely as the country's tourism industry is steadily growing, further boosted by the "Visit Malaysia Year 2026." "These connections will link economies that contribute more than two-thirds of global GDP growth and have 60% of the world's population with a median age lower than the global average. At this pace, we may very well see the economic integration of ASEAN+ 3 sooner than we think."[34]

[33] Businesswire, 2023, Juniper Research: QR Code Payments Forecast to Grow over 590% in Leading Southeast Asian Market by 2028, Spurred on by Pan-Asian Interoperability | Business Wire (https://www.businesswire.com/news/home/20230820288757/en/).

[34] Chinapress, 2023, https://www.chinapress.com.my/20231108/8%E5%9C%B0%E5%8C%BA%E7%94%B5%E5%AD%90%E9%92%B1%E5%8C%85-%E5%8F%AF%E9%80%8F%E8%BF%87duitnow%E6%94%AF%E4%BB%98/.

DuitNow QR cross-border linkages with Thailand, Singapore, and Indonesia (*Source* Soyacincau)

This strategic collaboration has opened up a world of seamless payment convenience for Asian travelers at DuitNow QR-affiliated merchants and for Malaysian electronic wallet holders making purchases abroad at Alipay+ merchants. It is a win-win that is not just promoting the development of local tourism industry and related services but also creating a wealth of revenue channels for local businesses.

On the consumer side, it is all about convenience, with more payment choices that are set to ignite spending in our post-pandemic world. By bridging DuitNow QR with the Alipay+ network, consumer's preference for digital lifestyle is spurring technological advancements and driving the uptake of digital services, all of which contribute to the flourishing of the broader digital ecosystem.

12.2.4 *Super Apps: Digital Construction and Industry Development*

On May 30th, Malaysian real estate company S P Setia (Setia Group) announced the official launch of its Super App, Setia Go, with the technical support of Ant Group's mobile development platform, mPaaS.[35]

[35] DoNews, 2024, https://www.donews.com/news/detail/4/4264043.html.

Setia Go is a lifestyle app designed to provide Malaysian users with a one-stop platform for managing properties and participating in promotions organized by developers. Setia Group will utilize it to optimize its membership system and offer digital services to local users. Prior to the launch of Setia Go, Ant's mPaaS provided S P Setia with a suite of modular tools and components, enabling their developers to modularize business applications. Previously, app version updates required a 1 to 2 weeks launch process waiting for the application platform, but the seamless update capability brought by the mini-program architecture has reduced the current app functional update time to half an hour, significantly enhancing the user experience.[36]

In the process of digital upgrading, the APP is an important entry point for corporate customer service. However, building an APP from scratch presents challenges such as technical talent, efficiency, and cost. To address these, Ant's mPaaS, a cloud-to-end one-stop mobile development platform, was created. The platform covers the entire lifecycle of APP design, development, testing, release, and operation. It also changes the traditional "code stacking" development model by providing a series of modular tools and components, allowing developers to create applications through drag-and-drop, configuration, and customization. Additionally, mPaaS brought China's leading "mini-program architecture" to Setia Go, changing the traditional model where "every functional change requires users to update the APP," achieving user-invisible optimization and greatly reducing the frequency of APP updates.[37]

"The launch of Setia Go marks the beginning of our strategic partnership with Ant Group," said Datuk Choong Kai Wai, Chairman and CEO of S P Setia, at the launch event. "For the past 50 years, the goal of building smart communities has inspired us and will continue to be at the core of all our business endeavors. We are committed to introducing the latest innovative technologies to enhance the consumer experience for homebuyers, which aligns with Setia Group's development philosophy of creating sustainable communities."[38]

[36] Huanqiu, 2024, https://finance.sina.cn/2024-05-31/detail-inaxakrn0891551.d.html?vt=4&cid=76524&node_id=76524.

[37] Haiwainet, 2024, https://finance.sina.cn/2024-06-06/detail-inaxuukv4369924.d.html?oid=5_nou&vt=4&cid=76524&node_id=76524.

[38] Huanqiu, 2024, https://finance.sina.cn/2024-05-31/detail-inaxakrn0891551.d.html?vt=4&cid=76524&node_id=76524.

Setia Group has hit the mark by leveraging the digital prowess of Setia Go, enhancing customer delight, and fostering the sustainable growth of communities. In the meantime, the synergy between Setia Group and Ant Group has culminated in a powerful alliance aimed at achieving common objectives.

12.3 Inclusiveness, Talent, and Sustainable Development: The Future of Ant International and Malaysia

12.3.1 Capital A × Ant Group: The Future of Digital Payments and Financial Technology

On March 29th, 2024, Ant Group and Capital A, the parent company of AirAsia, announced an expansion of their strategic cooperation in multiple international business areas. Solutions provided by Ant Group, such as cross-border mobile payments and international acquiring services, will join hands with the entire business ecosystem of Capital A, including AirAsia, the one-stop travel platform AirAsia MOVE, and the financial application BigPay. This collaboration aims to integrate more payment methods for the Capital A platform, provide acquiring services, promote business growth through digital marketing and business cooperation, explore inclusive and sustainable development opportunities, and achieve comprehensive collaboration.[39]

AirAsia MOVE will explore cooperation with Ant Group's international acquiring services to integrate the Alipay+ electronic wallet payment option into the AirAsia MOVE payment process. It will also explore the application of wallet technology, including super app-related solutions and the development of mini-programs within the Alipay+ ecosystem. AirAsia MOVE will also work with Ant to explore joint marketing and business sponsorship opportunities based on events such as the 2024 European Championship in Germany. Additionally, under this cooperation framework, BigPay will become the latest cooperative wallet of Alipay+, allowing its 1.5 million users to make seamless payments when

[39] China Daily, 2024, https://cn.chinadaily.com.cn/a/202403/29/WS66067f85a3109f7860dd78bc.html#:~:text=%E5%90%8C%E6%97%B6%EF%BC%8CCapital%20A%E6%97%97%E4%B8%8B%E4%BA%9A%E6%B4%B2,%E5%91%88%E7%8E%B0%E5%87%BA%E5%BC%BA%E5%8A%B2%E5%A2%9E%E9%95%BF%E6%80%81%E5%8A%BF%E3%80%82.

traveling abroad. BigPay will also explore the use of anti-fraud and other innovative technologies, including the development of a stronger payment platform and the creation of a more comprehensive super app. AirAsia, a subsidiary of Capital A, will use Ant Group's exclusive integrated payment solutions for the aviation industry to improve payment efficiency and achieve online, offline, and in-flight payment scenarios through cooperation with Ant Group.[40]

Tony Fernandes, Chief Executive Officer of Capital A, said, "We are thrilled to embark on this transformative journey with Ant International. Their global recognition as a fintech giant, coupled with their exceptional expertise, will undoubtedly fuel the rapid growth of our fintech enterprise BigPay and our online travel application (OTA) AirAsia MOVE. Both entities are committed to agile expansion, and this partnership is expected to accelerate our collective mission of providing seamless financial services and affordable travel experiences to customers worldwide."[41]

This strategic alliance spans various realms, delving into the integration of diverse payment solutions and offering acquiring services for Capital A's digital platform. Capital A has thrown its weight behind Ant International's latest venture, SIRIUS, which is aimed at extending sustainable financing opportunities to the small and microenterprise sector, thereby empowering and uplifting this often-overlooked community.

12.3.2 Talent and Employment Opportunities: Establishing a Digital Business Hub in Malaysia

Ant International is committed to accelerating global business and inclusive development through technological innovation. The majority of positions at the digital business hub are related to technology and innovation, including technology research and development, development, product, and design, as well as data technology and risk management. The company will invest resources to develop and cultivate these technical talents through comprehensive training, guidance, and a wealth of global business practice opportunities.

[40] Ibid.

[41] Airaisa, 2024, https://newsroom.airasia.com/news/sc-capital-as-digital-arm-to-form-strategic-partnership-with-ant-international-to-enhance-collaboration-on-digital-payments-and-financial-technologies#gsc.tab=0.

On April 18, 2024, Ant International announced the establishment of a digital business hub in Malaysia to attract local tech talent, promote global business development, and actively contribute to the economic development of Malaysia and the Southeast Asia region. It is reported that Ant will create more than 500 jobs in 2024, with continuous recruitment plans to be launched thereafter.[42]

At the media communication meeting on the 18th, Ant International President Douglas Feagin disclosed that the company's new digital business hub will be housed in Exchange Tower 106 by 2025. The Tun Razak Exchange Financial Central Business District is where the site is situated.

"The establishment of a new digital business hub by Ant International in Malaysia will play a significant role in helping local talents grow in the globalization of their digital industry businesses. This move brings more employment and investment opportunities, thereby helping to establish Malaysia's position as a digital hub in ASEAN. This is a clear testament to the close cooperation between Ant International and Malaysia to drive the development of the digital economy, and we will continue to support it," said Mahadhir Aziz, Chief Executive Officer of the Malaysian Digital Economy Corporation (MDEC).

With initiatives like the 10 × 1000 Tech Inclusion project, which teaches and inspires tech talent and leaders of the digital economy, Ant International also keeps working with regional partners to develop tech talent in Malaysia. More than 120 top representatives from Malaysia's fintech industry have finished the 10 × 1000 training, with over one-third of them being women, with the assistance of partners including MDEC, Touch 'n Go Digital, and the Malaysian Fintech Association.

12.3.3 Efforts of the National Bank and Boost's Inclusive Advocacy

Bank Negara Malaysia recognizes the gap in access to digital payments for certain segments of society and has introduced several initiatives to ensure these groups can benefit from digitalization.[43] How have they taken steps to promote a more inclusive digital society and ensure that payment solutions and financial services continue to reach all market segments?

[42] The Page, 2024, https://thepage.asia/ant-international/.

[43] Pathma Subramaniam, 2022, https://theedgemalaysia.com/article/cover-story-cashless-quandary.

12.3.3.1 Enhancing Public Awareness of Electronic Payments Through Advocacy

Practices include raising awareness and trust in electronic payments through media channels and outreach programs such as roadshows, to educate the public (underserved populations included) about the benefits and steps of transitioning to electronic payments.

"As awareness of electronic payments continues to grow, the elderly or disabled can now independently pay bills, purchase prepaid top-ups, or shop online comfortably from home without having to travel long distances to withdraw cash and run errands."

"From this perspective, electronic payments offer greater power and convenience for the elderly and disabled compared to cash. Programs targeted at seniors and retirees will be strengthened in the near future," stated Bank Negara Malaysia.

Promoting safe practices for electronic payments is also an ongoing effort, it notes. "As the adoption of electronic payments increases, financial fraud and cyber-attacks against users at all levels are also on the rise."

"It is everyone's shared responsibility to protect themselves from financial fraud and scams. While law enforcement and the financial industry will do their part, users also need to stay informed about the latest threats through programs provided by Bank Negara Malaysia and the financial industry. For instance, the public can visit the Amaran Scam Facebook page, which has more than 38,000 followers and is regularly updated by various agencies involved in combating financial fraud," stated Bank Negara Malaysia.

Besides, the regulatory authority is working to encourage employers to shift from cash to electronic wallets for wage payments, especially for migrant workers. "Electronic wallet providers will directly enable migrant workers without bank accounts to save and make payments and remittances to their home countries using electronic wallets. Notably, these migrant workers can remit money regardless of their immigration status, without any barriers, as long as the relevant details are recorded."

Bank Negara Malaysia is also committed to strengthening the role of agent banks to promote digital payments in rural areas, especially where ATM networks are limited. Since the majority of agent banks are located in rural areas, their function is further expanded to help rural communities with financial activities and payments, including opening savings accounts, transferring funds, paying bills, and withdrawing cash. Bank

Negara Malaysia said that "we are also taking measures to promote the use and acceptance of payment cards and electronic wallets in schools, colleges, and universities through initiatives taken by industry players, introducing electronic payments to students from a young age to change their behavior."

12.3.3.2 Bridging the Gap of Regional Development Differences Through Infrastructure

The central bank acknowledges that digital connectivity is another challenge in rolling out electronic payment facilities in remote areas. "Over the years, internet penetration has continuously increased, from 66.6% in 2014 to 87.4% in 2018. However, there are still areas lacking stable and reliable digital connectivity, especially in rural areas."

The Jendela Plan, a comprehensive five-year digital infrastructure project worth 21 billion ringgit, was introduced by the government to address the wider issues with digital infrastructure. The objective of the plan is to help Malaysia achieve greater digital connectivity by increasing the effectiveness of the country's infrastructure and optimizing spectrum usage. Improving total user connectivity and high-quality communication services is the main goal of this infrastructure. Along with attaining economic continuity towards a digital economy and preparing the nation for 5G connectivity, it is dedicated to enhancing the standard of living for Malaysians. It is expected that digital payments will reach remote areas in Malaysia through the implementation of the Jendela plan, bringing convenience to merchants and the public.

To further encourage the public to adopt smart mobile devices, the government has also put in place a number of measures. For example, the Jaringan Prihatin offers financial aid to low-income families so they can buy mobile devices and sign up for mobile broadband services.

12.3.3.3 Boost's Inclusive Advocacy

Financial technology companies like Boost, which has nearly 10 million electronic wallet users to date, are one of the few companies serving the unbanked and underbanked population.

Access to the basic functions of the Boost wallet only requires an internet connection, email address, and phone number. However, to access its advanced features (such as peer-to-peer transfers), a Malaysian identification card or passport is needed.

Boost's Group Chief Executive Officer Sheyantha Abeykoon said, "although we see a thriving fintech industry providing digital financial services in Malaysia, we still have significant potential and room for growth to develop and meet the underserved population, especially in meeting their unmet financing needs."

"Our future plans include expanding micro-financing services, using alternative data and artificial intelligence to reach more thin-file customers, expanding the buy now pay later sector, launching merchant solutions that cater to the needs of micro, small, and medium-sized enterprises, and bringing them into the digital economy. For our users, we will continue to build more convenient solutions to complement and simplify their growing digital lifestyle," said Sheyantha.

12.4 Conclusion

The 8th edition of the "e-Conomy Southeast Asia Report—Peaking Again: The Road to Profitable Growth," released in 2023 by Google, Temasek, and Bain & Company, indicated that the Southeast Asian digital economy would achieve approximately $100 billion in revenue that year. Since 2021, the Southeast Asian digital economy has grown at a compound annual growth rate (CAGR) of 27%.[44] In a significant shift, digital payments have overtaken cash, now constituting more than 50% of all transactions in the region. QR code payments are gaining widespread acceptance among Southeast Asian users, with an increasing array of online payment methods being seamlessly integrated into the checkout flows of popular apps. The merchant discount rate (MDR) is on a downward trend and is anticipated to continue its decline, as merchants and platforms look to cut costs and encourage the uptake of account-to-account (A2A) or e-wallet payment methods by reducing the MDR.

For Malaysia, the digital ecosystem's evolution has catalyzed transformative changes in local industries and daily routines. In 2021, a remarkable surge in cashless transactions propelled Malaysia to the forefront of Southeast Asia's payment markets, making it the region's second most cashless payment-adopting country, just after Singapore. Data from the same year indicated an influx of over 3 million new digital consumers

[44] Google, Temasek, and Bain & Company, 2023, https://www.bain.com/insights/e-conomy-sea-2023/.

in Malaysia, with an impressive 98% of Malaysian businesses already on board with digital payments. A Visa study corroborated this trend, noting a 64% drop in cash usage in Malaysia post-pandemic, contrasted with an 18% rise in electronic payment usage. According to PayNet, electronic payments have dethroned cash as the preferred payment method. While 78% of Malaysians now possess multiple e-wallets, the ecosystem, which includes more than 50 banks and non-bank e-wallets, is still in the midst of integration.

CHAPTER 13

Sustainable FinTech for Inclusive Partnership and Regional Development: Akulaku PayLater, BillEase, BPI, Gcash, HelloMoney, Maya and Alipay+ Ecosystem in the Philippines

13.1 The Digital Economy Ecosystem in the Philippines

13.1.1 *The Popularization and Development of Local Digital Payment Services*

In recent years, Southeast Asia has demonstrated impressive economic development, standing out relative to other regions in the world in withstanding global macroeconomic headwinds. The GDP growth for Southeast Asia has stabilized at over four percent while inflation rates have gone down to three percent.[1] In tandem with this, digital financial services in the region have expanded of late, notably in mobile and e-wallet services. The Philippines has been a little behind the curve in this regard but is catching up fast, developing its own local digital financial ecosystem and working with Alipay+ to promote sustainable tourism services for Asian tourists. Not only does it foster inclusive growth of the

[1] Southeast Asia e-Conomy Report, 2023, https://www.sohu.com/a/752446514_121640652.

J. Sun, *Sustainable Financialization*, Sustainable Development Goals Series, https://doi.org/10.1007/978-981-96-3580-1_13

economy within the country, but it also offers the potential of benefiting other economies in the Southeast Asian countries.

As reported by the World Bank in 2023, the population in the Philippines stands at more than 117 million people with almost 49 million of those being part of the labor force. With this, the Philippines becomes the second most populous country in Southeast Asia that has a GDP per capita equating to USD 3,725.[2] The vast size of the country's young-population structure also provides a strong potential market for consumer spending as well as the internet economy. Market boundaries will continue to be dominated by some traditional commerce, especially retail, and the Philippines is an appealing new market for international reputable organizations. Therefore, it can be said that the Philippines still has a long way to go in terms of adopting digital payment platforms as most citizens still use cash as a primary mean of exchange. But due to global trends including state backing of digitalization, technology innovation, and the global COVID-19 crisis, the Philippines has become more diverse in terms of modes of payment. During the last couple of years, there has been a drastic change in the financial landscape of the country.

In the past, the provision of financial services in the Philippines was left to the traditional banks, but the infrastructure was underdeveloped and there were issues of limited coverage. In comparison to many developed nations, the number of people with a bank card or a credit card is low in the Philippines. Getting a credit card is too hard for many consumers since they have no credit history, and opening an account with a bank is often associated with a lot of hassle and long procedures, which puts many people off from utilizing the banking services. On top of that, the narrow size of the market for the acceptance of cards has also prevented many people from enjoying the benefits of using bank cards. Consequently, this has left a significant proportion of the population as unbanked or underbanked and therefore limits the access to a wide range of financial services. A 2019 Leverage Financial Cushion Report by Bangko Sentral ng Pilipinas (BSP), a financial account was only possessed by 29% of Filipino adults as of 2019, with only 8% of them having accounts of e-money, and 12% of them holding bank accounts.[3] This lack of financial integration

[2] World Bank, 2023, https://data.worldbank.org.cn/indicator/SP.POP.TOTL?locations=PH.

[3] Xu Gao, 2022, https://www.mpaypass.com.cn/news/202208/24095844.html.

directly stifles economic activity of individuals and small enterprises, but it also demonstrates the vast opportunities available for the development of the country's financial sector.

Due to the progress made towards the improvement of digital payment systems, the pandemic providing heightened needs to go digital, and constant support for digital businesses, the Philippines has seen a growing digital financial ecosystem that has now become a trend. Systems that make electronic transactions possible are primarily at the forefront of this change. It caught on with the use of e-wallets such as Gcash and Maya which easily served the needs of both individuals and hierarchy in organizations at the time. The number jumped to 50 million in the year 2022 as Gcash's registered users increased from 20 million in 2019. More than 30 million people now use Maya (formerly PayMaya), showing that the company is expanding and growth has been stable since 2021. However, these platforms do not only offer services that are meant to be payments and transfers, but they have also been able to integrate loans, investments, and insurance, expanding the user's array of financial products. For example, an account holder can use the Gcredit service from Gcash and execute a payment even when the account balance is insufficient. On the contrary, Maya launched savings accounts with all high-interest rates called Maya Savings. These services have been beneficial in promoting financial literacy and as a result, have improved many households' access to modern-day banking services.

Buying goods or services can also be done through buy now pay later services (BNPL) and the same scenario applies in the Philippines as this market is growing quite popular and with a lot of demand. Many customers like the Akulaku PayLater and BillEase platforms, as they allow purchases with deferred payments or loans. This is in most cases adopted by the younger generation who have more e-commerce businesses and purchase activities making the sector grow even more.

On the other hand, the growth of the digital financial ecosystem has also improved the working of small and micro enterprises substantially. A lot of businesses, which were unbanked earlier and did not have a facility of digital payment acceptance, can now accept digital payments and get microloans for business purposes. This has not only improved their functionality but also affected the economy positively.

This progress is consistent with the recent trajectories undertaken by the Philippines in regard to the expansion of its financial inclusion policies. The National Strategy for Financial Inclusion 2022–2028 or NSFI which was launched in 2022 is an improvement from that introduction in July 2015 and serves as a framework for the provision of financial inclusive services for the next six years.[4] The NSFI's previous version established the principles of inclusive finance and coordinated actions between the public and private sectors for the effective provision of financial inclusion policies and projects. The new NSFI concentrates on four specific objectives: Closing the gap in financial inclusion, enhancing the viability and stability of the financial sector, increasing consumer awareness and empowerment, and widening access to finance for small and startup businesses and the agricultural sector. It also stresses that financial resilience is one of the indicators of financial inclusion, that the majority of the population especially the poor Filipinos are under stress brought about health emergencies, environmental change, and the dynamics of the economy and job market.

The roadmap for the transition to digital payment entails that by the year 2023, the Bangko Sentral ng Pilipinas (BSP) will have raised the share of adults with accounts or electronic accounts to 70% and moved 50% of total retail payments to electronic channels.[5] The onset of the pandemic also helped in some way as the proportion of the population aged 15 years and above who held a bank account in the Philippines increased from 29% in 2019 to 53% in the first quarter of 2021. Still, the quantum of total number of digital payment transactions as a percentage of total transactions in the economy only advanced by 1% from 2013 to stand at 20.1%.[6]

In the aspect about the Philippine's plan on digital finance development, former BSP Governor Benjamin E. Diokno did discuss a major component saying, "Digital banks will be key players in the digital finance ecosystem. These partners will help increase market's efficiency and provide a wider range of financial services enable us to reach our financial inclusion targets faster which include having at least half of retail payment transactions in Philippines to be electronic by the year 2030 and nearly seven in ten adult Filipinos to have transaction accounts."

[4] Bangko Sentral ng Pilipinas, 2022, https://www.bsp.gov.ph/SitePages/InclusiveFinance/InclusiveFinance.aspx.

[5] Bangko Sentral ng Pilipinas, 2023, https://www.bsp.gov.ph/PaymentAndSettlement/2023_Report_on_E-payments_Measurement.pdf.

[6] Ibid.

The last few years have however seen a watershed development in the area of digital payment advancement in the Philippines. Bangko Sentral ng Pilipinas (BSP)'s report for the year 2023 states that the use of digital payments among retail dealings increased up to 52.8% of the total value pushed over the target of 50% which was under the digital payment transformation roadmap launched from 2020–2023.[7] Such an increase is a considerable improvement from the 42.1% measured in the year 2022. Governor Eli M. Remolona of BSP viewed this as one of the achievements that pushes the country closer to its goal of greater cash-lite economy where the usage of cash is minimal, further emphasizing the role of digital payments in promoting financial inclusion and economic efficiency.

In relation to BSP, from the 1st to the 4th month of the year 2024, the overall volume of the electronic fund transfers in the Philippines increased by almost 70% recording 42.04 billion pesos after an increase from 24.78 billion pesos recorded during the same period last year. More specifically, values transacted using PESONet increased by 32% amounting to 3.12 trillion pesos, with the number of transactions growing by 8.5% amounting to 31.9 million transactions. At the same time, the value of InstaPay transactions increased by 45% amounting to 2.1 trillion pesos and the number of transactions surged by 78% to 388.5 million transactions.[8]

PESONet and InstaPay are mechanisms of the BSP's National Retail Payment System (NRPS) which has had a focus of upholding a secure, reliable, efficient, affordable, and inclusive payment system since its inception in December 2015.[9] PESONet's primary purpose is to facilitate high-value transactions and is considered an electronic replacement for paper check. As for InstaPay work instantly and has a limit amount of up to 50,000 pesos which are recommended for remittances and e-commerce situations.

This rapid transition to digital or mobile is transforming the financial ecosystem, empowering both individuals and small-scale enterprises with better access to finance and in the long run encouraging broader economic growth.

[7] Ibid.

[8] Business Times, 2024, https://businesstimescn.com/source/btzonghe.

[9] Fintech News Philippines, 2024, https://fintechnews.ph/63116/payments/pesonet-and-instapay-see-70-volume-surge-in-electronic-fund-transfers/.

13.1.2 Blue Ocean Opportunities in the Philippines' Digital Financial Ecosystem

However, the digital and the retail finance system of the Philippines has its own share of concerns and barriers to development notwithstanding the rapid transformation taking place.

First of all, although e-wallets have gained increased acceptance throughout the Philippines, security remains an issue for many consumers. Such fears are largely based on concern over the security of their money and the privacy of their data. Due to a surge in cyber-crime and data breaches in the recent past, public confidence has been eroded even further. A large number of users are uncomfortable utilizing Banks' websites due to fears that their personal information may be compromised during a transaction, preventing them from fully embracing the use of digital payment channels.

Filipinos can transact electronically using various payment platforms. Filipinos can use Gcash, HelloMoney, or Maya to transact electronically. However, the fragmented nature of this payment system results in a poor user experience. There is a difficulty switching platforms, so customers often stick to a certain platform which should not be the case in a good payment ecosystem. There are also major users of the platforms such as retailers who sell their goods and services to consumers. This fragmentation makes it hard for both consumers and providers of the services to use these platforms because there is no standard to follow. A payment system has different components that must work as one system in order to provide benefits to the users of the platform.

Urban dominance in mobile payments is quite evident, but the situation is different in the view of rural and remote areas. Still, a large part of the population in these areas has insufficient access to universal financial services. The lower penetration of such digital payment systems can be explained by the lack of needed infrastructure in these areas as well. According to the statistical data available, the country is lagging behind in incorporating modern technology in many aspects, as depicted in Table 13.1, there is potential for improvement in the digital structures within the country. These infrastructural gaps point out the reason why urban inhabitants of the country are able to utilize digital financial services more than those of the rural regions of the country and similarly provide a case for continued expansion of such services.

Table 13.1 Mobile phone and internet access in ASEAN-5 and Vietnam[10]

	Indonesia	*Malaysia*	*Philippines*	*Singapore*	*Thailand*	*Vietnam*
Mobile phone ownership (%)	66	97	79 (2019)	89	87	77
Mobile-cellular subscriptions per 100 inhabitants	134	141	146 (2022)	147	169	139
Mobile-cellular network coverage (%)	98	98	99	100	99	100
Individuals using the internet (%)	62	97	89 (2022)	97 (2022)	85	74
Households with internet access at home (%)	82	95	77 (2022)	99 (2022)	89	81
Active mobile-broadband subscriptions per 100 inhabitants	115	125	141 (2021)	147	112	88

Note All the numbers in the table are for 2021 unless stated otherwise
Source Digital Development Dashboard of International Telecommunication Union, OOKLA, PSA, National Telecommunications Commission

If these challenges are met, the Philippines might be able to leverage more opportunities in the country's digital financial ecosystem as there is still a very wide blue ocean in terms of potential in infrastructure development and financial inclusion.

Gcash is among the newest payment solutions in the Philippines and has grown over the years hand in hand with changes in the country's finance quadrant. This specific service belongs to Mynt which is a local digital finance company that was set up in the past as a department of globe, the biggest provider of telecommunications services within the Philippines. Gradually as the business of Gcash began to gain traction, Mynt began to separate itself from the parent company. It is worth mentioning that one of the investors of Globe is Ayala which is the second-largest conglomerate of the Philippines.

The Philippines' economy is more or less controlled by several key players, business families in which market share is mostly distributed. With

[10] Andrew Tsang, 2024, https://www.bworldonline.com/opinion/2024/04/11/587198/challenges-and-opportunities-of-philippine-economic-digitalization/.

this kind of competition, credibility comes with the size of the enterprise group, so the bigger the business group the more trust it has on the consumer level. Such credibility goes as far as the user trust level and fosters better communication with the regulatory bodies. As for Gcash, getting assistance from a big corporate name like Ayala came with many benefits such as getting relevant payment licenses and other operations required for the company's market entry.

Mobile transactions, however, are very much about technology, and the lack of infrastructure and the restrictive technological capability in the Philippines meant that it was possible to work with more advanced partners. In February 2017, mynt entered into a strategic deal with Ant Group, which took a huge but non-controlling 45% share of Gcash owner. Ant Group being active in micro payments and mobile lending—businesses that Jnt has a lot of overlap with—became a great assest in wographing to boost Gcash's offerings.

Beyond financial investment, Ant Group leveraged its technical expertise to enhance and optimize Gcash's platform. Following the strategic agreement, Ant Group immediately deployed technical specialists to work alongside Mynt's Gcash team, driving substantial improvements in both functionality and user experience. According to Anthony Thomas, former CEO of Mynt, Ant Group's involvement gave Gcash access to a larger global network of merchants. He emphasized that the partnership created a network effect, greatly increasing Gcash's appeal to merchants by linking it to the Ant Group's expansive global ecosystem.

Belonging to Ant Group's global mobile payment and lifestyle service provider Alipay+, a strategic outlook and broad market presence has been established by Alipay+ in the Philippine digital payment landscape today as the company has local partnerships with Akulaku PayLater, BillEase, BPI, Gcash, HelloMoney, Maya to create a fuller and more diverse ecosystem of digital financial services.

In the Philippines bilaterally, has transacted with Pagos and other services like Gcash, Maya, and HelloMoney wherein their technology integrates with the local digital payment ecosystem in order to extend support to Filipino's that utilizes their services when traveling abroad. Filipinos are now capable to transact overseas via enabling the cross-border payments with their wallets domestically and abroad because of this ability. In addition, policies involving the acquisition of local payments have simplified the scenarios of performance and attracted foreign clients to the Philippines in pursuit of simplicity. Additionally, platforms such as Akulaku PayLater or BillEase that use BNPL technology allow for increasing the payment flexibility for Filipino consumers. This demographic of consumers is often younger which values such payment options enhancing their experience while strengthening local e-commerce and the retail market.

On a different note, Alipay+ expands the concept of responsible tourism through "Sustainable Tourism Payment" or STP initiative. By working alongside local partners, Alipay+ helps Alipay products consumers bring in Asia, have an all-rounded eco-friendly trip to the Philippines. It makes up for a variety of activity, which spans from shopping, food, accommodation and transport around the area, and does also assist in the betterment of the local economy, while satisfying the needs of the tourists.

In the following paragraphs, we shall focus on two areas in particular: building the local digital payment system in the Philippines and also the role of Alipay+ in addressing the issue of sustainable tourism to demonstrate the impacts of an inclusive digital finance system that the Philippines and Alipay are building together.

13.2 Alipay+ Empowering the Development of the Philippine Digital Financial Ecosystem

13.2.1 The Rise of Buy Now, Pay Later (BNPL) Platforms: Akulaku PayLater and BillEase

As an open digital ecosystem platform, Alipay+ has strategically partnered with local digital payment platforms to foster the growth of the Philippine digital ecosystem, providing higher-quality digital financial services to local users.

13.2.1.1 Akulaku PayLater: Expanding Indonesian Expertise into the Philippine Market

For a long period in the Philippines, the traditional banking facilities for consumers and businesses have not worked satisfactorily, thereby leaving many people unbanked or underbanked. As shown in the e-Conomy SEA 2020 report, out of 400 million adults in the Southeast Asia region able to afford such services, only 104 million enjoy the full package of these services to include credit cards, investments, and insurance.[11] Similarly, small enterprises in most markets are also constrained to access affordable credit and this problem is more pronounced in Southeast Asia. Numerous small businesses require loans, yet are unable to obtain them at reasonable terms. In the Philippines, the proportion of people who are unbanked or underbanked is particularly high and the general banking coverage rate in the country is among the least in the region.

People who cannot easily reach conventional banks are at a disadvantage when it comes to utilizing regular financial services. On the other hand, technology-empowered business models are better able to target such a segment, thus opening up new market areas. Digital Financial Services (DFS) have been seen for a long time as a valid solution to the unserved populations' challenges in gaps within traditional banking. For the vast populations of the unbanked, the pure fintech and consumer tech platforms will not fulfill the huge need for digital financial services because such needs are linked to the developmental progress of national infrastructure capacities and underwriting needs.

This segment is where institutions will deploy digital financial services the most and is the real growth engine. With a large, growing, and active customer base in the consumer technology platforms, these platforms are able to capture the future contribution of customers within the underbanked side. Knowing this, Buy Now Pay Later (BNPL) services have floated in order to address the lack of credit card installment services that a lot of Filipinos currently have allowing consumers to purchase with the option to pay later. In the Philippines, Akulaku PayLater and BillEase are the two major users of BNPL services enabling Filipinos to make purchases and pay for them in parts. These services encourage consumers to buy on credit, with low interest rates, and pay later for the purchase made.

Akulaku Group is one of the fast-growing consumer finance service providers with a focus on technology, creativity, and innovation in Jakarta, Indonesia, where it was established in 2014. For its clients in the emerging markets, it provides a variety of services that include e-commerce, consumer finance loans, digital banking, online investments, and investment insurance. The release of the Akulaku PayLater app in 2016 has enabled the company to extend its credit limits to people with

[11] Google, Temasek, Bain Company, 2020, https://www.bain.com/insights/e-conomy-sea-2020/.

little or no credit histories and manageable installment plans across Southeast Asia, including the Philippines. Users can make repayments through e-wallets, bank transfers, and various other options.

In September 2022, Akulaku partnered with Alipay+, making Akulaku PayLater the first BNPL wallet that Alipay+ ever worked within Southeast Asia. Already, more than 4,000 merchants are already in the system. This allows the consumers in the Philippines to utilize Akulaku PayLater for BNPL services through the merchants of Alipay+. In the words of Akulaku's CFO Zhang Fan, "This partnership will bring more advanced financial services to the customers in Indonesia and Southeast Asia. We will rely on Alipay+ to quickly multiply the merchants who accept Akulaku PayLater, thus enhancing the user experience of our customers and the competitiveness of our BNPL services. Such combined assets will offer on-demand financial services to unbanked populations, stimulating growth in many industries within these markets."[12]

Jing Guoming, General Manager of Global Partnerships at Alipay+, noted that: "Since BNPL is a new but a very fast-growing digital payment option, it has made financial services available to the underbanked population." Most importantly, Mr. Guoming views BNPL as part of a digital payment ecosystem and in collaboration with Akulaku Paylater he is able to enable Indonesians and other targeted markets to provide seamless cross-border payments for their purchase enhancing their overall shopping experience.[13]

13.2.1.2 BillEase: The Growth of a Local BNPL Platform

Among those BNPL services provided by Akulaku PayLater that has pan-Southeast Asia influence, BillEase, a domestic BNPL platform based in the Philippines, has also shown considerable competitive advantage. BillEase is Philippines-based "Buy Now, Pay Later" service that allows qualified users to make installment payments sans a credit card and on a credit line. Since its launch in 2017, BillEase has connected with more than 2000 merchants such as Philippine airlines, Power Mac Center, Anson's, Vivo, Kimstore, and Focus Global Inc.

[12] Akulaku, 2022, https://www.akulaku.com/blog/akulaku-establishes-partnership-with-alipay-to-make-cross-border-shopping-more-accessible-for-consumers/

[13] Ibid.

For small and medium-size purchases, for instance ticket purchases, BillEase provides users with payment alternatives. The options include making a single full payment after 10 or 20 days of purchase, or payments broken down into four equal amounts. For more significant shopping requirements like buying cars, BillEase assists in purchasing through monthly install plans for 3, 6, 9, or 12 months. With such plans, paying interest is necessary, and in most instances, rates will be between 0% and 3.49%, which changes with the duration of the plan, and promotion.

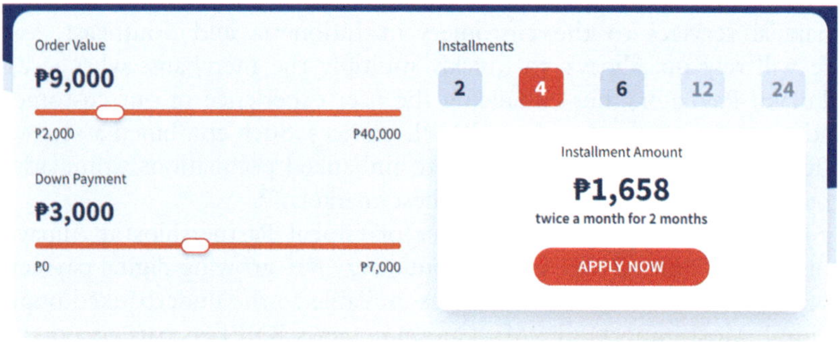

Screenshot of a BNPL Payment Plan Interface from BillEase (*Source* BillEase)

Apart from making payments online purchases, the BillEase app has grown to be useful in various daily transaction cases. The application itself allows users to recharge their electronic wallets that contain a number of local E-wallet services. Gamers can also use BillEase for the acquisition of in-game currencies or using it for mobile phone load recharges. So, BillEase serves as an efficient means to buy in-game credits for gamers. Additionally, those using BillEase pay utility bills including water, electricity, and gas, making life easier.

BillEase has partnered with a wide range of online merchants and service providers creating a strong affiliate network. Users shopping with these partner websites can easily select BillEase's installment plans during checkout. Given the variety of users and their requirements, BillEase also has several repayment options to suit the users. They include but are not limited payments made through InstaPay or PESONet platform, other e-wallets, or even bank wires. This model improves the way consumers shop and also improves the sales conversion rate and average order value for merchants especially for those consumers who do not have access to standard credit facilities.

In January 2023, the Philippine-based fintech company BillEase announced a collaboration with Alipay+ which is controlled by Ant Group, and now local customers can use the BNPL and cardless installment options for their purchases on international sites. Thanks to this partnership, BillEase's customers gain access to uncomplicated and effective cross-border online payments and improve the quality of their lives through an increased ability to shop around the world. Moreover, it has great implications for advancing financial inclusion in the Philippines, since it allows millions of Filipinos access to financial tools more easily and in turn provides global online retailers with a wider and more varied customer base.

In addition, this cooperation is crucial for enhancing the level of economic activity within the region as well as expanding the scope of financing activities. It also reduces the obstacles to entry into the electronic commerce industry, allowing more people especially the youth and lower-income earners to engage in the digital economy. This partnership also aids increase the competitiveness of businesses in the Philippines, enhances local innovations as the regional economic growth is further stimulated by creating local Fintech companies. As more and more people are integrated into the organized system of finance, the personal economic conditions of the people develop further aiding in poverty alleviation and the overall development. For international eCommerce business, this collaboration provides access to millions of Filipino customers who otherwise had been unable to engage in cross-border ecommerce because of a lack of efficient payment systems. This in essence not only enhances prospective sales for merchants, but also provides access to new markets.

Apart from border shopping, BillEase is taking advantage of the renewed interest in travel after the pandemic by integrating Alipay+ to extend payment methods at Agoda, an international online travel agency. This enables Filipino customers to book hotels, make flight reservations, or buy local tours through Agoda and use BillEase's Buy Now Pay Later or installment services even without a credit card. Tourists are able to divide the total payment into payments every two weeks or every month at a low interest rate of 0%, or at some of the lowest in the market rates for up to 12 months. This has very seriously has made many Filipino people to take up traveling.[14]

13.2.2 The Bank of the Philippine Islands: Traditional Bank's Digital Transformation

The Bank of the Philippine Islands (BPI), a pioneer bank in south east Asia that has been operating for almost 170 years, provides its customers with the most comprehensive services which include not only traditional banking and loans but also credit investments and modern online banking solutions. With such a broad area of operation, BPI's offerings extend beyond just mobile money, covering a host of banking services. With the constant advancements in financial services technology (fintech) and changing customer demand, BPI realized that there is potential to lose competitive edge unless there is a repositioning of finacial services.

This shift is not only an answer to the existing trends of the market but also an action that has been taken to make sure the bank moves forward in the future.

Fintech has profoundly altered how people make payments and access financial services in Asia's developing countries. A survey conducted by the World Bank indicates that over the last ten years, the proportion of accounts has improved, but approximately 1.7 billion adults are still unbanked as of 2014. In the Philippines, financial inclusion has advanced in the past few years; however, a segment of society has not properly availed the benefit of facilities that modern-day financial institutions offer. Moreover, young people, especially, have a preference for mobile devices for regular purchases and seek rapid, easy, and tailored services. It has thus become essential for traditional financial establishments such as BPI to embrace transformative strategies in this digital era to draw and cater for this fast-growing client base.

[14] Staff Writer, 2023, https://www.marketing-interactive.com/billease-partners-with-agoda-and-alipay-to-provide-flexible-travel-experiences.

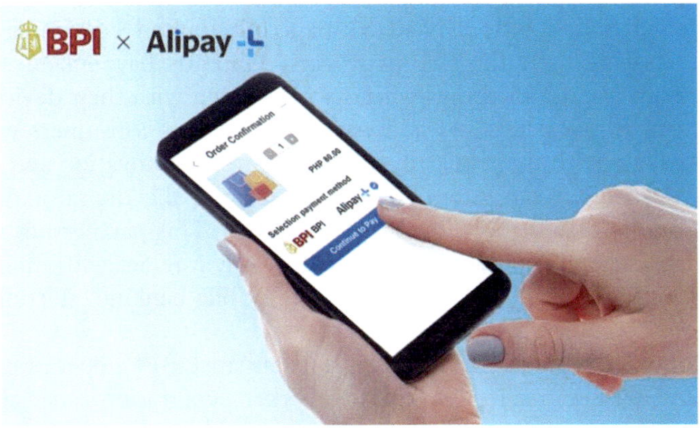

BPI and Alipay+ Payment Solution (*Source* Alipay+)

Previously, the Bank of the Philippine Islands (BPI) operated through physical branches. Although this was a good source of sustainability and guarantee, this had its limitations in terms of growth and improvement of its services. As more competitors began adopting internet technology and mobile application technology in providing better user experience, BPI faced ruthless competition. In response, BPI dramatically improved its digital technology investment and developed quite a number of novel initiatives.

In 2022, BPI partnered with Ant Group's Alipay+ to leverage its innovative cross-border payment solutions, enabling over 4 million Filipino consumers to enjoy a seamless "one wallet for global purchases" experience. By integrating Alipay+ solutions, BPI allows its users to conduct various online and offline transactions domestically and access services from millions of global merchants seamlessly. By 2021, BPI's digital customer base grew to 4.9 million, with active users increasing by 20% compared to the previous year.[15] Additionally, 91% of BPI's transactions were digital in 2021, up from 85% in 2020, demonstrating the bank's success in attracting more customers to its digital platforms for daily financial management.[16]

[15] BPI, 2022, https://www.bpi.com.ph/about-bpi/news/BPI-and-Alipay-Enables-Digital-Banking-Customers-to-Shop-Global.

[16] China Daily, 2022, https://caijing.chinadaily.com.cn/a/202206/07/WS629ef1f6a3101c3ee7ad9457.html.

BPI chief digital officer Noel Santiago pinpointed consumer digital evolution saying, "In the last two years, Filipinos have made a quick change from paying for things with cash to paying with their devices. For this reason, Alipay+ helps us address this market of consumers who are used to paying with their smartphones. With more innovation and expansion of our digital ecosystem, we are enhancing the digital experience for consumers that is more than just banking." This partnership further enables BPI's total 4 million customers to easily interact with millions of Alipay+ supported global merchants using mobile banking, starting from cross-border online payment options.

The digital transformation did not only enhance BPI's operational efficiency and service quality but also its market penetration especially into new areas. Furthermore, the transformation enables the advancement of a more equitable society by helping to extend the delivery of secure and dependable financial services to wider audiences. It promotes transactions and development in particular areas, establishing a good platform for more equitable and liberalized global economic setting.

Noel Santiago further commented, "The pandemic has rapidly accelerated the shift of payment behavior and preference of Filipinos from cash to digital. We aim to better serve and engage the now digitally savvy customers, by continuously innovating and expanding our digital ecosystem of partners and merchants, for an enhanced digital experience that goes beyond banking".[17]

13.2.3 E-Wallets and Mobile Payment Platforms: GCash, HelloMoney, and Maya

The Philippines also has a vibrant mobile payment culture with lots of e-wallets and BNPL services helping consumers shop online. GCash, HelloMoney, and Maya are among the most successful platforms. These platforms focus on the provision of digital wallets and mobile payments with functions like bill payments, online payments, fund transfers, and balance replenishment. Most of these services aim to simplify and enhance the user experience when making digital transactions. Furthermore, they are an important aspect of the larger digital finance ecosystem.

[17] BPI, 2022, https://www.bpi.com.ph/about-bpi/news/BPI-and-Alipay-Enables-Digital-Banking-Customers-to-Shop-Global.

13.2.3.1 Gcash: The Leading E-Wallet in the Philippines

GCash, a top e-wallet in the Philippines, is owned by Globe Telecom and benefits significantly from the technology and experience provided by Ant Financial (the parent company of Alipay). Early in their partnership, Alipay sent a 30-member team to assist GCash, bringing advanced technology like Ant's data platform, Alibaba Cloud's infrastructure, and Alipay's security and risk management models. Even Jack Ma himself promoted GCash during a visit to the Philippines, demonstrating how easy it was to pay using a QR code.

As of 2023, GCash boasts 66 million registered users, meaning that roughly one in two Filipinos use the platform, making it the country's largest digital wallet.[18] Its features closely resemble those of Alipay, providing services like:

Bill payments: Users can pay for mobile phone top-ups, utilities, credit card bills, government payments, and even some school fees.
P2P lending: GCash offers microloans, a critical service in a country where over 80% of adults have borrowed money at some point, and 47% are considered repeat borrowers.
Money transfers: Users can transfer funds to other GCash users or bank accounts for free, without minimum balance requirements.
Merchant payments: GCash is widely accepted for online shopping, such as through Lazada, ticket bookings, and at thousands of offline merchants, with QR codes that also support Alipay.
Payroll services: GCash helps businesses manage payroll, disbursing salaries, incentives, and allowances.

Meanwhile, GCash users can apply for GCredit, Mastercard, and American Express virtual cards, as well as link their accounts to PayPal for seamless access to funds. With its American Express virtual card, users can shop internationally.

In 2022, GCash has teamed up with Alipay+ to enhance cross-border payment options that give real-time currency exchange rates and remove the normal overheads that are associated with credit cards. This enables

[18] China Daily, 2023, http://cn.chinadaily.com.cn/a/202308/10/WS64d47216a3109d7585e48b0b.html#:~:text=%E7%9B%AE%E5%89%8D%EF%BC%8CGCash%E6%B3%A8%E5%86%8C%E7%94%A8%E6%88%B7%E6%95%B0%E9%87%8F,%E6%88%90%E6%9C%AC%E4%B9%9F%E5%B1%85%E9%AB%98%E4%B8%8D%E4%B8%8B%E3%80%82.

overseas Filipinos and travelers to save more when making overseas purchases. As stated by Gallegos Michelle, Head of Growth Gcash, it has become unnecessary for users to hunt around for exchange shops or shop for the best exchange rates as the app does this automatically.

A major edge over credit cards is that GCash provides receipts which have all relevant information. These receipts contain the amount in pesos and the local currency of Singapore for example the amount in Singapore dollars as well as the applicable exchange rate and discounts. This function enables users to monitor their foreign expenses effortlessly.

Gallegos pointed out how valuable this function is given the large presence of Filipinos in Singapore today and the number of Filipino tourists in the country. She pointed out that this latest tool is consistent with the boom in "revenge travel" after the pandemic where consumers could be able to travel after a long period of travel restrictions.

GCash actively works towards offering financial service solutions to Filipinos globally or those who are still in the Philippines. The vision of the platform is for all its users to never be without the e-wallet as they can perform tasks such as sending money to the 71 million users of GCash, purchasing goods, and dealing with their daily finances from anywhere.

13.2.3.2 HelloMoney: A Player in the Cross-Border Market

Prior to the global COVID-19 pandemic, Asia United Bank (AUB) has launched HelloMoney enabling users to register and conduct different kinds of banking activities without the need to go to the actual physical branches. These include transferring funds between banks, purchasing reloads for prepaid phones, PeraPadala remittances, making payments via QR codes, paying bills, and making ATM withdrawals. What is more, HelloMoney provides its users with a virtual Mastercard to enable them to go for online shopping. This type of remote service model greatly improved the customers' experiences as they hardly had to visit their service outlets physically and this was important during the times of the global pandemic when there was a focus on safety while ensuring that business activities do continue.

After this, Hello Pag-IBIG was launched in consideration of Pag-IBIG Loyalty Card Plus holders to be their complementary mobile app. Pag-IBIG is provided in the Philippines in a form of a housing fund program with the objective of enabling members to save in anticipation of future housing requirements. With Hello Pag-IBIG, account holders are able to do account management, balance checking, taking of loans and other transactions associated with banking. Services that used to require going to a physical branch are now easily performed in any location and at any time, which improved better the level of comfort and accessibility.

In 2021, AUB began issuing HelloMoney Mastercard cards for their customers, intended for online shopping. Even after the lockdowns were loosened, neither HelloMoney nor the other segments was idling, quite on the contrary, HelloMoney experienced a double figure growth in terms of transaction volume. Now that number 30 million transaction of HelloMoney is captivating. In contrast to last year's 19 million, the number has grown by 65%. Transaction value increased, reaching amounts of Republic pesos 115 billion which signifies an increase of 82% in values from 63 billion pesos a year earlier.[19] The strength of the HelloMoney Mastercard has helped users making online purchases quickly and easily which stimulated e-commerce participation further.

AUB even became the first Philippine bank to adopt a cross-border payment e-wallet in 2024. The bank is currently aiming to expand the global footprint of HelloMoney. And because of AUB's tie up with Alipay+, whose cross-border payment, marketing and digital solution platform is operated by Ant Group, users of HelloMoney may now pay for goods in e-wallets in the Korean, Malaysian, and Hong Kong markets by entering stores and displaying the QR code which the store has put on display. That growth facilitates cross-border payments for HelloMoney clients in four nations and regions with payment capabilities including Japan.

13.2.3.3 Maya: Digital Finance Facilitator

Under the umbrella of the PayMaya Philippines Group, Maya Business (previously known as PayMaya) has also gained popularity as a digital wallet platform. The platform is simple to use and free, allowing for credible and convenient transactions at any time and in any place, thereby assisting users in managing their finances. Furthermore, with the provision of an all-encompassing financial platform, Maya enhances access to finance as more businesses, government units, and social services can be integrated smoothly into the digital economy.

Maya's e-wallet application provides such users with the ability to perform secure and contactless transaction increasing interest among

[19] Kuafu, 2022, https://www.kuajinzhifu.com/bulletin/7982.html.

people for online shopping where one can pay, send remittances, do inter-bank transfers, pay bills, and do QR code payments. This service is in fact assists individual users alone but also in providing aid to governmental and social services agencies. And industries like e-commerce, gaming, and live-streaming platforms can avail from its payment services.

Maya offers its users a PayMaya physical card which is tied with its digital wallet enabling its users to make payments through stores that accept card payments, or at ATMs both locally and internationally. Maya offers a wide coverage of domestic remittances throughout the Philippine local area and provides a number of other financial services which include international and domestic remittance, bill payments, prepaid top-up, and cash in and out services.

Maya is a pioneer in providing businesses with a fully managed payment system that helps them receive and perform transactions using multiple sources such as debit cards, credit cards, prepaid cards, electronic wallets, and QR codes across an ecosystem of devices. Supplementary components of its digital payments landscape consist of:

1. E-wallet Payments

Users can pay bills, shop online, and purchase entertainment services (such as games and concert tickets) using the Maya app, which supports both Philippine pesos and several international currencies.

2. QR Code Payments

Maya allows users to scan QR codes in stores, restaurants, and public service locations throughout the Philippines, promoting the convenience of cashless transactions.

3. Top-up and Cash Withdrawal

Users can load funds into their Maya accounts through 250,000+ cash in points or bank transfers and withdraw funds from designated locations.

4. International Payments

Maya users can purchase international goods and services using their PayMaya account, facilitating global transactions.

Maya is active and trying to aid Filipino enterprises by extending its digital financial solutions to them, allowing them to offer better and more integrated internet customer services while contributing to the development of a more computerized Philippines.

Promoting Financial Inclusion: The Philippines being an archipelago with several islands which are underdeveloped and lack financial services, constitutes its target market. Maya extends the reach of formal financial service provision to people who do not have bank accounts. Users engage in normal activities such as buying goods, paying their utility bills, and transferring money to other people with no requirement for a normal bank account. This service facilitates even people who live in the most inaccessible places to use financial services thereby bringing down the financial disparity between urban and rural people.

Supporting Small Businesses Growth: Maya also seeks to enable small businesses by providing them with a wide range of payment solutions that are important when accepting payments in an online environment. For both e-commerce websites and physical stores, businesses can utilize Maya's QR code payment system to simplify their point of sale and improve overall business processes. Also, through strategic partnerships with different e-commerce sites, Maya provides SMEs with more ways to generate sales. In this way, Maya enables SMEs to leverage the opportunities of digitalization in growing their businesses.

Maya Business provides a suite of digital cash services which translates every business into a cashless organization. In the views of Shailesh Baidwan, the company president and a co-founder, the Philippines needs to completely embrace a digitalized financial system in order to become one of the most advanced nations. "Maya Business has provided a digital payment method for various businesses for close to a decade. Our intimacy with businesses grows stronger as more of them embrace the idea of combining every aspect of their operations with various digital banking services," Baidwan said.[20]

With large companies as well as micro small and medium enterprises (MSMEs) relocating their business, the Philippines' consumer landscape has become even more online-oriented. Maya has considerably enhanced its strategy to become mobile through major sectors such as retail, e-commerce, and service industries. Among these, the most notable families of fast food in the Philippines, Jollibee Group and McDonald's, have successfully paved ways for their businesses to jump into the future by implementing digital payments and other financial features. Other Companies in the Jollibee Foods Corporation such as Chowking, Mang Inasal, Red Ribbon, Greenwich, Burger King, Panda Express, and Yoshinoya that are part of the Golden Arches have received the 2022 Maya Business Trailblazer Awards for being key players in the financial services ecosystem.

Suyen Corporation, a well-known Philippine conglomerate famed for its popular clothing brand Bench, is a pioneering partner in Maya's digital financial solutions suite. With an emphasis on convenience, more choice to consumers, well-structured marketing strategies, and offers during sales, integration of payment during sales, Suyen Corporation was able to almost double its cashless payment volume processed through Maya Business compared to the previous year.[21]

Maya has developed into an important player in changing the financial landscape of the Philippines. Maya voted the best digital bank in South East Asia at the World Digital Bank Awards and featured by Forbes Magazine as one of the World's Best Banks—has established itself as one of the region's leaders.

In April 2022, Maya changed its name from PayMaya, its original brand, and promised to transform the mundane nature of traditional everyday transactions by introducing rewards. Today, users can earn high interest on their deposits without extra effort, with funds transferred through the Maya platform and interest accruing every day now. Moreover, Maya has applied its strong AI-based credit scoring capability and

[20] MB Business, 2022, https://mb.com.ph/2022/09/08/maya-business-honors-digital-payments-ecosystem-builders/.

[21] Inquirer Net, 2022, https://business.inquirer.net/357005/maya-business-recognizes-growth-leaders-in-business-payments.

provided loans in the app itself, which simplified the whole borrowing experience.

A year after the launch of Maya's operations there, the users were already 2.6 million. It is evidence that the recent trends of consumer preferences are changing toward more digital banking solutions and services. This growth has demonstrated to us that Maya is able to keep up with changes in the economy and demands of today's consumers who are seeking a fast, easy, and enjoyable banking experience.

13.3 Alipay+ Partners with Philippine E-Wallets to Build Sustainable Tourism Scenarios

The Philippines is situated in the western Pacific region and is comprised of numerous islands which makes it perfectly positioned to be successful in tourism development. The Philippines has over five million people who work in the tourism industry. Even so, the Philippines which is the second most populous nation in Southeast Asia, has yet to be a major player in tourism in the region. Despite having more than 30 million people more than its neighbor Thailand, the Philippines receives only one-fifth the number of tourists that travel to Thailand in a year.

The context provided in the above paragraph suggests that the tourism industry in the Philippines is under development. As indicated by the statistics published by the Philippine Statistics Authority, the tourism sector contributed 8.6% toward the economy of the nation with regards to GDP in the year 2023 and accounted for 6.21 million jobs or 12.9% of the national employment.[22] At the 2023–2028 National Tourism Development Plan National Summit, Tourism Secretary Christina Frasco said that since more than one million international people had already visited the Philippines in the first three months of 2023, the goal of luring 4.8 million international tourists over the course of the year was realistic, indicating that the country's tourism and economy was in the process of fully recovering.[23] Recently, the summit revealed the ambitious 2023–2028 National Tourism Development Plan, which is aimed at "building a tourism industry that is culturally, historically and nationally Filipino," in

[22] Philippine Statistics Authority, 2023, https://psa.gov.ph/.

[23] China Trade News, 2023, https://www.ccpit.org/a/20230410/202304104fbw.html.

order to enhance the Philippines's position as one of Asia's top tourism destinations.

As for the previous performance of the country in terms of foreign visitors, the Department of Tourism (DOT) had previously targeted 7.7 million international arrivals for the duration of the entire year 2024. As per the DOT's statistics today, 3.1737 million foreign tourists have come to the Philippines between January and July 2024. The highest foreign tourists' number came from South Korea with 932,000 visitors making up 26.24% of the total, followed by the US (580,000), China (219,000), Japan (213,000), and Australia (150,000).[24]

The 1990s saw the beginning of a surge in both trade and investment between South Korea and Southeast Asian countries including the Philippines, and this explains why South Korea remains the biggest market for those who travel to the Philippines for tourism. In addition, English language resolutely and university courses are to be found at far more lower cost in the Philippines relative to Western countries. Hence many Korean students find it attractive. Bureau of Immigration statistics in 2014 indicated that Koreans were the largest nationality group of foreign students in the Philippines. Due to geographical factors as well as cheap airfares, the Philippines has become a favorable travel spot for Koreans.

As a blue ocean market, the Philippines has become a key region for Alipay+'s global strategy. By partnering with local digital payment platforms, Alipay+ has played a crucial role in enhancing financial inclusion and promoting sustainable tourism. Through these partnerships, Alipay+ has enabled a variety of payment methods, providing greater convenience for a wider range of international visitors to the Philippines. To significantly enhance the travel experience for visitors, Ant Group, the operator of Alipay+, announced that four mobile wallets—AlipayHK, Kakao Pay, Touch 'n Go e-Wallet, and Alipay from mainland China—are now accepted in the Philippines. These mobile wallets have been integrated into the local payment ecosystem, covering various use cases such as shopping, dining, and transportation, allowing visitors from different regions to enjoy a seamless and cashless payment experience, making their travels more convenient and secure. Since its initial acceptance in 2017, Alipay+ has been widely embraced by merchants in the Philippines. It is

[24] Republic of The Philippines Department of Tourism, 2024, http://www.tourism.gov.ph/files/2024/tourism_demand/08/08-05/JAN-JULY.pdf.

now widely accepted in Manila and Cebu, the two cities that were the most visited by tourists before the 2019 pandemic.[25]

Mactan-Cebu International Airport (MCIA), the second-largest airport in the Philippines, has taken a pioneering step to meet the needs of digitally savvy travelers.[26] MCIA integrated Alipay+ in around 90% of its merchants, which means that the moment a visitor sets foot in Cebu, he or she can already expect to go around the island without cash. Aboitiz InfraCapital, which runs MCIA, CEO Athanasios Titonis said it is critical to ensure traveling passengers' ease and that incorporating Alipay+ was consistent with MCIA's principles of providing world-class services. Due to its unique geographic position and state-of-the-art infrastructure, MCIA forms an important link in connecting the world with the Philippines. Furthermore, MCIA is the first and still the only Philippine airport to receive the Airport Customer Experience (ACX) certification from Airports Council International.[27]

Fueled by Alipay+, Philippine tourism is rebounding strongly. More than 3.5 million visitors were received in the first seven months of 2024, according to the Department of Tourism (DOT), over 3.2 million of whom were international visitors. Overall visitor arrivals increased by 10.43% as compared to June a year ago. In addition to South Korea, Japan, Australia, United States, Canada, United Kingdom, and Singapore are in the top ten countries of international visitor arrivals.

Also, Dr. Cherry Huang—General Manager Alipay+ Offline Merchant Services at Ant Group speaks about the collaboration and its effects on the travelers with optimism. She observed that the digital applications and solutions provided by Alipay+ will set the standard of expectations and interact with the businesses that are related to tourism. Presently, the aim is to work with local partners in order to bring more merchants, particularly small businesses, in order to advance the objectives of digital tourism and provide the necessary support for the growth and recovery of the industry in the Philippines.[28]

[25] DigitalCFO Newsroom, 2023, https://digitalcfoasia.com/2023/09/11/alipay-expands-mobile-wallet-acceptance-in-the-philippines-enhancing-tourist-experience/.

[26] Ibid.

[27] Ibid.

[28] Ibid.

13.4 The Positive Impact of the Philippines' Open Digital Financial Ecosystem and Sustainable Tourism

As the Philippines faces an influx of tourists and as its tourism sector regains strength, the potential for greater use of mobile wallets will revolutionize payments and the manner of travel in the Philippines. Alipay+ reiterates its dedication towards advancing digitalization for the benefit of travel in the Philippines and other locations alike.

13.4.1 *Social Impact: Inclusive Finance, Decent Work, and Economic Development*

The digital payment market in the Philippines is on the rise as more people start using local payment gateways as well as the efforts seeking to enhance the operations of Alipay+. This competition seeks to provide more people in the Philippines access to inclusive finance services through Alipay+'s open ecosystem. This includes small businesses and individuals, especially those who had difficulty accessing finance services because they did not have traditional bank accounts.

In supporting small businesses, Alipay+ collaborates with local financial institutions to provide loans and other financial products, helping them expand their businesses and increase their revenue. For example, through its partnership with BPI, Alipay+ offers more accessible financing channels, lowering the barriers to financing for small businesses and fostering the prosperity of local business activities.

For those lacking comprehensive financial services, Alipay+'s partnerships with local e-wallets like GCash, HelloMoney, and Maya allow even unbanked individuals to enjoy convenient payment services, further enhancing the inclusivity of the digital financial ecosystem. This model contributes in alleviating poverty and enhancing economic growth expansion. Alipay+'s payments also ease everyday transactions for users and allow them to accumulate credit histories to possibly gain access to more leverage in the sorts of financial services offered in the future.

For the management and promotion of sustainable tourism in the Philippines, the partnerships of Alipay+ with local tourist sites, hotels, and retailers allow international visitors to be able to pay with ease and thus increasing the number of tourists who visit the country. As cited in the report of the Department of Tourism, the effort has brought and significantly enhanced the experience of the visitors and the income from the tourism activities. Over 100,000 Chinese tourists were brought into the local economy in 2023 through a special promotional program of Alipay+ and the Department of Tourism. Such efforts as increasing the number of foreign visitors to the Philippines help not only to stimulate the economy but also to create additional jobs. The advent of digital payments and many other tasks such as those covering technical support, marketing, and customer care, among others. It is reported that in the year 2023 alone, the Ecosystem Alipay+ employed several thousands of direct and indirect employment opportunities in the country of the Philippines. Such new posts give more and better opportunities of employment to the host people so improving the skill acquisition and growth of the active group in the country and hence making this practice sustainable.

13.4.2 Enhanced Governance: Industry Innovation and Upgrading

From a case study perspective, Alipay+'s expansion in the Philippines' digital payment and financial ecosystem is an innovative practice that empowers the country's governance capabilities through robust technological infrastructure. Alipay+ actively collaborates with the Bangko Sentral ng Pilipinas (BSP) to ensure its operations comply with local regulations and standards. This collaboration helps establish a stable and transparent digital payment environment, enhancing consumer trust and promoting the healthy development and upgrading of the entire industry.

Additionally, Alipay+ exports its marketing experience from the Chinese market, providing technology and business training to local partners to help them better utilize digital tools to serve customers. These training programs not only improve the competitiveness of local small businesses but also drive innovation and development across the industry. For example, through a series of online webinars and in-person training sessions, Alipay+ helps local small businesses understand the latest payment technologies and marketing strategies.

Moreover, Alipay+ facilitates the growth of local communities through engagement to outreach programs such as provision of educational scholarships and technical assistance to startups and young business folks. These initiatives help in moving towards the creation of a more cohesive and sustainable socioeconomic structure which would create more decent jobs and workplaces for the people. For example, Alipay+ works with local

Non-Governmental Organizations to provide technological teaching aids and access to the internet to schools in far flung localities and help reduce the information inequality between urban and rural areas.

To develop viable tourism scenarios, Alipay+'s integration with local e-wallets is aimed at proving deeper and more sustainable economy. Alipay+ created open ecosystems that enhance financial inclusion and decent work opportunities, bolster industry advancements, and upgrades governance structures. This cooperation between the Philippines and Alipay+ is a good benchmark to harness for future development in digital finance and economic growth that is sustainable.

CHAPTER 14

Sustainable Fintech for Inclusive Development in the Digital Economy and Underbanked/Unbanked Social Groups: Doku, DANA, Akulaku PayLater, Kredivo, and Alipay+ Ecosystem in Indonesia

14.1 The Digital Ecosystem of Indonesia

14.1.1 The Thousand Island Nation with Potential: The Rapid Growth of Indonesia's Digital Economy

Data from the World Bank indicates that in 2017, only 48.9% of the adult population aged 15 and above in Indonesia had bank accounts, which, despite significant growth, was still below the global average of 69%. At that time, 60% of Indonesia's GDP was contributed by 60 million small and medium-sized enterprises (SMEs), most of which could not afford the installation and maintenance costs of transaction terminals such as

J. Sun, *Sustainable Financialization*, Sustainable Development Goals Series, https://doi.org/10.1007/978-981-96-3580-1_14

POS and had to resort to cash transactions. The more cash transactions there were, the less information banks could obtain, preventing them from providing services based on the financial levels and credit status of merchants and consumers, thereby further exacerbating Indonesia's economic dependence on cash.

Indonesia's digital economy development started relatively late and is still in its infancy and germination stage, but its rapid growth rate in recent years has demonstrated the potential it may possess in the future.

The pandemic has objectively accelerated the digital transformation in Southeast Asia, and Indonesia is no exception. In 2021, when Satu Kahkonen, the World Bank's Chief Representative in Indonesia, presented the World Bank's Indonesia Digital Report, he said, "We see that the daily activities of households and communities are increasingly reliant on digital technology, whether for working from home, online shopping, remote education, and other activities... For companies, digital technology has helped them continue operations; this has become a warning and a reminder to prepare for the future, where digital innovation is one of the most important factors in the future business landscape."[1] During the pandemic, contactless digital payments were favored by consumers, merchants, and the government, and this preference has persisted after the pandemic. The 2021 Southeast Asia Internet Economy Report shows that since the start of the pandemic, Southeast Asia has added 60 million digital consumers, with 20 million people becoming new digital consumers in the first half of 2021 alone. Looking at the Indonesian market, the adoption rate of digital payments by merchants has also increased significantly. From December 2020 to October 2021, the number of merchants adopting Indonesia's QRIS system increased from 5.8 million to 12 million.

[1] MetroXinwen, 2021, https://www.medcom.id/cn/business/read/2021/07/29/21651.

Indonesia's digital landscape is a testament to its thriving online community. As of 2022, the country boasted over 200 million internet users, a figure that is nearly three-quarters of its total population. This digital revolution has not gone unnoticed, with its youthful demographic structure being a key catalyst in this surge. Moreover, Indonesia is home to approximately 97 million middle and upper-class consumers, a cohort whose demand is a powerful force in the nation's digital economy. This tech-savvy and increasingly affluent segment has been instrumental in pushing Indonesia's digital economic frontiers forward.

Furthermore, the digital payment infrastructure has been gradually improved with government support. Nezar Patria, the Deputy Minister for Information and Communication in Indonesia, pointed out that nearly 90% of Indonesia's territory is covered by 2G–4G networks. At the same time, according to data from the Association of Indonesian Internet Service Providers (APJII), the coverage rate of internet facilities has reached 80% of the population.[2]

As illustrated in the figure below, the digital economy outlook for Indonesia/ASEAN is presented. It can be observed that from 2017 to 2025 (projected), Indonesia's digital economy has developed rapidly and holds a dominant position in the Southeast Asian region. The proportion of Indonesia's digital economy in GDP has remained higher than the ASEAN average since 2018.[3]

[2] Antara, 2024, https://en.antaranews.com/news/311577/indonesia-has-great-potential-in-digital-economy-deputy-minister.

[3] Statista, 2022, https://www.statista.com/outlook/dmo (accessed 21 October 2022); and Economist Intelligence Unit, 2022, https://viewpoint.eiu.com/data/.

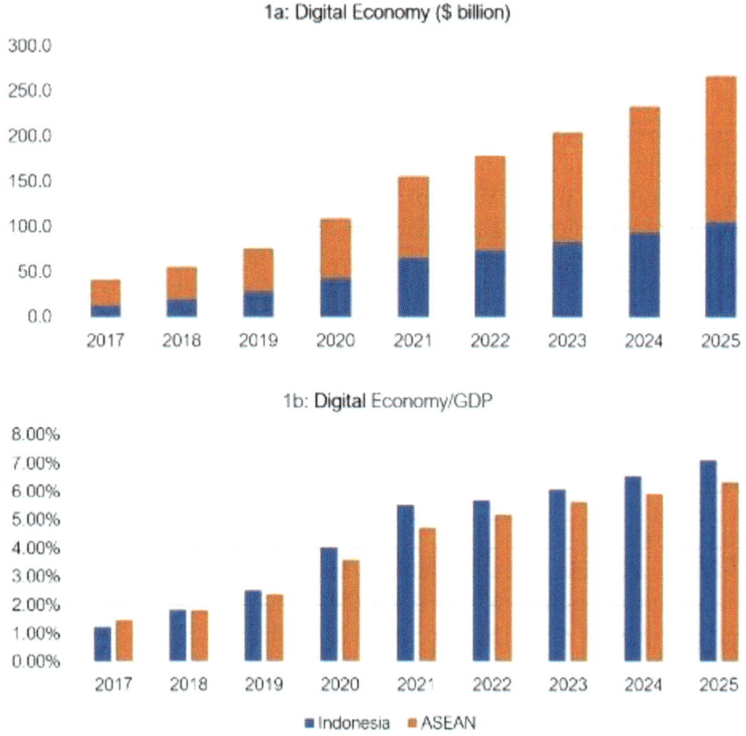

Digital economy outlook for Indonesia/ASEAN (*Sources* Statista and Economist Intelligence Unit)

14.1.2 The Development of E-commerce and Analysis of Indonesia's Online Retail E-commerce Platforms

Analysis indicates that Southeast Asia's demographic dividend, the rising economy, substantial investment in the digital economy, and widespread adoption of mobile devices have all laid a massive market volume foundation for its cross-border e-commerce.[4] Under the pandemic, the number of e-commerce transactions and patent applications has seen a significant

[4] TMO Group, 2024, https://www.tmogroup.com.cn/insights/southeast-asia-ecommerce-sales/.

increase. Asia currently accounts for nearly 60% of the global online retail sales. In 2020, e-commerce revenue in Southeast Asian countries such as Vietnam and Indonesia grew by 40–60%, surpassing most countries worldwide. The driving force behind this rapid growth is the disappearance of cash payments and the subsequent rise of digital payment alternatives (particularly e-wallets and prepaid cards).[5]

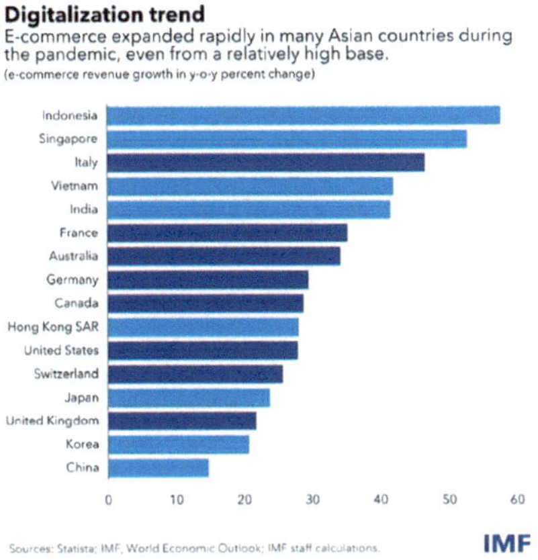

Indonesia's e-commerce growth is nearing 60%, topping the rank (*Source* Statista; IMF)

In Indonesia, e-commerce is a vital component of the local digital economy. A recent report by Singapore-based payment company 2C2P, in collaboration with US research firm IDC, forecasts a significant leap in Southeast Asia's e-commerce spending, predicting it to hit $179.8 billion by 2025. This growth is set to add 188.6 million new e-commerce users to the region's digital economy by the same year. With this surge, the adoption of digital payments is also anticipated to soar, with e-commerce transactions through digital payments estimated to claim 91% of the market share by 2025.

Analysts at China International Capital Corporation, including Hou Liwei and Wu Weijia, are particularly bullish on Indonesia, anticipating its B2C and C2C e-commerce market to balloon to $95.4 billion by 2026,

[5] IMF blog, 2023, https://www.imf.org/en/Blogs/Articles/2023/01/09/asias-productivity-needs-a-boost-that-digitalization-can-provide.

up from $53 billion in 2021, at a compound annual growth rate of 12%. This outpaces other Southeast Asian nations significantly. Tech in Asia estimates place Indonesia's e-commerce market size at $133 billion in 2020, highlighting that e-commerce's contribution to Indonesia's GDP has already reached 12.6%.[6]

Shopee, established in Singapore in 2015, is estimated to have a monthly traffic volume of 132,776,700 in Indonesia. As a subsidiary of Tencent, Shopee has seen rapid growth in Indonesia, supported by the strong backing of Chinese conglomerate Tencent. It has evolved from a C2C specialist to a B2C platform, competing with Singapore-based platform Lazada across multiple Southeast Asian markets, including the Philippines. In recent years, Shopee ID has quickly become the undisputed leading e-commerce platform in the country.[7] A report by U.S. research firm IDC shows that ShopeePay has reached over 500 cities in Indonesia, with significant benefits seen in second and third-tier cities—transactions using ShopeePay outside the Greater Jakarta area have increased by more than eight times.

Tokopedia, which came onto the scene in 2009, was once the go-to online marketplace in Indonesia and has now grown into the country's largest digital marketplace. It has made it a breeze for individuals and businesses to set up and manage their online shops, boasting an impressive monthly traffic volume of 157,233,300. The platform is now home to over 8,000 official brand stores and has a logistics network that spans 93% of Indonesia's regions, with nearly a quarter of consumers enjoying the convenience of same-day delivery services. Starting as a C2C platform, Tokopedia has recently expanded into the B2C sector, offering business models like "flagship stores" for brands. Alibaba sealed the deal with a Series G lead investment in Tokopedia in December 2018, further solidifying its foothold in the Indonesian market following its involvement with Lazada, which is also part of Alibaba's ecosystem.

Beyond the top tier, Lazada stands out as a leader in Indonesia's e-commerce scene, which is part of Alibaba's expansive ecosystem. Lazada operates as both an online department store and a marketplace, enabling retailers to sell their wares and drawing a monthly web visitation of

[6] https://finance.sina.cn/tech/2021-04-25/detail-ikmyaawc1657308.d.html?fromtech=1.

[7] Tmo Group, 2024, https://www.tmogroup.com.cn/insights/indonesia-marketplace-ranking/.

43.6 million. The platform has a successful track record in e-commerce across six Southeast Asian countries, including Indonesia, Malaysia, the Philippines, Singapore, Thailand, and Vietnam.

In the region, TikTok Shop has seen an impressive annual growth rate of 60%, allowing it to quickly gain ground on its competitors. However, the Indonesian government's ban on in-app transactions in October posed a challenge. In response, a few months post-ban, TikTok announced a strategic move by acquiring a 75% stake in Tokopedia, Indonesia's e-commerce giant, to navigate around the restrictions and resume operations in the country. This merger is anticipated to place PT Tokopedia as a strong second contender with a market share of 28.1%, significantly narrowing the gap with Shopee and potentially disrupting the original e-commerce duopoly of Shopee and Lazada.[8]

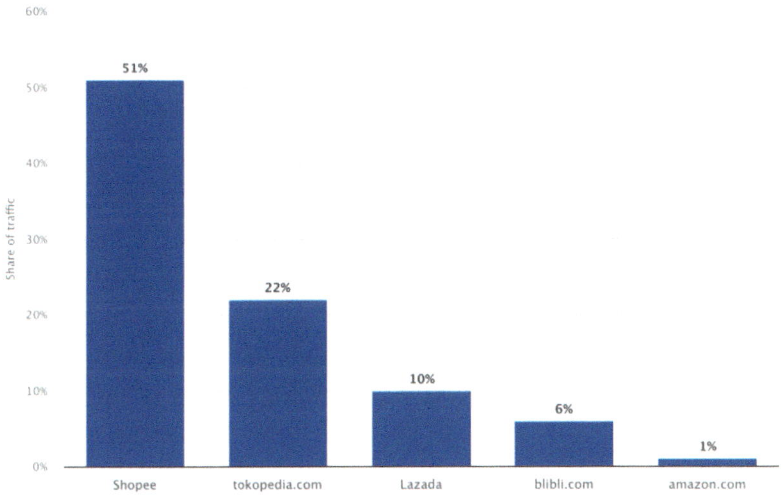

Market Share of E-commerce Platform Traffic in Indonesia from February 2024 to April 2024 (*Source* Statista)

[8] Tmo Group, 2024, https://www.tmogroup.com.cn/downloads/sea-outlook/.

14.1.3 *Digital Payment System in Indonesia*

14.1.3.1 *Fragmented Payments in Indonesian Market*

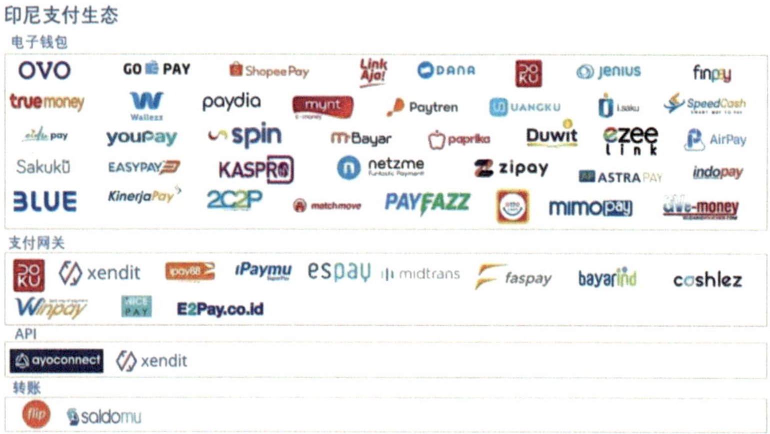

Fragmented Payments in Indonesia (*Source* Fintechnews Indonesia)

Taking Indonesia's to To-Consumer payment wallet market as an example, Indonesia is known to have the "Big Five Wallets," which are GoPay, DANA, OVO, ShopeePay, and LinkAja. Locally frequently used and larger local service online payment methods include Gopay, OVO, and DANA, while a common e-commerce to C-end online payment method is ShopeePay. According to InsightAsia data, GoPay is the most widely used digital wallet in Indonesia, with 71% of users having used GoPay before.[9] The second-ranked digital wallet brand in Indonesia is OVO, which has been used by 70% of local users, slightly lagging behind GoPay.

In the Indonesian market, GoPay reigns supreme among the top five payment platforms in terms of recognition, with an impressive 97.2% of respondents being familiar with it. Hot on its heels are DANA, OVO, LinkAja, and ShopeePay, each boasting recognition rates of 96.3%, 95.6%, 83.5%, and 64.6%, respectively.

9 InsightAsia, 2024, https://insightasia.com/gopay-largest-market-share-and-the-strongest-consumer-loyalty/.

When it comes to the usage rates of the big five payment platforms in Indonesia, GoPay leads the pack with 87%, followed closely by OVO at 80.4%, DANA with 75.6%, ShopeePay at 53.2%, and LinkAja bringing up the rear at 47.5%. Moving to popularity, DANA captures the crown with 36.8% of the vote, with OVO not far behind at 30.8%, and GoPay nabbing a solid 20.6%. But when frequency of use enters the picture, ShopeePay surges to the front. The following figure offers a more thorough view of the transaction volumes.

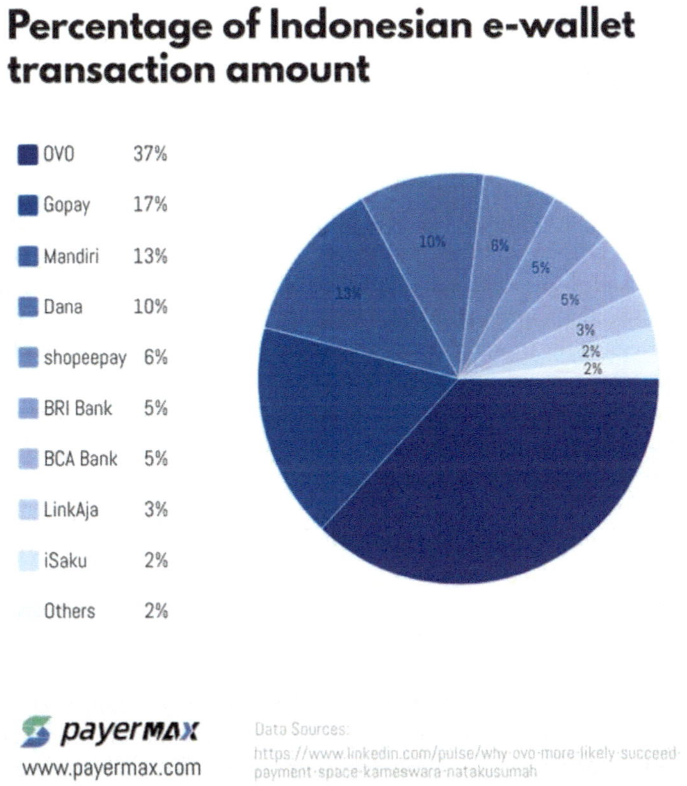

Percentage of Indonesian E-wallet Transaction Amount (*Source* Internet)

14.1.3.2 Connection Between E-commerce and Digital Payments

Sea, the majority shareholder of which is Tencent, launched the e-commerce platform Shopee in 2015, focusing on a C2C model similar to Taobao.[10] ShopeePay is an e-wallet introduced by the e-commerce platform Shopee, which was first launched in Indonesia in 2018 and has since served consumers in Singapore, Malaysia, Thailand, the Philippines, and other countries. ShopeePay has over 35 banking and e-wallet partners (including BDO, Bank of the Philippine Islands (BPI), RCBC, Metrobank, Chinabank, UnionBank, etc.), which can be used for online transactions within the Shopee app, offline (QR) transactions with merchants accepting Shopee payments, and for receiving and making payments.[11] In addition to paying for e-commerce orders on the Shopee platform, it can now also be used for game top-ups, mobile recharges, and purchasing movie tickets, among other things.

DANA, known as "Indonesia's Alipay," has been part of Ant Group's financial technology overseas partnership since 2016. In 2016, Ant partnered with Thai payment company Ascend Money, invested in the Philippines' largest e-wallet GCash in 2017, acquired Singapore's payment service helloPay, and jointly established DANA with Emtek in Indonesia. Through the DANA wallet, Indonesian users can conveniently perform common e-wallet services such as mobile recharges, bill payments, utility payments, social security contributions, and bill installments, similar to Alipay.[12] Regarding the connection to the e-commerce, DANA is deeply integrated into two of Indonesia's top e-commerce platforms, Lazada and Bukalapak. Lazada, an e-commerce platform invested by Alibaba in Indonesia, has received a total investment of 4 billion USD from Alibaba to date, making it the fourth-largest e-commerce platform in Indonesia.[13] In 2022, Indonesian e-wallet DANA secured 250 million USD in financing, invested by local conglomerate Sinar Mas and Lazada. DANA will use these funds to invest in new technologies and launch more financial services.

[10] Geekpark, 2021, https://www.163.com/dy/article/GAJP7EK405119FMA.html.

[11] Antomdocs, 2023, https://global.alipay.com/docs/ac/Payment_method_desc/shopeepay.

[12] 7.5 Degree, 2022, https://new.qq.com/rain/a/20220812A0901500.

[13] PayerMax, 2020, https://www.mpaypass.com.cn/news/202012/28111212.html.

In Tokopedia's shareholder composition, SoftBank is its largest shareholder, and Alibaba has participated in multiple rounds of investment in Tokopedia through Taobao, currently holding a 28.25% stake. Lazada, controlled by Alibaba, attempts to take a branded and high-end route in the Indonesian market, which seems to face challenges in Indonesia, where the average consumption level is currently low. This may lead Alibaba to continue increasing its stake in Tokopedia to improve its strategy in the Indonesian market. Alibaba Cloud is also helping Tokopedia complete its annual Indonesian version of "Double 11 Shopping Festival".[14]

14.2 Development of DOKU and DANA Digital Wallets in Indonesia

14.2.1 Challenges in the Development of Digital Wallets

14.2.1.1 Localization

"Localization involves local social relations. In terms of company compliance, it is necessary to achieve tax compliance and structural compliance. In terms of product compliance, it is necessary to ensure formal importation and local halal certification. It is only on this basis that the product and brand itself can be considered. For example, Chinese teams have strong operational capabilities and brand-building capabilities, but they have a weak understanding of local cultures, requiring long-term adaptation in the face of cultural conflicts."[15]

In September 2023, the Indonesian Ministry of Trade issued Regulation No. 31 of 2023, prohibiting social media from being used as a platform for selling goods, which also led to the closure of TikTok Shop (hereinafter referred to as TTS) in Indonesia. In fact, for several years before the introduction of Regulation No. 31, the Indonesian government has been tightening cross-border e-commerce policies and adopting various measures to protect local small and micro enterprises. However, in this situation, Yu Deshui expressed that even if Indonesia introduces new regulations on cross-border business supervision, as long as local company registration is properly handled, certification is completed, and

[14] https://finance.sina.cn/tech/2021-04-25/detail-ikmyaawc1657308.d.html?fromtech=1.

[15] Ebrun, 2023, https://m.huxiu.com/article/2046308.html.

operations are legal and compliant, with patient long-term investment and even becoming a truly local Indonesian company, it is still possible to gain the recognition of the Indonesian government and consumers, and there is still a great opportunity to tap into the Indonesian e-commerce market.[16] Therefore, the key lies in localization.

In fact, behind many well-known Indonesian local brands are Chinese cross-border e-commerce teams. This also means that Chinese sellers who want to enter the Indonesian market in the future must do a good job in localization, develop local supply chains, recruit local teams, and place brand building, content operations, channels, and warehousing in Indonesia in exchange for long-term development. Attempts to gain a foothold in the Indonesian or even Southeast Asian market through short-term heavy investment and relying on low prices to increase volume will be difficult to tell a good story.[17]

14.2.1.2 Inclusiveness

Indonesia consists of more than 17,000 islands, and the establishment and maintenance of logistics and distribution networks are costly and limited in efficiency. Therefore, Chinese companies need to cooperate with local logistics service providers to optimize distribution networks and develop application versions adapted to low-bandwidth environments to ensure a good user experience in remote areas. At the same time, considering the low level of financial inclusion in Indonesia, digital wallet services also need to work closely with local banks and financial institutions to build a widely covered payment ecosystem to increase user trust and usage frequency.

Not only in Indonesia but also in the entire Southeast Asian region, the terms "underbanking" and "fintech" have become increasingly common in recent years. "Underbanking" refers to the group of people who cannot normally access or have bank services (and even financial products), while the latter mainly refers to businesses that provide lower-threshold solutions for this situation. In the entire Southeast Asian region, which currently has a scale of over 680 million, nearly 70% of the group is still in this state. Among them, the countries with larger populations in Southeast Asia—Indonesia, the Philippines, and Vietnam—also occupy

[16] Globe, 2023, http://www.news.cn/globe/2023-12/07/c_1310752409.htm.

[17] Globe, 2023, http://www.news.cn/globe/2023-12/07/c_1310752409.htm.

the forefront of this scale due to their large population sizes, with the underbanked population reaching 77%, 78%, and 79%, respectively.[18]

On the other hand, there are local small and micro enterprises. Including the warungs and sari-sari stores that are common on the streets and corners of the region, small and micro enterprises account for 97% of the total number of enterprises in Southeast Asia. Among them, more than 60% cannot timely access this service when they encounter loan needs. Under the influence of this factor, when considering subsequent expansion or business growth, these enterprises obviously need to make more preparations. At the same time, due to the limitations of this channel, these enterprises often need to prepare a certain amount of cash to deal with operations and salary payments, not to mention the lack of convenient digital payment services at the consumer and their own levels—all of which greatly restrict their development.

14.2.1.3 Other Challenges

Table 14.1 outlines some other challenges and opportunities within the Southeast Asian e-commerce sector, providing a comprehensive view of the current business landscape from the perspectives of market maturity, entry strategy, infrastructure, and consumer base.[19]

14.2.2 Collaboration and Optimization: Digital Wallet DOKU Forges Local Payment Services

Established in 2007 with electronic payment technology services as its business inception, DOKU is the first electronic payment service provider in Indonesia. In December 2012, it obtained an electronic money license, being one of the few non-bank institutions to acquire such a permit. In April 2013, DOKU launched its electronic wallet. As a quintessential enterprise of the current mainstream development model of third-party payments in Indonesia, which is the "payment gateway + electronic wallet" model, DOKU has evolved over the years into a well-known online payment solution provider.[20] Currently, DOKU primarily

[18] TechNode, 2024, https://cn.technode.com/post/2024-03-10/now-for-seas-unerbanking/.

[19] Tmo Group, COMMENCE DATA.

[20] Weiyang, 2018, https://www.weiyangx.com/273261.html.

Table 14.1 Other challenges and opportunities of Southeast Asian e-commerce market

Aspect	*Challenges*	*Opportunities*
Market Maturity	Underdeveloped markets suggest that existing structures are not yet stable, requiring brands to pave new paths in the region without the benefit of numerous successful precedents to learn from	The e-commerce market is maturing, with varying development speeds across countries. None have reached the stable and saturated market conditions seen in neighboring China and Singapore, offering ample room for foreign brands to enter
Entry Strategy	Creating a local legal body usually involves requirements for job development and requires a significant financial investment	Due to the lack of substantial entry barriers and high import duties, cross-border e-commerce is still largely free and loosely controlled
Infrastructure	Transportation from a national hub to another area of the country can take more than a week in certain areas due to the uneven logistical infrastructure	In impoverished regions, regional marketplaces like Lazada and Shopee can support the growth of infrastructure and logistics
Consumer Base	Southeast Asia still has a high rate of poverty, which makes some regions temporarily inaccessible to e-commerce	Younger demographics and the middle class's rapid expansion offer exceptional growth potential

offers three types of services: payment gateway channels for businesses, cross-border transfers, and collaborative commercial payments (electronic wallets and community payment apps).[21]

14.2.2.1 *Creating Partnerships: Compliance and Reliable Connectivity*

Under the watchful eye of the Central Bank of Indonesia (Bank Indonesia), DOKU is tasked with complying with stringent financial regulations, including the essential requirement of having a robust disaster recovery plan in place. The company is searching for a trustworthy and reputable cloud partner that can provide a range of services to address issues with IT infrastructure. The disaster recovery site must be located

[21] 7.5 Degree, 2021, https://www.baijing.cn/article/36071.

inside Indonesia's borders in accordance with banking regulations, which requires DOKU to use first-rate data centers that satisfy both domestic and international standards. Moreover, DOKU is in quest of a cloud provider with deep knowledge and technical acumen to address the complexities faced by fintech enterprises in Indonesia.[22]

Regarding the promotion of technological infrastructure, DOKU intends to transition from a local IT infrastructure model to a cloud infrastructure in order to assist the company in moving from a capital expenditure model to an operational expenditure model and reap significant cost optimization benefits. By making this adjustment, the company intends to maintain a consistent and dependable production environment while complying with regulatory standards and effectively controlling infrastructure costs.

Besides, DOKU has implemented a hybrid cloud architecture by connecting its local infrastructure with Alibaba Cloud. It utilizes Alibaba Cloud's Express Connect service to establish a secure and reliable dedicated connection between its local and cloud environments, achieving high network throughput and low latency. For database management, the company has adopted ApsaraDB RDS for Redis, MongoDB, and PolarDB. These have benefited the company with increased stability and built-in high availability options, liberating staff from many typical database management activities and aiding in the implementation of a robust disaster recovery strategy.

Currently, DOKU is gearing up to migrate the majority of its operations to Alibaba Cloud, while gradually decommissioning its local infrastructure. This strategic shift marks a key milestone in the company's mission to focus on its core payment solutions and contribute to Indonesia's transition towards a paperless economy.

14.2.2.2 Optimizing Services: Upgrade of Local Community Payments in Indonesia

DOKU offers users the convenience of accessing financial services directly from their iOS and Android devices without the need for a traditional bank account. DOKU users can make bill payments, online shopping purchases, interbank transfers, and mobile bill payments through DOKU's supported banks, utilizing the DOKU third-party gateway

[22] Alibaba Cloud, https://www.alibabacloud.com/en/customers/doku?_p_lc=1.

payment feature funded by their bank accounts. Moreover, consumers can use DOKU (electronic wallet) for consumption by registering for a DOKU account via the web, obtaining a DOKU ID; topping up their DOKU account (through ATM, bank/online banking transfers, convenience stores, etc.), and logging into DOKU to conduct transactions and make purchases at DOKU partner merchants. The following chart illustrates the process of making payments with DOKU.[23]

Easy Payment Process of DOKU

In terms of To-Businesses, DOKU provides customized solutions based on needs, such as offering payment methods for digital content within the Google Play store. Speaking about serving B-end enterprise clients, DOKU's Chief International Business Cooperation Officer, Alison, believes there are no shortcuts, "We have dedicated sales teams in Jakarta and Bali, always responsible for follow-ups and communications. For our clients, our sales team acts more as a business/demand consultant."

Since 2020, the pandemic has affected various industries in Indonesia to varying degrees. In terms of the payment sector, DOKU has discovered numerous new business opportunities during this exceptional period. Alison explained that before the pandemic, many small and medium-sized enterprise merchants primarily conducted sales activities through offline physical stores, but the pandemic largely severed the path of physical sales, forcing these merchants to start using WhatsApp or other online channels for business activities. Specifically, there has been more than a 50% increase in transaction volumes in areas such as online retail platforms/

[23] Antom docs, https://global.alipay.com/docs/ac/antomop/hhl8gy0s.

markets, online streaming platforms, and digital gaming.[24] To meet the needs of these merchants to simplify the payment process, DOKU has launched a targeted WhatsApp payment link. This link does not require merchants to upload any company documents or support technical compatibility; merchants can register and use it when communicating payment matters with users. DOKU has a comprehensive KYC process for its served merchants. "We have always placed great emphasis on KYC, and we carefully verify all types of merchants. From the category of goods sold, to the source of funds, to cross-checks on different websites, we have a professional team to handle these matters, ensuring the quality of merchants' products and services." DOKU has also launched the "Juragan Doku" service, targeted at SMEs, aiming to enhance the efficiency and performance of these businesses through digital payments. By July 2023, over 15,000 SMEs in Indonesia had adopted this service.[25]

In addition, DOKU has also added payment functions to various community applications to facilitate merchant sales. "For these community applications, technology is scarce, and our service method is mainly technological output." For instance, in May, DOKU created the E-Istiqlal app for the largest Muslim (Hajj) community at Istiqlal Mosque in Indonesia. E-Istiqlal adds payment and transfer functions on the basis of traditional social applications, not only meeting the needs of the Muslim community to view religious content and participate in the Hajj in a digital way but also facilitating donations, payment of various fees, and transfers. It can be said to be the Indonesian Muslim version of "WeChat/Alipay." Alison clearly stated that applications targeting specific communities like E-Istiqlal are the focus of subsequent efforts and will continue to use technological methods to promote digital upgrades.

[24] 7.5 Degree, 2021, https://www.baijing.cn/article/36071.

[25] Antara, 2023, https://www.antaranews.com/berita/3791862/doku-proses-200-juta-transaksi-digital-genjot-kinerja-umkm.

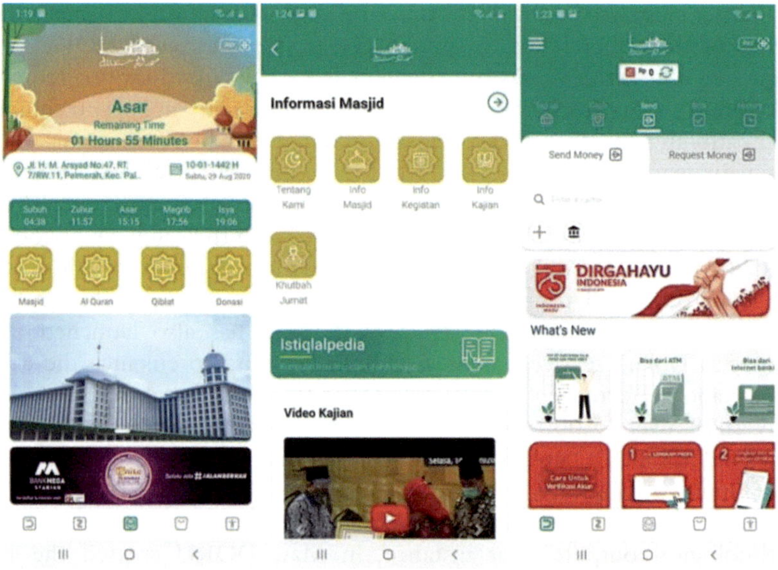

E-Istiqlal App Screenshots (*Source* E-Istiqlal)

Notably, in 2021, DOKU obtained a $32 million investment from Apis Growth Fund II, which was used to accelerate the development of new products and services and to increase Indonesians' access to digital payments.[26] The growth of DOKU is consistent with the Indonesian government's long-term goals to build digital infrastructure, which includes telecommunications, digital payments, cybersecurity, and Know Your Customer. The digital economy of Indonesia is predicted to triple to 5,800 trillion Indonesian rupiah by 2030, according to President Joko Widodo, with digital payments increasing 2.5 times to 12,300 trillion Indonesian rupiah. DOKU's innovation strategy is centered on financial inclusion and community empowerment, both of which are aided by its growth and innovation.[27]

[26] Apis, 2021, https://setkab.go.id/en/president-jokowi-indonesias-digital-economy-has-potential-to-reach-rp5800-trillion-by-2030/.

[27] Cabinet Secretary of the Republic of Indonesia, 2024, https://setkab.go.id/en/president-jokowi-indonesias-digital-economy-has-potential-to-reach-rp5800-trillion-by-2030/.

14.2.3 Trust and Convenience: Digital Wallet DANA Creates Financial Inclusion

DANA has over 150 million users and 5 million daily transactions in Indonesia. DANA is an integrated application that enables users to safely conduct cashless and card-free payments in their daily usage. The extensive functionality of DANA provides users with faster and more practical transaction convenience. DANA aims to provide a trustworthy and convenient mobile payment system for Indonesia to meet customer needs and serve as a bridge to a cashless economic ecosystem.[28]

14.2.3.1 Reliable Cloud Network: Modernizing Infrastructure Locally

As DANA's digital operations continue to scale with the rising tide of transactions, the pressure is on to address the mounting challenges. Key questions arise: How to efficiently manage the costs associated with local infrastructure? How to ensure compliance with local digital payment regulations? In light of these demands, DANA is setting its sights on constructing a robust, reliable, and agile cloud-based infrastructure for its mobile app.

DANA has deployed a variety of Alibaba Cloud Infrastructure as a Service (IaaS) and Platform-as-a-Service (PaaS) products and services within its architecture. A suite of Alibaba Cloud network solutions has also fine-tuned DANA's network to achieve high-performance and robust security. These solutions include Alibaba Cloud Web Application Firewall (WAF) for real-time server monitoring to ensure high availability and DDoS protection to defend against Distributed Denial of Service (DDoS) attacks and reduce other potential cyber threats. Alibaba Cloud's DataWorks big data platform, MaxCompute large-scale data warehousing solution, and Elasticsearch analytics solution have enhanced DANA's data analysis capabilities, supporting DANA in understanding market trends and shifts in user behavior patterns. DANA also extensively uses Alibaba Cloud Data Management Services (DMS) to manage multiple environment databases, track data changes in a visual data management platform, and improve database development efficiency.

Alibaba Cloud has played a pivotal role in streamlining and upgrading DANA's IT infrastructure, slashing operational burdens and forging a

[28] Alibaba Cloud, https://www.alibabacloud.com/zh/customers/dana?_p_lc=1.

secure, swift transaction network. DANA hasn't rested on its laurels; it's continued to partner with Alibaba Cloud in the quest for further modernization opportunities. As Norman Sasono, DANA's Chief Technology Officer, puts it, "We've rolled out a suite of innovative software development solutions to keep modernizing our IT infrastructure on Alibaba Cloud. This collaboration has not only simplified and modernized our IT infrastructure but also lightened our operational load and established a secure, high-speed transaction network."

DANA's development team stands to gain the most from these upcoming enhancements. They are now free to concentrate on driving innovation and delivering a smooth, secure mobile payment platform, unburdened by management and scalability concerns. This platform, powered by world-class technology, connects everyone and supports Indonesians in their pursuit of a better life.

14.2.3.2 Convenient Digital Payments: Electronic Wallet Inclusive of Rural and Remote Areas

In Indonesia, where over 273 million people are scattered across thousands of islands, almost half of all adults remain unbanked or underbanked. Bank service points and ATMs are mostly located in Java, with limited presence on other islands.[29] Therefore, "the country's geography requires a unique approach to creating an inclusive and prudent fintech innovation ecosystem to provide daily financial services to the population." For these unbanked populations with few options, digital financial services are a crucial interface. "Fintech (such as digital wallets) helps people access finance by equipping them for their daily transactions. Digital wallets provide easy access through smartphones and have changed the way people receive and transfer money." Various digital wallets, including DANA, have brought the possibility of accessing financial services to local rural and other remote areas with their convenience and growing security performance.

DANA focuses on catering to the needs of rural areas in Indonesia, especially the small and medium-sized enterprises (SME) segment. While big companies compete in big cities (77% of OVO's user base in Greater

[29] The World Economic Forum, 2024, https://www.weforum.org/agenda/2024/02/indonesia-financial-inclusion-digital-wallets/.

Jakarta), DANA finds that entering the rural market and enabling residents in rural areas to use mobile payments is a wise next step for the company.

Chief Executive Officer Iswara said, "The vast majority of people in big cities already have many options, but most people outside big cities have few choices. Therefore, we sincerely hope that these people can benefit from electronic payment technology. After all, the most important value of technology is to bring solutions to the public." The company has begun to establish offices in rural areas and has formed partnerships with village businesses and local bank agents, ready to go.[30]

DANA's Chairman Rudiantara said, "In the past six years, I have witnessed DANA's journey to create an inclusive payment and financial service platform with the vision of achieving a cashless society in Indonesia. In addition to growing to 1.8 billion users, DANA's inclusiveness has also increased, as 60% of users live in third and fourth-tier cities in Indonesia." He stated that considering the number of independent cellular service users in Indonesia exceeds 240 million, DANA's growth potential remains very large. "Therefore, everyone can play a role in improving Indonesia's financial inclusiveness."[31]

Now, with just a mobile phone and internet access, many Indonesians can access a world full of financial possibilities. These digital wallets are catalysts for enhanced financial literacy.

"Digital payment and financial service platform DANA conducted a survey of its Micro, Small, and Medium Enterprise (MSME) merchants. This survey revealed that our digital wallets significantly impact MSMEs by shifting payments from cash to cashless."[32]

14.2.3.3 *Convenient Digital Payments: DANA Bisnis Inclusively Benefits SMEs and Merchants*

As part of DANA's additional vision, the construction of DANA Bisnis provides Indonesian MSMEs with digital business solutions. Merchants receive payments via electronic payments methods, improving business

[30] Maypass, 2019, https://www.mpaypass.com.cn/news/201907/04090549.html.

[31] Indonesia Shangbao, 2024, https://indonesiashangbao.com/Index/News_info/id/31380.html.

[32] World Economic Forum, 2024, https://www.weforum.org/agenda/2024/02/indonesia-financial-inclusion-digital-wallets/.

processes and productivity, and gaining opportunities to promote their businesses through DANA Bisnis.

The pandemic has sparked a significant increase in merchants joining DANA Bisnis, with numbers skyrocketing from 130,000 to 200,000 this year. There is a noticeable shift as merchants pivot from traditional to digital business models, steering transactions online. This move not only sidesteps the risks of offline gatherings but also empowers businesses to rapidly scale their operations digitally, mitigating the pandemic's hit on their activities and aligning with the government's orientation for a cashless society.

Looking ahead to 2021, DANA is set to expand its electronic transaction horizons by teaming up with a variety of sectors. In a nifty partnership with Parkee, for instance, DANA users can now settle parking fees with just a few taps in the app. DANA Delivery is another innovation born from synergistic ties with logistics aggregators, simplifying the process for the public and MSMEs to ship goods. It offers a practical, all-in-one delivery solution through DANA's platform, making life easier for all parties involved.

Moreover, support for the implementation of artificial intelligence technology and other innovations has strengthened DANA's optimistic attitude towards the development of digital payments and financial services in Indonesia. DANA has utilized artificial intelligence technology to assist in solving user issues through the DANA customer service digital assistant, personalize user financial solutions, and enhance the security of digital transactions through risk engine technology. DANA's Chief Technology Officer, Norman Sasono, stated that the use of artificial intelligence not only makes the daily work of internal technology engineers easier but also enriches DANA's digital financial functions and services, making them more reliable.[33]

Iswara stated that DANA's ultimate goal is to play a significant role in promoting Indonesia's digital economy and to become one of the main electronic payment participants in the Indonesian market. However, he pointed out that he is not worried about competition because the market opportunities are very obvious. He said, "Therefore, the company's goal is not to be first, second, or third, but to help Indonesia transition

[33] Indonesia Shangbao, 2024, https://indonesiashangbao.com/Index/News_info/id/31380.html.

to a digital economy. If we can change the way the vast majority of Indonesians pay, then we will be very happy."[34]

DANA's business model shines with its inclusive approach, underscored by a dedication to boosting financial literacy and lending a hand to communities that are often overlooked by traditional financial systems. By casting its net wide to include rural areas and those on the periphery of banking services, DANA has played a key role in leveling the economic playing field. The platform's intuitive design and commitment to local language support have been game-changers, particularly in encouraging uptake among business owners for whom English is not their first language.

Indonesia's fintech scene has been a powerhouse for economic growth, providing a lifeline to MSMEs and merchants in particular. DANA, with its all-embracing business strategy, has been a bridge between conventional finance and the evolving demands of the digital economy. By offering financial tools that are not just accessible but also affordable, DANA has given businesses the firepower they need to not just survive, but flourish in a cutthroat market.

14.3 Buy Now Pay Later (BNPL) Platforms Promote Financial Inclusion in Indonesia: Akulaku PayLater

14.3.1 Popularity of BNPL in Indonesia

The shopping experience is changing quickly due to the rise of BNPL players. At the same time, it is changing how companies process payments. Customers can choose to divide the cost of high-value items into several interest-free installments, making them more affordable.

While the flexible payment solutions bring clear benefits to consumers, retail merchants are also riding the BNPL wave, as it offers a range of commercial advantages. These include increased consumer spending due to lower barriers to purchase and accelerated sales conversion rates, as well as opportunities to penetrate the younger demographic.[35]

[34] Ardi Wirdana, 2019, https://www.mpaypass.com.cn/news/201907/04090549.html.

[35] Zachary Pestana, 2023, https://aspireapp.com/blog/top-10-buy-now-pay-later-bnpl-players-in-southeast-asia.

Top 10 BNPL Players in Southeast Asia (*Source* aspire)

Indonesia has a population of 181 million individuals without bank accounts, which is more than the population of most countries, and the number of underbanked individuals is significantly higher. Interest-free installments (if paid on time) seamlessly integrated into e-wallets could become the de facto form of credit in the country. Japanese mega-banks have been steadily increasing their investments in Indonesia's fintech industry, with BNPL being a key focus area. In December 2022, Mitsubishi UFJ Financial Group announced it would invest $200 million in Alibaba-backed Akulaku, where BNPL is becoming an increasingly important part of its business. By mid-2024, Akulaku aims to provide BNPL services to 10,000 online merchants and over 100,000 offline stores. The use of BNPL as a payment method is becoming increasingly mainstream in Indonesia, especially for e-commerce. A recent survey and online study by local media Tirto, involving 1,500 respondents, showed that over 60% of consumers have used BNPL at least once, while 30% of consumers stated they frequently use BNPL. Electronics and gadgets are the most popular categories for BNPL purchases, followed by home appliances, internet and data top-ups, and fashion and accessories. Indonesians primarily use BNPL for convenience.[36]

[36] Kapronasia, 2023, https://www.kapronasia.com/asia-payments-research-category/why-is-bnpl-resilient-in-indonesia.html.

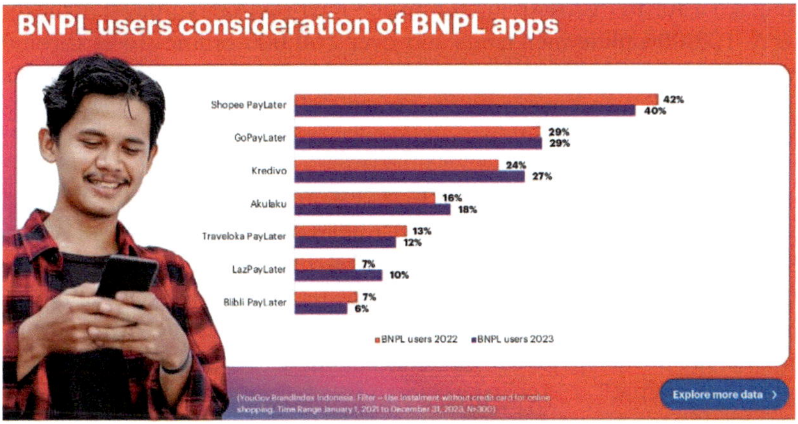

Application usage preferences of Indonesian BNPL users (*Source* YouGov)

14.3.2 Akulaku PayLater and the Collaboration with Alipay+

Established in Jakarta in 2014, Akulaku Group is a technology-driven mobile finance platform with a substantial user base in Southeast Asia. Launched in 2016, the Akulaku PayLater App is dedicated to providing virtual credit card installment loan services for individuals in Southeast Asia without credit cards. Users can repay through e-wallets, bank transfers, and other methods.[37] It supports interest-free installments and allows for three payments within three months, with customers able to pay using debit or credit cards, and offering early repayment options.

Alipay+ has incorporated Indonesian Akulaku's BNPL solution into its supported digital payment methods, marking its first foray into Southeast Asia. This partnership will enable Akulaku PayLater users to use their BNPL solution for all Alipay+ merchants globally.

The Indonesian BNPL solution has been able to meet the growing demand for flexible payment plans and installment purchases, thereby boosting consumer purchasing power. Akulaku PayLater also aims to ensure that previously underserved markets can participate in convenient digital financial services. To date, there are over 4,000 merchants in the

[37] Antom, https://global.alipay.com/docs/ac/antomop/akulaku?version=1.0.105&pageVersion=14.

Akulaku PayLater ecosystem, and in the next two years, Akulaku aims to serve 10,000 online merchants and over 100,000 offline stores through its BNPL solution. Furthermore, according to an IDC InfoBrief commissioned by 2C2P, Indonesia is expected to welcome the most new e-wallet users by 2025—130 million, making the archipelago the largest BNPL market in Southeast Asia.[38]

"As an emerging yet rapidly adopted digital payment method, BNPL brings inclusive financial services to underserved communities. We see BNPL as an integral part of the digital payment ecosystem and are delighted to partner with Akulaku PayLater to enable communities in Indonesia and other potential markets to enjoy seamless and convenient cross-border payment services, enriching their daily lives through global shopping," said Cheng Guoming, General Manager of Global Partnerships at Alipay+.[39]

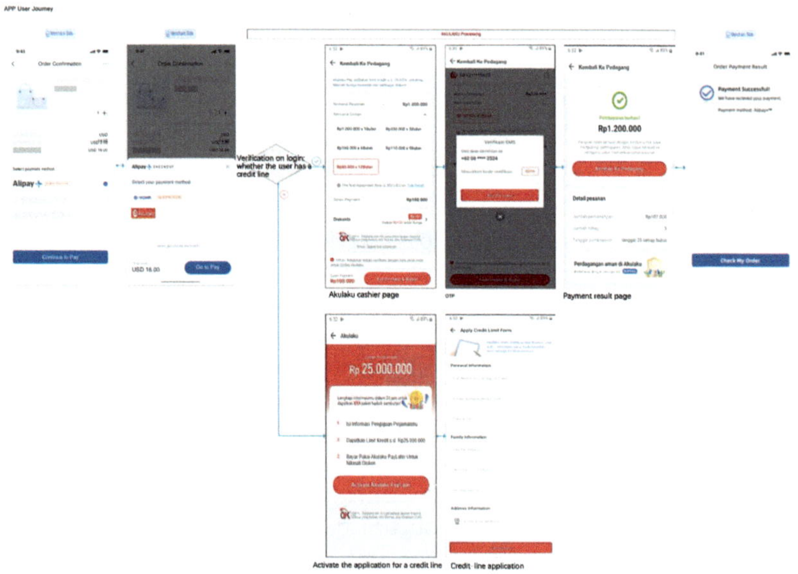

Akulaku Payment Procedure (*Source* Ant International)[40]

[38] Techwire Asia, 2022, https://techwireasia.com/2022/09/akulaku-paylater-first-bnpl-in-southeast-asia-to-partner-with-chinas-alipay/.

[39] https://www.kuajinzhifu.com/bulletin/3706.html.

[40] https://global.alipay.com/docs/ac/antomop/akulaku?version=1.0.105&pageVersion=14.

Indonesian consumers stand to gain a great deal from the partnership between Alipay+ and Akulaku PayLater in the field of cross-border payments. This collaboration is especially advantageous since it provides credit lines and installment payment plans to consumers, increasing their purchasing power and promoting more consumption by branching out into both online and offline consumption scenarios. Moreover, the convenience of cross-border payments is anticipated to generate more employment opportunities in related sectors. As international tourists return, this alliance is poised to enhance the payment experience within Indonesia's tourism industry, enticing more foreign visitors to spend locally. Most importantly, the partnership enables markets with limited financial services to tap into convenient digital financial services, fostering greater financial inclusion and regional economic integration.

Kredivo, which was established in 2016 and is headquartered in Jakarta, Indonesia, is committed to offering local users low interest, instant credit financing services. Its parent company, Singapore-based financial technology firm FinAccel, has been supported by multiple high-quality investment institutions since its inception, reflecting its strong financial backing and credibility in the industry.

The partnership between Alipay+ and Akulaku PayLater, which is growing in both digital and physical retail spaces, has the potential to greatly increase growth for Indonesian customers, particularly in the field of cross-border payments. Akulaku PayLater is successfully increasing consumers' purchasing power by offering credit lines and flexible installment plans. It is anticipated to encourage even more spending in the long run. Besides, the ease of making cross-border transactions could pave the way for more job openings in associated sectors. With the revitalization of global travel, this joint effort is expected to contribute more to the payment landscape within local tourism industry, encouraging more international tourists to get involved in local society and market while visiting. Perhaps most crucially, this alliance opens a window for domains lacking robust financial services to join the digital financial transition, fostering broader financial inclusion and make regional economic more prosperous.

14.3.3 Kredivo Promotes Financial Inclusion by Providing Loans to Over One Million Indonesians

Kredivo, established in 2016 and headquartered in Jakarta, Indonesia, is dedicated to providing local users with low interest, instant credit financing services. Its parent company is Singapore-based financial technology firm FinAccel, which has been supported by multiple high-quality investment institutions since its inception.

Leveraging an artificial intelligence-powered real-time decision-making system, Kredivo enables users to apply for and be approved for a credit limit within Kredivo online within just two minutes. Users can enjoy interest-free buy now, pay later services for up to 30 days and various monthly installment credit services when shopping at partner merchants. Subsequent repayments can be made through various methods, including electronic wallets, online banking, or cash.

Kredivo and Alipay+'s partnership is based on Alipay+'s cutting-edge cross-border payment technology, which aims to support international payment integration and encourage financial technology innovation. With a solid presence in Indonesia's BNPL sector, Kredivo teams up with Alipay+'s inclusive network and technological prowess to streamline the payment process for Indonesian consumers and promote the expansion of cross-border transactions.

Kredivo sets itself apart from the Alipay+ ecosystem by offering its clients the convenience of deferred payments and installment options. With this credit-funded payment option, customers can select Kredivo as their payment method at participating retailers. Regarding the consumer side, Alipay+ boasts a strong worldwide network that spans numerous nations and regions and includes AlipaySG, AlipayEU, AlipayUS, AlipayUK, and other regional behemoths.

According to Kredivo's data, through its collaboration with Alipay+, it had surpassed over one million transactions by October 2023, with a transaction value of at least 300 billion Indonesian rupiah. Kredivo has served more than 6 million users through its BNPL service, and its user base has doubled in the last 10 months.[41]

Kredivo has obtained a significant increase in its funding ceiling by obtaining a staggering 70 trillion Indonesian rupiah from Hana Bank in South Korea. This financial leap is expected to improve retail credit accessibility and inclusivity. In the meantime, Kredivo is actively embracing digital transformation, launching Krom Digital Bank through its subsidiary, Krom Bank. This move is a game-changer for the younger generation, empowering them to achieve financial independence. Besides,

[41] Kredivo Group, https://kredivocorp.com/about/.

Kredivo is strategically expanding its market reach. With intentions to expand its services through a strong cooperation with Alipay+, it is keeping its sights set on regional countries such as Vietnam and Thailand. Apart from these efforts, this expansion links more accessible digital financial services to underdeveloped markets, increasing financial inclusion and fostering regional economic integration.

On December 20, 2023, DBS Bank Indonesia joined forces with fintech innovator Kredivo to extend loans to a milestone of over one million accounts, all under a collaborative financing scheme. This financial boost is reaching recipients in diverse regions throughout Indonesia, aligning with DBS Bank Indonesia's dedication to elevating the financial inclusion index for those who traditionally fall outside the scope of mainstream banking services.

With operations in 19 markets, DBS is one of Asia's top financial services groups. DBS, which has its headquarters in Singapore, is well-represented across the three Asian growth axes of South Asia, Southeast Asia, and Greater China. The bank has "AA-" and "Aa1" credit ratings, which are among the highest in the world. DBS has won awards for its global leadership, such as "Global Bank of the Year" from The Banker and "World's Best Bank" from Global Finance magazine. DBS is in the forefront of using digital technology to influence the future of banking, as evidenced by its designation as the "World's Best Digital Bank" by Euromoney and the "Most Innovative Digital Bank" in the world by The Banker.

DBS offers a comprehensive range of consumer. Rooted in Asia, it has a keen grasp of the intricacies of thriving in the region's most vibrant markets. Beyond the realm of traditional banking, the DBS Foundation is at the forefront of creating meaningful change. It backs enterprises that chase a dual mandate—chasing profits while also making a positive dent in society and/or the environment. The DBS Foundation's dedication to giving is demonstrated by a variety of programs, such as strengthening food security and resilience and equipping communities with skills that will prepare them for the future.

In 2023, the Central Bank of Indonesia shed light on a pressing issue: approximately 97.7 million Indonesians, roughly 48% of the nation's population, were grappling with inadequate access to financial services, a state often referred to as being underbanked. This disconnect is attributed to a tapestry of challenges, including a lack of financial literacy that hinders effective financial planning and management. Factors such as fluctuating incomes, limited loan accessibility, and a dearth of credit history or an insufficient track record also erect barriers, preventing many from tapping into traditional financial institutions for loans.[42]

These figures demonstrate how much potential there is to use technology to increase financial inclusion. Kredivo has been offering technology-based financial services since 2021 after realizing this opportunity. Umang Rustagi, CEO of Kredivo Indonesia, stated, "Our vision is to become a financing company that successfully aligns economic and social aspects with its business activities. Through our collaboration with DBS Bank Indonesia, we are committed to not only providing financial access but also helping the underbanked population across Indonesia to become bankable in the future. With this innovative joint financing, our aim is to empower more people to achieve their financial goals and build an inclusive financial ecosystem, in line with our goal to serve tens of millions of users in the coming years".[43]

Besides, Melfrida Gultom, Deputy Head of Consumer Banking at PT Bank DBS Indonesia, said, "Our partnership with Kredivo is a tangible manifestation of DBS Indonesia's first sustainability pillar, Responsible Banking. We focus on delivering innovative and responsible banking products to accelerate financial inclusion in Indonesia. In this digital era, we believe that technology is not only a solution but also the key to bringing a positive impact to the broader community, especially the underbanked. To benefit even more people, DBS Indonesia and Kredivo periodically increase the limit of the joint financing, from Rp300 billion at the initial stage to Rp2 trillion by mid-2022. We hope that this collaboration will be a concrete step towards economic empowerment and the well-being of the community."[44]

[42] DBS, 2023, https://www.dbs.com/newsroom/Bank_DBS_Indonesia_and_Kredivo_promote_financial_inclusion_give_more_than_1_million_Indonesians_access_to_loan.

[43] Ibid.

[44] Ibid.

14.4 Conclusion

Indonesia offers a distinctive case study. Unlike other nations reliant on tourism or those with established fintech ecosystems, Indonesia has forged its own digital financial ecosystem by partnering with global behemoths, bypassing the traditional banking phase and directly transitioning into a model encompassing the digital finance era and the digital economy.

Indonesia, with its robust domestic economy and sizable local population, does not rely heavily on tourism for financial growth, which only makes up a small fraction of its GDP. This means that attracting foreign visitors is not necessarily the main driver of the financial sector evolution in Indonesia. Meanwhile, the local market has nurtured a sophisticated digital payment ecosystem, with homegrown players like GoPay and OVO securing dominant market positions. Even state-backed LinkAja and Ant Group-backed DANA are finding it challenging to compete with these established players in the market.

The fintech landscape in Indonesia is witnessing a rapid evolution, with a spotlight on e-commerce, mobile payments, and digital wallets, echoing the global trend of leapfrogging from fixed telephone networks to mobile technology. This shift is particularly focused on bringing financial services to the previously unbanked and underserved populations, allowing them to tap into financial products through smartphones and mobile apps. Indonesia's fintech success story is built on its ability to innovate according to local needs, rather than relying on traditional banking models. This innovative spirit has created a fertile ground for local fintech companies to flourish and has also opened doors for international players like Alipay+ to contribute to the ecosystem and bolster the growth of the local digital economy.

When it comes to technological infrastructure, Indonesia's development hinges on collaborations with local partners and Chinese FinTech giants such as Ant Group. The aim is to establish a comprehensive digital financial ecosystem that nurtures the digital economy and facilitates a smooth transition to a digital economy paradigm. This approach is characterized by the development of indigenous digital wallets and payment solutions tailored to the specific needs of the Indonesian market. As a result, Ant Group and other international titans play a supportive role in Indonesia, enhancing the overall digital financial infrastructure by offering expertise in science and technology, risk management, data analysis, and cross-border payment capabilities.

In Indonesia, Alipay+ and its partners are dedicated to constructing an inclusive digital financial ecosystem that supports the inclusive growth of the digital economy. Here, the Sustainable Development Goals are centered on cultivating a robust and innovative local fintech industry to address the needs of the unbanked and underbanked populations, while also fostering local economic growth and social development. Should Alipay+ and its local partners play a pivotal role in this domain, contributing to the cultivation of Indonesia's digital ecosystem and the incubation of its digital economy, they could potentially exert a more significant influence on the country's future digital industry development.

PART VI

Sustainable Travel Program—Innovating Best Practices for Sustainable Development Goals

CHAPTER 15

FinTech for Sustainable Travel, Mega Events, and Best Practice Partners in Hong Kong and Inclusive Development in the Greater Bay Area (GBA)

15.1 Dual-City Life: Economic Integration Prompted by the Stable Consumption Demand of Millions

15.1.1 Hong Kong Residents Heading North: Post-pandemic Economic Recovery Driven by Consumptions

This section will analyze the trend of Hong Kong residents moving to Shenzhen and other mainland cities in the GBA due to economic benefits and the implications for the recovery of consumption in the post-pandemic era. It will evaluate the effects on local economies as well as the usefulness of the digital payment systems towards this movement practice.

Since the full resumption of border crossings between the mainland and Hong Kong in February 2023, exchanges between Shenzhen and Hong Kong have become increasingly close, with Hong Kong residents' spending on food and leisure activities in Shenzhen now a part of their daily routine.

According to statistics from the Hong Kong-Zhuhai-Macao Bridge inspection station, the number of passengers crossing the bridge exceeded 16.3 million in 2023, with more than 2.68 million coach trips. This figure represents an increase of 1.29 times and 2.45 times compared to

J. Sun, *Sustainable Financialization*, Sustainable Development Goals Series, https://doi.org/10.1007/978-981-96-3580-1_15

2019, respectively. Notably, the number of single-license vehicles from Hong Kong and Macau has been steadily rising, reaching 1.4 million trips throughout the year, accounting for over 50% of the total coach traffic at the port.[1] By July 2024, the bridge is expected to witness another surge in northbound traffic. According to statistics from the Zhuhai Border Inspection Station and the Hong Kong-Zhuhai-Macao Bridge inspection station, from early July to early August, over 3 million passengers and more than 600,000 vehicles crossed the Zhuhai road port, representing year-on-year increases of 62.15% and 74.27%, respectively, both setting historical records for the same period.[2]

In a "Secretary's Essay" on July 30, 2023, Hong Kong's Financial Secretary Paul Chan wrote, "After three years of the pandemic, many citizens' lifestyles have changed, and the spending of Hong Kong residents in the mainland has significantly increased." Data from Hong Kong's AlipayHK shows that in the second quarter of 2023, the amount and number of transactions made by Hong Kong residents in the mainland surged more than three times compared to the previous quarter, indicating a trend of increased cross-border travel. In just six months, travel and spending data have shown a stepwise increase.[3] Since the resumption of border crossings in January 2023, millions of people have been traveling across cities in the Guangdong-Hong Kong-Macao Greater Bay Area daily, with residents' consumption returning to 140% of the amount spent during the same period in 2019.

According to statistics from the Hong Kong Immigration Department, approximately 53 million Hong Kong residents traveled north in 2023, averaging about 7.2 trips per resident. Notably, during the period overlapping the Easter holiday and the Qingming Festival (from March 28 to April 7), around 540,000 Hong Kong citizens traveled north for consumption.[4]

The Luohu Port is currently one of the busiest entry and exit ports in Shenzhen and is the preferred hub for various travel groups. Among the

[1] GBA Review, 2024, https://mp.weixin.qq.com/s/X_iNl54xZkYgEAgxzWf9yg.

[2] Radio The Greater Bay, 2024, https://mp.weixin.qq.com/s/5-9vjSMZl5QikT0K6ekgPA.

[3] AlipayHK, 2024, https://www.alipayhk.com/en/shoppers/. See also, https://mp.weixin.qq.com/s/TB7ZcJajh9fkwlIQorzG1Q.

[4] Immigration Department of HKSAR, 2024, https://www.immd.gov.hk/hkt/facts/passenger-statistics-menu.html.

seven land Class I ports currently open to the outside world in Shenzhen, three are located in the Luohu District: Luohu Port, Wenjindu Port, and Liantang Port. In the six months following the reopening after the pandemic in 2023, the Shenzhen Border Inspection Station processed approximately 62.347 million entry and exit passengers. The combined traffic at these three major ports accounted for 22.82 million trips, representing 44.7% of the total cross-border traffic between Shenzhen and Hong Kong, with Hong Kong residents making 17.61 million trips, comprising 77% of that volume.[5]

Luohu Port before and after the pandemic (*Source* Internet)

Luohu District has always been a popular shopping destination for Hong Kong residents visiting Shenzhen. Located near the terminus of the East Rail Line, a large number of citizens of Hong Kong enjoy spending their weekends in Luohu, engaging in activities like eating, shopping for clothes, and attending some shows. Most especially the youth tend to show such behavior. Jinguanghua Plaza, located about 1.5 kilometers away from the Luohu port, has become a place very popular for the people of Hong Kong to shop. Summer holiday is a season of highest sales and many of the popular chain stores will be thought to be fully reserved at dinner periods. At peak periods, some beverage outlets in the Jinguanghua Plaza may have waits of up to 1.5 hours and several residents of Hong Kong would bring some of the drinks home before returning to their home.

The Shenzhen government has been quick to follow consumption trends, organizing events to attract crowds. In May 2023, a highly successful "Hong Kong·Trendy" shopping festival was held in Futian District, featuring nearly 80 exhibitors from Hong Kong and over 130

[5] Luohu Release, 2023, https://mp.weixin.qq.com/s/g9EzP6xbXhkc2TaNNLOV_w.

Hong Kong brands that achieved great success during the event. In June of the same year, Shenzhen launched its first night market district, "Little Hong Kong Night," which fully aligns with the consumption habits and cultural characteristics of Hong Kong residents in Yantian District. Additionally, business circles in both Shenzhen and Hong Kong collaborated to offer exclusive promotions. Deeply City and the popular Hong Kong brand "Xidorsi" jointly introduced special discount events, with nearly 30 merchants providing dedicated services for Hong Kong residents.

July and August, being the summer vacation period, are the prime season for youth study tours. In the first summer following the pandemic reopening in 2023, Shenzhen held the "Luohu Shenzhen-Hong Kong Cultural Month" cultural and tourism event, featuring a wide array of entertainment, food, and market activities, attracting numerous tourists from both regions. During the 2024 Spring Festival, Shenzhen ports collaborated with the Zijing Cultural Group to organize six themed flash cultural events titled "Dragon Soars Celebrating the New Year—Elegance Flows Through All the Port" at various land ports between Shenzhen and Hong Kong. These activities not only enriched the holiday experiences of Shenzhen-Hong Kong travelers but also promoted a deeper cultural integration between the two regions.

2023 Luohu Shenzhen-Hong Kong cultural month (*Source* Luohu Government)

The strong purchasing power of Hong Kong residents is particularly evident on weekends and holidays. With a considerable number of public holidays in Hong Kong, every weekend and holiday becomes a time for Hong Kong residents to "invade" major shopping malls in Shenzhen. During the first Christmas holiday after the pandemic at the end of 2023, over a million people crossed the border daily, with the number of people leaving Hong Kong significantly higher than those entering. The Executive Director of the Tourism Promotion Association, Coi Ding-bong, noted that the mainland has developed rapidly in recent years, and many

Hong Kong residents, after crossing the border, experienced the luxury and comfort of local hotels, finding the value for money to be very high, which naturally encouraged more people to head north for consumption.

During the 2024 Spring Festival, the Hong Kong Special Administrative Region optimized its border crossing policies, leading to a significant 45% increase in the number of transactions made by Alipay users in Hong Kong compared to the same period in 2019. The coverage rate of Alipay among merchants in Hong Kong and Macau exceeded 90%.[6] Tourists are not only keen on shopping but also adept at utilizing local transportation for an immersive travel experience. Alipay has been widely integrated with the MTR, busses, Star Ferry, and trams, becoming the most popular travel options for tourists, with each user averaging over three scans to use transportation.[7]

The interconnectedness of the cultural and tourism industries in the Greater Bay Area is deepening, with the governments of Hong Kong and Macau actively collaborating with various sectors to host a rich array of Spring Festival activities, making Hong Kong and Macau favorites for short trips during the festival.

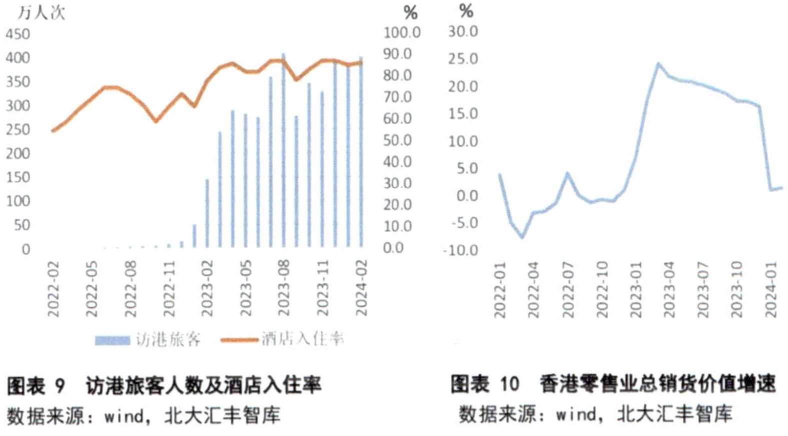

Number of Hong Kong travelers and retail sales situation as of early 2024 (*Source* Wind, Peking University HSBC Business School)

[6] Alipay, 2024, https://www.alipay.com/index.html.

[7] IT Home, 2024, https://www.ithome.com/0/750/510.htm.

Subsequently, on the first day of the Hong Kong Easter holiday on March 29, 2024, the number of Hong Kong residents exiting the territory nearly reached 800,000, setting a new single-day high since the pandemic. Among them, nearly 500,000 people crossed into Shenzhen via the six land ports along the Hong Kong-Shenzhen border, while another approximately 170,000 traveled to the mainland via the West Kowloon Station of the high-speed rail and the Hong Kong-Zhuhai-Macao Bridge.[8]

During the Qingming Festival in 2024, statistics from the Hong Kong Immigration Department indicated that approximately 362,000 mainland travelers entered Hong Kong over the three-day holiday, while Hong Kong residents exiting the territory surpassed 838,000. The number of Hong Kong residents traveling north for consumption has clearly exceeded the number of tourists heading south.[9] On the day of the Qingming Festival itself, over 1.01 million people crossed the borders into and out of Hong Kong, representing an increase of about 52% compared to the same period in 2022, with around 70% of those being Hong Kong residents, nearly reaching 10% of the city's population.

Hong Kong residents waiting in line at the Port to Head to Shenzhen (*Source* Sing Tao Daily)

In January 2024, the Shenzhen Retail Commercial Industry Association released the "Shenzhen-Hong Kong Consumer Satisfaction Survey Report." The report indicated that one year after the resumption of

[8] Immigration Department of HKSAR, 2024, https://www.immd.gov.hk/hkt/facts/passenger-statistics-menu.html.

[9] Ibid.

border crossings between the mainland and Hong Kong post-pandemic, 95.24% of Hong Kong residents reported visiting Shenzhen multiple times for consumption. Many typically come to Shenzhen with family or friends, with 27.62% of high-frequency visitors traveling more than once a week, while 73.33% visit at least once a month. Most prefer to make same-day round trips.[10]

Overall, the main reasons attracting a large number of Hong Kong consumers to Shenzhen for shopping are the excellent value for money, a wide variety of products, and convenient consumption experiences. Areas like the Dongmen-Renmin South business circle in Luohu District and the Futian CBD, which are close to the ports and have convenient public transportation like the subway, are particularly favored by Hong Kong residents. Many Hong Kong visitors find the border crossing and local transportation in Shenzhen to be very convenient, noting that Shenzhen offers a rich selection of products and services at great prices, with over 40% of consumers highlighting these aspects.

Hong Kong residents' preferred shopping locations in Shenzhen are primarily shopping centers and large supermarkets, accounting for 78.10% and 60.00%, respectively.[11] Moreover, dining consumption accounts for over 90% of their spending, followed by shopping and leisure activities.

What began as a "revenge spending" spree for a late-night snack has now evolved into comprehensive shopping for daily necessities, making "heading north" a hot topic in the daily lives of Hong Kong residents. Young people are eager to queue for trendy brands, middle-aged individuals flock to warehouse-style membership stores for bulk purchases, parents with children are drawn to spacious indoor playgrounds, while older adults enjoy affordable restaurants and massage shops.

Hong Kong residents' shopping scenes include bulk purchases at membership-based warehouse stores like Sam's Club and Costco, as well as affordable chain stores like Hema Fresh. They enjoy tasting popular mainland dining brands and also explore electronics, wellness, beauty, and medical services. A project manager for a major shopping center under the Excellence Commercial Group stated in a media interview that, compared to local Shenzhen customers, Hong Kong residents have clear

[10] Shenzhen Smart Retail Association, Shenzhen Chain Operation Association, & Shenzhen Retail Trade Association, 2024, https://www.szrba.org/news/page-74/2021/%E6%B7%B1%E6%B8%AF%E4%B8%A4%E5%9C%B0%E6%B6%88%E8%B4%B9%E6%BB%A1%E6%84%8F%E5%BA%A6%E8%B0%83%E6%9F%A5%E6%8A%A5%E5%91%8A%E5%8F%91%E5%B8%83.html.

[11] Ibid.

goals when shopping in Shenzhen, focusing on "eating, drinking, playing, and bringing back."[12]

When it comes to dining, Hong Kong residents focus on freshness, affordability, and good service, frequently visiting charcoal-grilled, barbecue, and hot pot restaurants. For entertainment, they look for fun and exciting activities which don't take too much time, with high cost-performance ratios like massages and hair treatments. While Hong Kong residents enjoy high-quality chain brands, they seek novelty as well, which contribute to a growing popularity of independent brands created by individual entrepreneurs.

One Hong Kong resident mentioned that spending 1,000 HKD in Hong Kong is only enough for a proper dinner, but in Shenzhen, it can cover a massage, dinner, and several cups of freshly made lemon tea, and with some savings, they might even include hotel costs. "Our family's expenses are equivalent to what one person spends in Hong Kong."

Compared to Hong Kong's limited-scale chain supermarkets, the affordable and value for money warehouse-style stores in mainland China continue to attract Hong Kong residents. On the opening day of Shenzhen's first Costco in early 2024, there was a significant crowd by 8 a.m. A couple from Hong Kong shared that they arrived in the Longhua District of Shenzhen the night before specifically to shop at the new store and also to enjoy a short trip. Similarly, Wu Fan, a dedicated fan of Sam's Club, has been making bulk purchases every two weeks for the past two months. His repurchase list primarily includes Sam's private label products like American roast chicken, coconut-flavored yogurt, scallion pork floss bread, and Swiss rolls. "Even accounting for transportation and time costs, these products are still cheaper than similar ones in Hong Kong, and Sam's curated products come with quality assurance; I haven't had any bad experiences." He noted that the pace of life in Hong Kong is fast, and most daily necessities are purchased at the nearby Wellcome supermarket. "Usually, I just go when I run out of something; Hong Kong supermarkets aren't large, and the brands and products they offer are quite fixed, so I rarely browse in the supermarket."

Farmers' markets have also attracted a large number of Hong Kong residents for bulk shopping. Compared to the prices of agricultural products in Hong Kong, some residents have stated, "The same products can

[12] Caixin, 2024, https://weekly.caixin.com/2024-01-20/102158618.html.

be twice as expensive in Hong Kong as they are in Shenzhen." This trend has even spread to other farmers' markets far from the ports. The Huafu Market near the Longhua subway station in Shenzhen usually serves local residents; however, even on workdays, Hong Kong residents can be seen shopping there. Vendors at the market expressed surprise at the number of Hong Kong shoppers. "It's not difficult to identify them; they typically wear outdoor clothing, carry backpacks or pull small carts, and their bags are often equipped with hand sanitizer. Their Mandarin usually has an accent." Hong Kong shoppers tend to favor local specialties, such as Yunnan blueberries, Dandong strawberries, Yangshan honey peaches, and Guangxi sugar oranges.

A Farmers' market in Shenzhen (*Source* Internet)

Moreover, the immense consumer demand from Hong Kong residents has even given rise to a new phenomenon: Shenzhen-Hong Kong "reverse cross-city purchasing services." A quick search on China's social media platform Xiaohongshu for "Shenzhen-Hong Kong cross-city delivery" reveals a flood of promotional offers to transport goods from Shenzhen to Hong Kong. These goods include, but are not limited to, popular mainland snacks and foods, fresh fruits and vegetables, and flower gifts.

Since the full reopening of borders, the trend of reverse purchasing has surged. Businesses like Baoshifu, Zhou Hei Ya, milk tea chain "Yidian Dian" have become the hottest candidates for reverse shopping. A number of residents of Hong Kong prefer to make joint orders with their families and friends. They can get these for over 100 yuan with the delivery charge for an item less than 50 yuan at roughly 15 yuan. This new type of business activity has further addressed the consumption needs of other Hong Kong residents who on normal days, find it a hassle to shuttle back and forth between Shenzhen and Hong Kong.

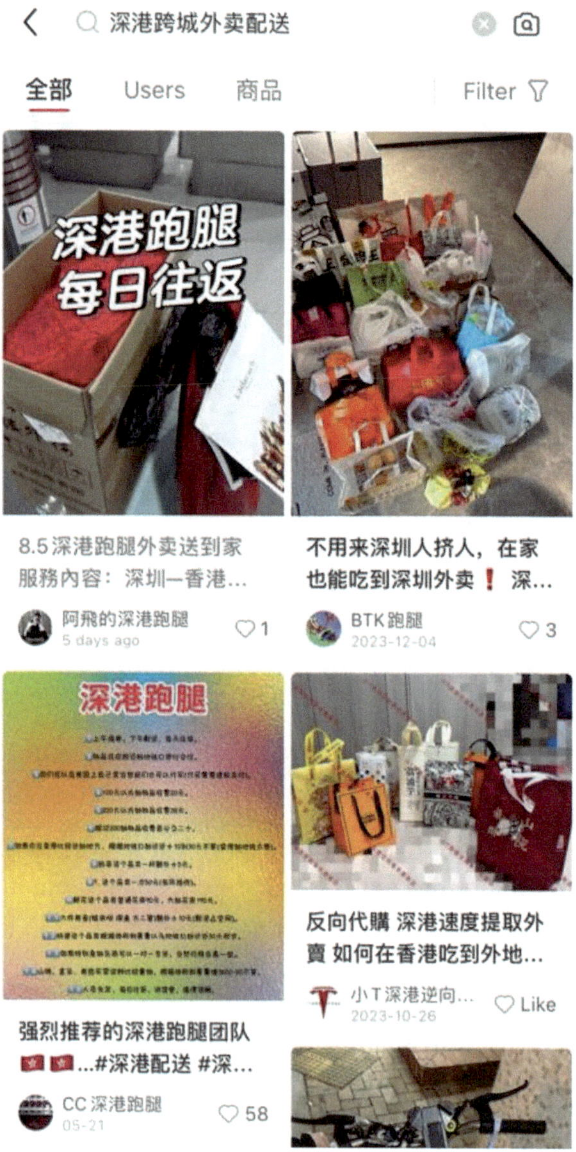

Content on Shenzhen-Hong Kong Cross-City delivery on Xiaohongshu (*Source* Xiaohongshu)

In addition to the surge in shopping trips to Shenzhen during holidays and even on regular days, healthcare has become another rapidly growing demand for Hong Kong residents heading north. Shenzhen's Luohu District People's Hospital, located near the border, is filled with numerous Hong Kong and Macau patients. The hospital is equipped with Cantonese-speaking medical staff to facilitate better communication between patients and doctors.

In terms of healthcare services, the most popular departments among Hong Kong patients are dentistry, ultrasound, and reproductive medicine. Xiong Yi, the executive director of Luohu District People's Hospital, noted that he treats a significant number of pregnant women from Hong Kong every week, as private hospitals in Hong Kong tend to charge high fees for medical examinations. He explained that mainland hospitals offer comprehensive prenatal examinations at a fraction of the cost, with prices less than one-third of those in Hong Kong's private hospitals, making it highly cost-effective.[13]

On the same note, elderly residents of Hong Kong looking for medical treatment from the mainland will also enjoy additional options. In February 2024, the Health Bureau of Hong Kong announced its intention to implement its plan of extending the "Elderly Health Care Voucher Greater Bay Area Pilot Scheme." By the beginning of Q3 of the same year, elderly people 65 years and over will be able to use health vouchers at more hospitals in Shenzhen, Guangzhou, Dongguan, and Zhongshan.

The newly added medical institutions include five general hospitals: The First Affiliated Hospital of Sun Yat-sen University and its Nansha branch in Guangzhou, Zhongshan Chen Xinghai Integrated Traditional Chinese and Western Medicine Hospital, Dongguan Tung Wah Hospital, and Shenzhen New Frontier United Family Hospital. Additionally, two dental clinics near Shenzhen's border will be included in the trial. Secretary for Health Lo Chung-mau stated that this pilot program aims to extend coverage to all core cities in the Greater Bay Area, showing care for the elderly who retire in Guangdong.

[13] Hong Kong Commercial Daily, 2024, https://mp.weixin.qq.com/s/aD5n3UxASVaRMHr8nlBAbQ.

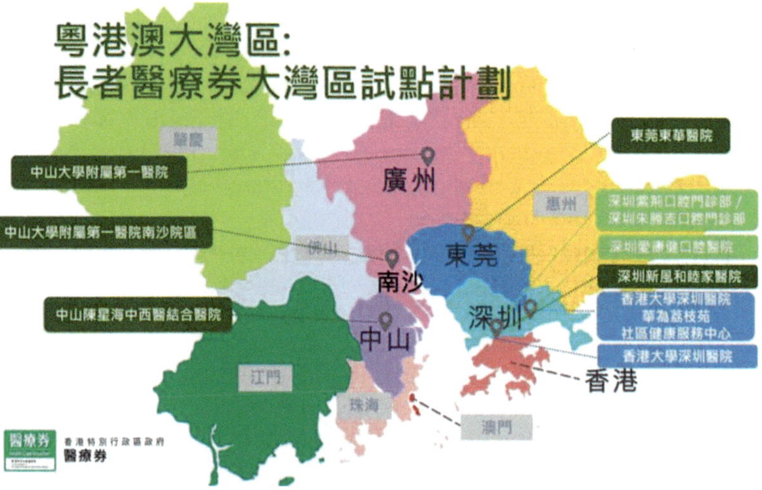

Coverage of the Elderly Health Care Voucher Greater Bay area pilot scheme (*Source* Hong Kong SAR Government's Guangdong Liaison Office)

If the initial wave of "Hong Kong residents heading north" was driven by "revenge spending" after border reopening, why has this trend lasted so long? This phenomenon has not been a fleeting trend; instead, it shows enduring vitality and has gradually become a normalized pattern. Some institutions refer to this as a "reverse siphon effect" of mainland commerce attracting Hong Kong and Macau tourists, or a "shopping migration."

However, regardless of how it is named, the deeper issues revealed by this large-scale cross-border consumption phenomenon go far beyond spending. It shows the growing economic, cultural, and social interdependencies of Hong Kong and the mainland—and more so within the Greater Bay Area. As transport networks improve and the integration process of the greater bay region advances, it is apparent that, the residents of the region are not only going over to shop or for short tours up north. Rather, there seems to be a shift as these residents are gradually more looking forward to a more sustainable "dual-city lifestyle" where cross-border living is normal. That is, residents' work, life, and shopping are all integrated across the two regions so much so that residents are never too far away from the resources they need.

Not only does the cross-border shopping culture adopt the spending habits of millions of Hong Kong shoppers heading north for the weekend but this also becomes a major space for economic integration and fintech. Throughout this process, the mainland market's more attractive pricing, diverse consumption options, and increasingly sophisticated service systems have become key factors driving Hong Kong residents to travel north more often.

The depreciation of the Hong Kong dollar (HKD) against the Chinese yuan (RMB) has led some residents of Hong Kong to increase their spending level in mainland China. There have been periods in time where the HKD has appreciated against the RMB, allowing consumables from the mainland to be cheaper for those in Hong Kong. For example, the exchange rate of the currencies increased from 1 HKD to 0.8 RMB in February 2022 to 1 HKD to 0.93 RMB in May 2024. Such exchange rates enabled consumers in Hong Kong to purchase more goods and services from the mainland as the prices were more competitive.

The expansion of the mainland's market economy with new cities like Shenzhen in the Greater Bay Area presenting new business and residential opportunities, has allowed for the satisfaction of Hong Kong residents' desires while also creating new opportunities for them. The significant improvement in the mainland's consumption environment and service quality has also started to change the perception Hong Kong residents have of the mainland's market. From diverse shopping, dining, and entertainment options to high-quality service experiences, these features have become highlights that attract Hong Kong consumers. Hong Kong residents' view towards the quality of mainland products has shifted with time leading to an increasing acceptance and appreciation of the mainland consumption environment.

More significantly, with the development of the Greater Bay Area and in particular, the construction of the high-speed rail and the Hong Kong-Zhuhai-Macau Bridge, it has become much easier for Hong Kong residents to travel to the mainland. Other initiatives and policies implemented by Alipay Hong Kong (AlipayHK), Octopus, and other platforms have also contributed in reducing the burden on Hong Kong residents traveling north and improved the ease of cross-border trips.

In summary, the pattern of Hong Kong residents moving north should no longer be associated with mere shopping. It has developed into a cross-border, deepened lifestyle, thus contributing to the regions becoming even more economically integrated.

15.1.2 *Expanding from Hong Kong: Linking a Broader International Market*

15.1.2.1 The Breakthrough of Mainland Catering in Hong Kong

This subsection will look at the proliferation of mainland China's catering industry in Hong Kong, providing an insight into the reasons for this trend and the impact on the local dining industry and economy.

Hong Kong residents have shown immense enthusiasm for emerging mainland chain restaurant brands, a trend that has encouraged more mainland brands to expand into the Hong Kong market. Beginning 2023, a number of big names in the restaurant business with chains in the mainland have had their outlets in Hong Kong, such as Meow BBQ, Xita Laotaitai, Chinese Sauerkraut Fish, Hefu-Noodle, Muwu BBQ, and Manner Coffee. These brands have either developed through direct ownership or regional franchises at a rapid rate.

Most of these brands have chosen to set up shop in Hong Kong's most bustling commercial districts, such as Mong Kok, Tsim Sha Tsui, and Causeway Bay. For example, Hefu-Noodles has secured a spot on Russell Street, once known as the "world's most expensive shopping street," with a monthly rent of HK$300,000. And another mainland brand—NongGengJi Hunan Cuisine opened in Hong Kong in September 2023, and after just ten months it already boasted six locations. Come mid-2024, Chagee, another well-known tea brand from the mainland, was already gaining traction, with a large placket close to Tsim Sha Tsui MTR. Chagee is set to launch in Hong Kong in September 2024, at the trendy K11 Art Mall, a favorite among young people. As one of the most popular tea brands among Hong Kong people, the news of its first Hong Kong store generated significant online buzz across various social media platforms.

In May 2023, Meituan launched a new food delivery platform in Hong Kong called KeeTa, with the slogan "10 Billion in Rewards." KeeTa's initial partners included popular chains like McDonald's, Maxim's, KFC, Yoshinoya, and Pacific Coffee, along with local favorites such as Hung Fook Tong, Nam Kee Noodles, Sum Congee, Ken Sushi, and Chicha San Chen. New users received HK$300 worth of vouchers and free delivery, and compensation vouchers were provided for delayed deliveries. With meals available for as low as HK$50–60, KeeTa's prices were generally cheaper than those offered by its main competitors, Foodpanda (a German platform) and Deliveroo (a UK platform). KeeTa also adopted some of Meituan's signature marketing strategies, such as daily promotions through the "Daily Deals" campaign, which randomly distributed coupons to users. By the end of 2023, more than 1.3 million users had registered for KeeTa. Among them, some users, particularly business professionals, showed high loyalty, with one customer purchasing meals 336 times within six months through the platform's "One-Person Canteen" section, which does not require a minimum order amount—averaging nearly 50 orders per month. This reflects KeeTa's high user retention and repeat purchase rate.[14]

Brands from mainland China continue their expansion into Hong Kong. Beyond popular chain restaurants, experiential and lifestyle brands are also planning to open branches in Hong Kong. For instance, the aesthetic brand OCE and the domestic makeup brand ToSummer are both penetrating the Hong Kong market.

While mainland restaurant brands are rapidly gaining ground, local Hong Kong dining establishments are shrinking. Many such long-established local restaurants are closing owing to high rents, or because there is no next generation to take over such family businesses. This is not to mention the local popular restaurant chain Tao Heung which confirmed it will shutter no fewer than four of its branches by July, 2024. The company's Chairman and CEO, Wong Kit Lung, advised other restaurant owners to "act within their means and not overextend themselves financially. Reopen after the business environment improves."

Currently, gross street level shop rents in Hong Kong are said to be the lowest in history. As noted by JLL's Senior Director of Research Cathie Chung, with the firm's rental index, prime street level shop rents resided at a 41% reduction during the 2020–2022 pandemic, which was a 72.5% reduction when set against 2014 when the rentals peaked. She has also observed the economic indicators and outlook, which tell her that rent for shops is unlikely to be nearing their peak regions either within the next five years.[15]

David Ji, Executive Director and Head of Retail Services for Cushman & Wakefield Hong Kong, mentioned that as more and more Mainland brands continue to penetrate the Hong Kong market, the level

[14] Investor, 2024, https://www.investorchina.cn/article/75716.

[15] Jones Lang LaSalle (Hong Kong), 2023, https://mp.weixin.qq.com/s/oRfMEBFCUjG35Al_ghz-AA.

of understanding by local residents of the services and products offered by such brands will greatly increase, which will see them gain ground in Hong Kong. This trend is seen as a way to comprehend the impact caused by the closure of stores in Europe, America, and Japan, and is predicted to succeed. Nonetheless, substantiating this, he cautioned that some of such mainland brands emerging here could substitute some street level shops in Hong Kong's retail market snags, possible hampers market recuperation.

15.1.2.2 Widespread Participation in Hong Kong's Local Consumption Market: Sasa's Best Practice

This subsection will present a case study of Sasa, the Hong Kong-based cosmetic retailer and how its management has sought to penetrate the local market and their level of success in doing so.

Sasa International Holdings Limited, or more commonly known as Sasa, was established in 1978 and has since become one of the most well-known and influential beauty retailing groups in Hong Kong and Asia. Among the strength of the group includes an expansive brand as well as a wide retail network. Sasa ranges over 600 brands of products in various categories which include skin care, perfumes, and general makeups.

With the continuous integration of economy between Shenzhen and Hong Kong and the social lifestyle network of the Greater Bay Area making its impacts, Sasa's business growth and level of market response exemplify an ideal model of a Hong Kong-based entity. Sasa enhances its presence within the limits of Hong Kong by constantly changing its product range, scaling up the geography of its activities, and improving the quality of services offered at its points of retail.

Sasa stores in Hong Kong (*Source* Internet)

Sasa's largest market, the Hong Kong and Macau region, is now on the path of recovery owing to the resurgence of consumer spending post-COVID-19 and arrival of mainland Chinese tourists, coupled with the opening of stores in prime tourist markets. As disclosed in the last financial report of Sasa international, for the half year ended September 30th, 2023, the company's turnover increased by 38.3% amounting to HK$2.144 billion where the offline sales in Hong Kong and Macau surged by 64.9% and reached HK$1.624 billion. During this period, HK$102 million were reported as a net income of the company. This is in stark contrast to a loss of HK$133 million the company reported during the same period of the previous year.[16] This marks an improvement, moving from approximately a loss of 100 million Hong Kong dollars to a profit of the same sum.

Sasa observed new post-pandemic tendencies among mainland tourists who used to resort to mass shopping rather than the sea and ocean park tours. Now, they are more likely to go on culturally enriching tours around Ocean Park, museums, and the Kowloon Cultural District. Sasa pointed out that this change has been good for the business as its stores near the West Kowloon station. Sasa continued to adjust its store portfolio and business model to better fit this new wave of the tourism trend.

Data from the 2023–2024 interim report shows that the Hong Kong and Macau Special Administrative Region (SAR) markets contributed the most to the group's revenue. The financial report reveals that offline sales in the Hong Kong SAR grew by 58.7% year-on-year, mainly benefiting from the return of tourists, exhibitions, and various consumption-boosting activities, as well as an increase in customer and VIP member loyalty. During the period, same-store sales grew by 64.9%, while the total number of transactions and the average transaction value per sale increased by 24.9% and 33.9%, respectively.[17]

[16] Sasa, 2024, https://doc.irasia.com/listco/hk/Sasa/interim/2024/intpresssc.pdf; Jiemian News, 2023, https://www.jiemian.com/article/10418071.html.

[17] Ibid.

The strategy of the Sasa business is still focusing on the Hong Kong and Macau area. However, although the total number of its brick-and-mortar shops in the mainland has shrunk, the average turnover per outlet increased. And Sasa is still rapidly entering service in the straights and many countries in Southeast Asia. This indicates that the improved situation in the tourism industry, in particular the Sasa Group revolution, has been beneficial.

In the dimension of enhancing sustainable tourism as well as promoting low-carbon emissions development, Sasa Group is a key player based in Hong Kong. Sustainability gives this company focus in all its operations. In 2021, Sasa launched a "Cosmetic Bottle Recycling" program that allows customers to return empty cosmetic containers to the stores with an incentive. Such containers are recycled into new plastic products, thus minimizing plastic wastes. In 2022, Sasa installed rooftop solar panel systems in some of its stores in Hong Kong as a way of utilizing clean energy and reducing their carbon footprint and power costs. The company reports that these systems, in particular, avoid around 150 tons of CO2 emissions per year.

Besides, Sasa partners with suppliers to package their products in recyclable or biodegradable materials like paper and plant-based bottles, an aspect that enhances the circular economy. A comprehensive waste mapping, segregation, and recycling system has been developed in stores, which includes recycling of paper, plastic, and metal, and reduces landfill and incineration impact. Such activities are complemented by Sasa's programs to engage staff in the restoration of nature such as cleanups of beaches and tree planting, increasing their environmental awareness and stewardship.

Furthermore, Sasa has been promoting paperless mobile payments, particularly in partnership with AlipayHK, to reduce paper waste and environmental burden. The widespread use of Alipay has facilitated cashless transactions, enhancing the payment experience for tourists while reducing the use of paper receipts, which has had a positive environmental impact.

15.2 Inclusive Finance: From E-wallet to One-Stop Service of Life Scenario (Alipay's Best Practice)

The three-year border closure due to the pandemic has brought the application and popularization of big data in mainland China into a new phase, particularly accelerating the use of mobile payments and digital economy scenarios. As the mobile payment market leader in the mainland, Alipay launched its Hong Kong version AlipayHK in 2017. As of 2019, AlipayHK had over 2 million users and was accepted by 50,000 offline merchants in Hong Kong. Thus, it allows us to say, roughly, that every third Hong Kong resident uses it, and every third local merchant accepts it.

As per a report from the research firm Cimigo, AlipayHK is the leading e-wallet in terms of market share at 33% in Hong Kong.[18]

15.2.1 AlipayHK's Convenient Cross-Border Payment Service for Shenzhen and Hong Kong

This part will focus on payments made through Alipay's cross-border payment service specifically in the GBA and how they have enabled transactions between Shenzhen and Hong Kong and thus aided in the region's economic integration.

In 2019, AlipayHK enabled seamless connectivity with all merchants in mainland China through the Alipay+ cross-border payment solution. After the reopening of the borders, many Hong Kong residents found that they could use AlipayHK to scan and pay directly in mainland China, making payments more convenient and encouraging more Hong Kong residents to take weekend trips to the mainland areas of the Greater Bay Area (GBA). As people resumed their "dual-city" and "multi-city" lifestyles in the GBA, Alipay data shows that the most popular intercity routes were between Guangzhou and Foshan, Shenzhen and Dongguan, as well as between Guangzhou and Shenzhen.

With the surge in shopping and consumption by Hong Kong residents traveling north after the pandemic, various initiatives tied to convenient payment scenarios have been introduced to stimulate consumption and

[18] Mobile Payment Network, 2019, https://www.mpaypass.com.cn/news/201903/06172651.html.

support economic recovery. Offering Hong Kong residents a seamless cross-border transaction and payment experience is part of the broader goal of inclusive finance. The Hong Kong Monetary Authority (HKMA) and the People's Bank of China are working to interconnect the two regions' fast payment systems—Hong Kong's Faster Payment System (FPS) and mainland China's Internet Banking Payment System (IBPS)—to facilitate real-time, small-value cross-border payments. Additionally, plans are underway to allow mainland branches of Hong Kong-registered banks to issue bank cards in the mainland.[19]

In the summer of 2023, AlipayHK and Alipay, in collaboration with the Hong Kong Retail Management Association, launched the "Hong Kong Happy Shopping Festival." AlipayHK and Alipay supported 130 brands in distributing shopping offers and experiences worth approximately HKD 1.5 billion. From the data provided by AlipayHK, in the last year, the Hong Kong merchants managed to gain the benefits of the increased presence created by the dual wallet system and more than 35 million of the coupons issued and redeemed by the residents of both regions were issued. This initiative aimed at rejuvenating retail activities in Hong Kong which created new impetus to the economy of the city.

Shenzhen was not left behind in this trend. Zhou Wei, deputy store manager at Hema Fresh Supermarket located at Huangting Plaza, Shenzen observes that, "After integrating AlipayHK, our transaction volume increased by 183% within a month. The store has also introduced various convenient new services, such as shuttle busses to border crossings and the launch of Cainiao delivery services. Customers can place an order by 5 PM, and have it delivered to their homes in Hong Kong the next day."

In June 2023, data from AlipayHK showed that in the second quarter, spending and transaction volume by Hong Kong customers in mainland China surged more than threefold compared to the previous quarter, indicating increasingly frequent cross-border travel. Travel and consumption data showed a steady upward trend over the first half of the year. Moreover, since the reopening of the border in January 2023 until the end of the first half, millions of people commuted across cities daily in the Guangdong-Hong Kong-Macao Greater Bay Area, with consumer

[19] HKSAR, 2024, https://www.policyaddress.gov.hk/2024/tc/policy.html.

spending recovering to 140% of the levels seen in the same period in 2019.[20]

There has been a remarkable boost in the need for travel-related expenditure as a result of the simpler mobile payment. Based on data held by Alipay's outland travel services, Ooi explained, that during the 2024 Chinese New Year, Alipay users exceeded transaction levels by 107% compared to the same time period in 2019, with 140% more spending in comparison with 2023.[21] These figures also indicate that along with the substantial recovery of outbound tourism, there is a shift in the global travel trends, and the use of mobile payments is growing and becoming more diverse. Hong Kong, Macau, other Southeast Asian markets, Japan, and South Korea were more favored. However, some large-distance countries such as Europe, North America, Australia, and even the Middle East started to show gradual recovery, meaning that the possibilities and reach of Chinese tourists' trip preferences are expanding as well.

In terms of spending, the top ten destinations favored by Alipay users were Hong Kong, Japan, Macau, Thailand, Malaysia, Singapore, South Korea, France, Australia, and Canada. These regions fit the main markets and mobile payment opportunities where Alipay+ has been focusing and developing into recently.

[20] AlipayHK, 2024, https://www.alipayhk.com/en/shoppers/; see also, Economic Observer, 2023, https://www.eeo.com.cn/2023/0812/601111.shtml.

[21] Alipay, 2024, https://www.alipay.com/index.html; see also, China News, 2024, https://www.chinanews.com.cn/cj/2024/02-17/10164591.shtml.

The rich ecosystem of Alipay's "Global Fun" and AlipayHK's "Mainland Deals Section" (*Source* Alipay and AlipayHK)

As for the transportation aspects, the electronic payment systems of Shenzhen and Hong Kong have started to last be integrated with each other. Promotional messages on Shenzhen Metro trains now advertise that "Hong Kong residents can use Alipay in Shenzhen and pay in Hong Kong dollars, whether for metro rides, taxis, supermarket shopping, or ordering food." These messages have also been translated into clear and simple Cantonese to make them easily understandable for Hong Kong residents. In Hong Kong, AlipayHK's advancements in smart travel reflect this growing cross-border collaboration. In 2021, MTR (Mass Transit

Railway) announced the launch of QR code ticketing services across all 93 stations, excluding the Airport Express, with an exclusive first-year partnership with AlipayHK. Passengers can use AlipayHK's "EasyGo" or MTR Mobile's "Ticket QR Code" to scan and ride the train. AlipayHK's "EasyGo" allows users to scan and board within just 0.3 seconds. In addition, it also provides offline as well as post-payment which are already part of business technology in mainland China, as elaborated by Alipay.

A localized team of AlipayHK has also designed a real-time refund model in line with the travel habits of people living in Hong Kong. Further, some island minibus routes have been linked up with AlipayHK. Also, AlipayHK has established a partnership with HopeGoo, which is a new travel booking platform that falls under Tong Cheng Travel, thus enabling AlipayHK users to buy train tickets for travel all over the mainland directly and in turn making it easier for people from Hong Kong who travel to the north.

On the eve of the 2024 New Year, MTR launched a 50% discount on adult tourist day passes for the first time through Alipay. Alipay and AlipayHK also teamed up again with the Hong Kong Retail Management Association to launch the "Hong Kong Festival of Splendor" following the success of the "Hong Kong Happy Shopping Festival." This event aimed to offer exclusive merchant benefits to Alipay and AlipayHK users, attracting both tourists and driving local consumption, injecting fresh momentum into Hong Kong's economy.

Additionally, customers who make payments through Alipay+ can earn green points, which can be redeemed for eco-friendly products or donated to environmental organizations. This motivates consumers to spend in a way that endorses "green spending" and allows them to engage in sustainable development.

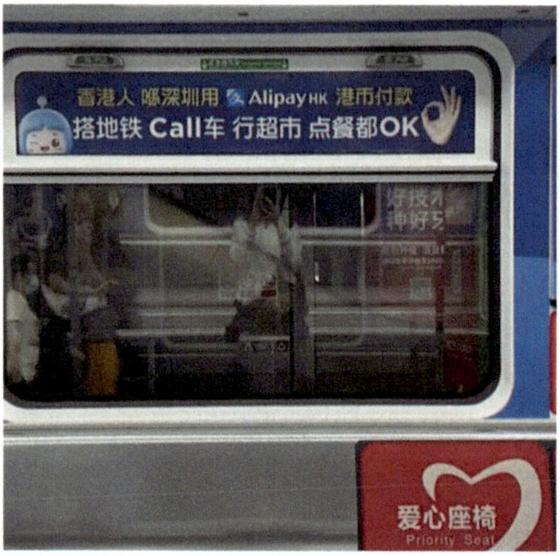

AlipayHK's Cantonese advertisements on Shenzhen Metro (*Source* Internet)

The number of cross-border vehicles between Shenzhen and Hong Kong keeps increasing. There has been an increase in the border crossings between the two regions now having 15 types of border crossings including 7 road ports and an increase in the ease of border clearance. The integration of ticketing based on QR codes for the public transport system has been achieved and the driving policy for Hong Kong residents has been provided by the Hong Kong vehicles going north.

Recently, Shenzhen announced a series of measures to optimize the border clearance environment, including the establishment of "temporary convenient financial service points" to assist Hong Kong travelers in using mobile payment applications such as Alipay, WeChat Pay, and UnionPay. These measures also detail the construction of rain and sun shelters in border areas. By the first half of 2024, data from Shenzhen's border clearance shows that the cumulative number of cross-border vehicles at land ports has reached 5.2441 million since the beginning of the year, marking a 64.31% increase compared to the same period last year (including the West Kowloon Station border crossing for the Guangzhou-Shenzhen-Hong Kong High-Speed Railway).[22]

[22] Shenzhen Ports, 2024, http://ka.sz.gov.cn/gkmlpt/content/11/11476/post_11476433.html#324.

Cross-Border land route between Shenzhen and Hong Kong (*Source* Internet)

In the field of culture and entertainment, we say that the Hong Kong Book Fair is a success after the lockdown due to the pandemic. Because of Hong Kong's unique background, he's able to sit in on exhibits behind the Taiwan Strait and the three regions every year. The occasion also drew various notable cultural figures including Yu Hua, Xu Zidong, Ma Boyong, Chen Chuncheng, and Xu Zhiyuan. Given the present times, when urban culture and the quest for self-fulfillment are constantly on the rise, the book fair stands out as one of the focal points in Guangdong-Hong Kong-Macao Greater Bay Area and to other foreign visitors as well.

The 2023 Hong Kong Book Fair was the first one held after the pandemic and it attracted 850,000 visitors and had 780 exhibitors. The 2024 fair is expected to follow suit and draw in 990,000 visitors of whom 40% to 50% are expected to be mainland tourists. According to Wen Liling, who is in charge of the Human Resources and Market Development Departments at the China Publishing Group, there was a higher turn up at the book fair during weekends than weekdays. She commented that this year‘s sales figures were not affected, and they attributed this to books associated to popular movies and TV Shows. She noted that approximately 40% of the attendees at their booth were China customers, who mainly bought local literature and history books, with a special interest towards the books of Ellen Cheung Yin, a local author. The mainland buyers are said to be one of the potential future book fair visitors.

At least 90% of the exhibitors were able to provide electronic payment systems, with 72% of the surveyed using electronic payment systems, during the period of the 2024 Hong Kong Book Fair. During the time of the sponsored event, Alipay donated somewhere between 4,000,000 and 5,000,000 HKD. This year alone, Hong Kong's Trade Development Council reported a yearly average of 912 HKD being spent per visitor, a 4.6% growth from last year, further indicating that the visitors spent over half their yearly budget on books over 15% of the people who attended the fair.

The gross consumption created is nearly worth 900 million HKD with almost million people having attended the fair, spending almost 900 HKD each. Such a large number of visitors and large consumptions in one region is a great addition to the economy of Hong Kong and is also a great step forward in terms of the cultural exchange between the 3 regions.

15.2.2 Diverse Localized One-Stop Services

This section will focus on the introduction of different localized services available with Alipay including the importance of the services in addressing the different demographics of the society in the promoting inclusivity.

The entry and expansion of Alipay in Hong Kong is driven by the close collaboration of two teams. Alipay has brought a wealth of knowledge and progressive ideas from its mainland operations, while the local team strives to work off its extensive understanding of the market as well as consumers to offer the most critical services to residents of Hong Kong. "Hong Kong is a key market for Alipay's global expansion. We will continue to fully support the development of AlipayHK by bringing the best products and technologies to Hong Kong and consistently providing unique value to users and merchants. In the future, Alipay will also introduce electronic wallets from around the world to Hong Kong, offering more business opportunities for local merchants," said Jing Xiandong, Chairman and CEO of Ant Group, during the first anniversary celebration of AlipayHK.

Apart from the simple payment services, Alipay extends its notable services in Hong Kong as followed[23]:

[23] AlipayHK, 2024, https://www.alipayhk.com/en/shoppers/.

Bill Payments: Users can pay for water, electricity, gas, and government bills through the AlipayHK app.[24]
Cross-Border Payments: AlipayHK supports cross-border payments in mainland China, Japan, Singapore, and other locations, making it convenient for Hong Kong users to shop in these places.
Transport Payments: AlipayHK is the first mobile wallet in Hong Kong to offer QR code pay and go transport fares to metro and other public transportation fares. AlipayHK is the first mobile wallet in Hong Kong to offer QR code pay and go transport fares to metro and other public transportation fares.
Remittance Services: Users can send international remittances through AlipayHK, currently supporting transfers to mainland China and the Philippines.
Consumer Rewards: AlipayHK has devised A. Point reward program which enables cash or discounts on all purchases made.
Investment and Insurance: Users can invest in mutual funds or purchase insurance through the AlipayHK app.

These services are integrated into one app, making AlipayHK not just a payment tool but a comprehensive financial services platform.

As a mobile electronic payment tool, AlipayHK collaborates with domestic banks in Hong Kong to provide financial services that include:

Collaboration with Ant Bank (Hong Kong): Users may use a mini-program and AlipayHK e-wallet to remotely register as an account holder of an Ant Bank (Hong Kong) and access eM + payment options. This effectively provides virtual bank accounts that are integrated with digital wallet accounts.

Collaboration with Hang Seng Bank: Users can register their Hang Seng credit cards on the AlipayHK platform to make local purchases that automatically deduct + FUN Dollars upon purchase. AlipayHK is the first in Hong Kong to offer credit card customers points redeeming services for purchases made through the platform.

Cross-Border Remittance Services: AlipayHK clients can conduct remittance transactions to the Philippines by sending money through Gcash wallet using blockchain technology while Standard Chartered

[24] Ibid.

Bank of Hong Kong does the end of day fund settlement and currency exchange so as to make remittances possible within a limited period.

Credit Card Linking Service: AlipayHK permits attaching credit cards from almost all regions and from Visa and MasterCard, allowing for fast and quick paying facilities for the clients. Credit cards can be used for various payment types on AlipayHK as provided by the clients.

These services not only improve the ease of payments for the residents of Hong Kong but also encourage the growth of the fintech sector in Hong Kong and the uptake of inclusive financial services.

Another interesting area where AlipayHK excels is where the city supports the ethnic minorities who use AlipayHK. It is particularly observed that the city is densely populated with the south and southeast Asian communities especially the Filipinos. For example, AlipayHK, which has entered the foreign domestic helpers' market, launched its remittance service to the Philippines over four years ago. Alipay claims that more than half of the people from the Filipino community who are present abroad mainly use AlipayHK to send remittances to the Philippines, six to eight times a month on average. For those who prefer Gcash e-wallets, Gcash allows them to save around 10 million Hong Kong dollars in transaction fees. This is good news for the Philippine community in Hong Kong, as it eases the pressure on their wallets on the whole.

According to Li Yongshi, General Manager of Ant Group Greater China International Business, Hong Kong AlipayHK's objective is to diversify and deepen the complexity of the financial lives of its residents through technology and other activities. This includes launching remittance services to the Filipino community from Hong Kong, working with local organizations to conduct mobile courses targeted at seniors who wish to use mobile payments, and providing students who register for specifically selected campuses with discounts, so as to lessen the financial pressures that they experience. The emergence of remittance services to Indonesia recently improves the ecosystem of inclusive finance that AlipayHK seeks to offer. Building on the success of its remittance services to the Philippines, AlipayHK now meets the needs of over half the foreign population in Hong Kong, further supporting the integration of immigrant communities.

15.2.3 Imagination of a Digital Society: Sustainable Hong Kong-Shenzhen Experiences

The final section will discuss the opportunities that exist within GBA's digital economy that will be vital in scaling sustainable development. The experiences of China will be relevant because Hong Kong and Shenzhen can provide a case study of how FinTech and digital economy can be used for sustainability in the region and what lessons can be learnt and applied in the broader GBA region.

With 11 cities, a total area of 56,000 square kilometers, a population of over 86 million, and an economic output surpassing 14 trillion yuan, the Guangdong-Hong Kong-Macao Greater Bay Area is gradually integrating to become a world-class bay area and a leading urban agglomeration. Horizontally, the economic total of the Greater Bay Area indicates that it has generated one-ninth of the country's economic output from less than 0.6% of the nation's land area. This economic output is comparable to that of Japan's Tokyo Bay Area, positioning it among the top four bay areas in the world.

In this region, there exists one country, two systems, three customs zones, three currencies, and three legal jurisdictions, making the construction of the Greater Bay Area unprecedented globally. A transfer from the MTR Admiralty Station to the East Rail Line allows for a journey to the Luohu or Futian border crossings in just about half an hour. When accounting for customs clearance time, the "one-hour living circle" between Shenzhen and Hong Kong is rapidly becoming a reality. As commuting and tourism infrastructure continue to be developed and improved, the integration of the Greater Bay Area is steadily advancing. This trend of integration further stimulates deeper cooperation between Guangdong and the Hong Kong-Macao regions in the digital economy.

Despite being highly developed, the transportation infrastructure in the Greater Bay Area is still undergoing enhancements aimed at further improving commuting efficiency. When unfolding the map of the Greater Bay Area, one can see the shape of the Lingdingyang Sea embedded into the land, dividing the coastline into eastern and western segments. Currently, the most challenging cross-sea engineering project in the world, the Shenzhen-Zhongshan Link, will open to traffic on June 30, 2024, effectively becoming the "artery" of the bay area. Spanning 24 kilometers from Shenzhen in the east to Zhongshan in the west, it will enable travel within 30 minutes in the future. This link is situated in the core area of the Greater Bay Area and integrates "bridges, islands, tunnels, and underwater interconnections," injecting robust momentum for the industrial upgrading and urban transformation of the cities on either side of the Pearl River Estuary.

Additionally, the Nanzhong Expressway's Wanqingsha Branch, currently under construction and expected to be completed within the year, will connect directly with the Shenzhen-Zhongshan Link. This will reduce the commuting time between Shenzhen and Guangzhou to just 20 minutes. Furthermore, the Greater Bay Area airport cluster has an annual passenger throughput capacity of over 280 million, while the port cluster handles over 85 million TEUs (twenty-foot equivalent units) annually.[25] The route network covers major cities around the world and key global trade ports.

While in respect to industrial clusters and technology reach GBA, Guangdong-Hong Kong-Macao Greater Bay Area is spear heading the development of a global technology innovation center and a national science center, a significant concentration of high-level talents as well as creating key laboratories, so at the same time few important laboratories are developed. The Greater Bay area has in excess of 75,000 national high-tech enterprises with several leading tech companies churning including Huawei, Tencent, BYD, Midea, and Gree.

The He Tao Shenzhen-Hong Kong Science and Technology Innovation Cooperation Zone, demarcated by the Shenzhen River, consists of a 0.87 square kilometer Hong Kong park and a 3.02 square kilometer Shenzhen park, symbolizing the continuous enhancement of the Greater Bay Area's technological innovation capabilities. On August 5, 2024, the first cross-border direct bus from Shenzhen Park in He Tao departed, passing through the "No. 1 Channel" and arriving at the Hong Kong Science Park in about half an hour, providing "point-to-point" convenient commuting services for Shenzhen and Hong Kong researchers.

> "The technologies developed in Hong Kong can find rich application scenarios in the mainland, and we also assist our mainland partners in expanding into overseas markets," a representative from the Hong Kong Applied Science and Technology Research Institute stated in a media interview. Five "world-class" universities, including the University of Hong Kong and the Hong Kong University of Science and Technology, have initiated over 13 key research projects at the He Tao Shenzhen Park. Many companies are achieving production value growth from zero to tens of

[25] Shenzhen Satellite TV News, 2024, https://www.sztv.com.cn/ysz/dsdb/szws/ssxw/.

millions through a collaborative innovation model of "Hong Kong experimentation + Shenzhen R&D + Pearl River Delta production + nationwide sales."[26]

In the ranking of technological clusters in the top 100 of the World Intellectual Property Organization's "2023 Global Innovation Index," the "Shenzhen-Hong Kong-Guangzhou" tech cluster has ranked second globally for four consecutive years.[27]

Table 6 Top S&T cluster by economy or cross-border region ranked among the top 100, 2023

Rank	Cluster name	Economy	Top applicant	Top organization
1	Tokyo–Yokohama	JP	Mitsubishi Electric	University of Tokyo
2	Shenzhen–Hong Kong–Guangzhou	CN/HK	Huawei	Sun Yat Sen University
3	Seoul	KR	Samsung Electronics	Seoul National University
4	Beijing	CN	BOE Technology	Tsinghua University
6	San Jose–San Francisco, CA	US	Google	Stanford University
12	Paris	FR	PSA Automobiles	Sorbonne Universite
20	London	GB	Nicoventures Trading	University College London
22	Munich	DE	BMW	Technical University of Munich
26	Amsterdam–Rotterdam	NL	TNO	Utrecht University
27	Taipei–Hsinchu	TW*	Hewlett-Packard	National Taiwan University
30	Tel Aviv–Jerusalem	IL	Yeda Research and Development	Hebrew University of Jerusalem
31	Moscow	RU	Samsung Electronics	Lomonosov Moscow State University
33	Singapore	SG/MY	A*Star	National University of Singapore
34	Tehran	IR	Ghanbari, Ahmad	University of Tehran
38	Stockholm	SE	LM Ericsson	Karolinska Institutet
42	Melbourne	AU	Monash University	University of Melbourne
48	Madrid	ES	LM Ericsson	Complutense University of Madrid
49	Zürich	CH	ETH Zurich	ETH Zürich
50	Milan	IT	Pirelli Tyre	University of Milan
51	Brussels–Antwerp	BE	Agfa	KU Leuven
52	Toronto, ON	CA	DH Technologies Development	University of Toronto
56	Bengaluru	IN	Samsung Electronics	IISC – Bangalore
59	Istanbul	TR	Sanovel Ilac Sanayi Ve Ticaret A.S.	Istanbul University
61	Copenhagen	DK	Novo Nordisk	University of Copenhagen
72	São Paulo	BR	Braskem	Universidade de São Paulo
73	Helsinki	FI	Nokia	University of Helsinki
76	Vienna	AT	Technische Universitat Wien	Medical University of Vienna
90	Warsaw	PL	Samsung Electronics	University of Warsaw
97	Basel	CH/DE/FR	DSM IP Assets	University of Basel

Source: WIPO Statistics Database, May 2023.

Notes: Tables in this section use ISO alpha-2 country codes, with the following additions: TW* = Taiwan, Province of China; IISC – Bangalore = Indian Institute of Science – Bangalore; TNO = Nederlandse Organisatie voor Toegepast Natuurwetenschappelijk Onderzoek. Economy labels were assigned to a cluster, when at least 1 percent of a cluster's output occurred in a given economy.

The "Shenzhen-Hong Kong-Guangzhou" technology cluster ranked second in 2023 (*Source* WIPO Statistical Database)

[26] Greater Bay Area Official Website, 2024, https://www.cnbayarea.org.cn/homepage/news/content/post_1255675.html.

[27] World Intellectual Property Organization (WIPO), 2023, wipo.int/export/sites/www/pressroom/en/documents/pr-2023–907-cluster-ranking.pdf.

Their cross-border movement and transactions has been made a lot more comfortable along with harnessing technology to the vast majority of Hong Kong residents. The Hong Kong residents are also attracted to other benefits in this city such as work, support, and opportunities for growth that are key to developing their businesses. This fascinating affirmation was confirmed by the General Office of Shenzhen Municipal Government in combination with the Bureau of Management Control. Not only do they actively do this, but they also organize to make online bookings as convenient as possible for their residents.

Their cross-border movement and transactions has been made a lot more comfortable along with harnessing technology to the vast majority of Hong Kong residents. The Hong Kong residents are also attracted to other benefits in this city such as work, support, and opportunities for growth that are key to developing their businesses. This fascinating affirmation was confirmed by the General Office of Shenzhen Municipal Government in combination with the Bureau of Management Control. Not only do they actively do this, but they also organize to make online bookings as convenient as possible for their residents.

Additionally, to facilitate the daily lives of Hong Kong and Macao residents in Shenzhen, the "iShenzhen" app's service area aggregates information on local cultural, sports, tourism, science education activities, and other cultural life updates, enabling users from Hong Kong and Macao to stay informed about relevant information in a timely manner. Moreover, with the capabilities of the Guangdong Province Unified Identity Authentication Platform, the "iShenzhen" app supports real-name authentication for two types of documents: the travel permit for entering the mainland and the residence permit for Hong Kong and Macao residents. Once authentication is completed, users can process related business online.

The Shenzhen municipal government has developed a comprehensive promotion plan covering various aspects, including consumer supply, customs facilitation, promotional activities, overall marketing, and rights protection. This initiative is based on the orientation aimed at promoting Shenzhen and Hong Kong cross-border consumption integration—enhancing the economic dynamism of the area of the two regions.

For the entire Greater Bay Area, the widespread adoption and innovative iterations of the digital economy and fintech are crucial elements driving rapid development and enhancing international competitiveness, both at the enterprise (B2B) level and the consumer (B2C) level.

This involves coordinated collaboration among multiple local governments and reflects the economic needs and daily choices of the broader population. On an international scale, this represents an unprecedented economic achievement. Currently, the active development momentum of the Greater Bay Area provides the world with a sustainable and actionable Chinese experience. The Greater Bay Area's "one-hour living circle" not only reflects daily life scenarios but also embodies a promising picture of technological innovation, economic integration, and complementary advantages among multiple regions.

The FinTech integration in the GBA, particularly between Hong Kong and Shenzhen, has profound implications for sustainable travel and inclusive development. It promotes a cashless economy, enhances consumer protection, and fosters financial inclusion. More impacts are expected such as reinforcing the economic vibrance of the region, enhancing the prospects for technology development, and advancing the sustainable development goals of the United Nations by 2030.

GBA global trade will consolidate the area's positioning enabling it to be a synergistic impact cumulatively creating stimulating factors for growth. GBA has already grown rapidly throughout the years and has become dynamic forcing the area in the forefront of integration for economy and society.

15.3 Conclusion

For the Hong Kong government, the expansion of fintech (such as Alipay) brings multiple benefits. Firstly, it promotes the development of fintech, further consolidating Hong Kong's position as an international financial center. Additionally, by introducing more payment methods and financial services, the government can effectively stimulate economic growth and enhance overall economic activity. This not only supports the government's digitalization policies and smart city initiatives but also encourages the promotion of cashless payments, thereby reducing the management costs associated with cash transactions. Moreover, as digital payments become more prevalent, the government can minimize paper transactions and lower carbon emissions, effectively aligning with green policies and contributing to the achievement of sustainable development goals. The expansion of fintech thus becomes a significant driving force for the Hong Kong government's digital transformation and green development, enhancing its global competitiveness and regional cooperation potential.

For Hong Kong merchants, adopting digital payments offers numerous advantages, particularly in expanding sales opportunities and attracting more customers. Through different e-payment systems, merchants can cater to a larger pool of both local and international buyers, especially mainland tourists and the younger generation. Also, electronic payments are usually quicker and cheaper than the traditional way of payments, therefore, lowering transaction costs and helping the merchants reduce their operational costs. In addition, these modern payment methods are able to reduce transaction times and improve the customer's payment experience, which results in a higher level of customer satisfaction and repeat business. For smaller businesses, digital payment systems have an added advantage of improving their market competitiveness while at the same time providing chances of working with bigger platforms, thus increasing foot traffic and creating vibrancy in the market.

For Hong Kong residents, AlipayHK has brought a more convenient and efficient payment experience, playing a vital role in daily local life and cross-border consumption. Cashless transactions help in making on the go shopping, travel, and entertainment less complicated and more resource-efficient for the residents. Moreover, consumers through loyalty rewards can enjoy more offers and benefits, adding pleasure and value to their spending. A green consumption reward program has also been initiated by AlipayHK where residents also get monetary rewards for using green products or services which helps the local tourism and greening activities. This not only improves the consumer experience but also helps ensure that sustainability and conscious consumption are pursued, thereby, enhancing the quality of life and environmental protection activities of residents. Therefore, it has been the case that digital payments have now transformed and upgraded the convenience level of the regions.

In summary, fintech plays a crucial role in promoting economic integration, sustainable development, and inclusive growth within the Guangdong-Hong Kong-Macao Greater Bay Area (GBA). Through cashless payments and digital services, fintech not only facilitates cross-border consumption and economic activities but also promotes the development of green tourism and cultural events. It supports the United Nations Sustainable Development Goals (SDGs), particularly those related to responsible consumption and production (Goal 12) and sustainable cities and communities (Goal 11). Hong Kong's best practices demonstrate the potential of fintech to enhance financial inclusivity and strengthen economic resilience. Moreover, fintech helps achieve environmental objectives in the GBA by reducing paper transactions and optimizing resource usage, aligning with environmental, social, and governance (ESG) standards. As the digital economy in the Greater Bay Area continues to grow, fintech will remain a vital driving force propelling the region toward a more sustainable, interconnected, and inclusive future.

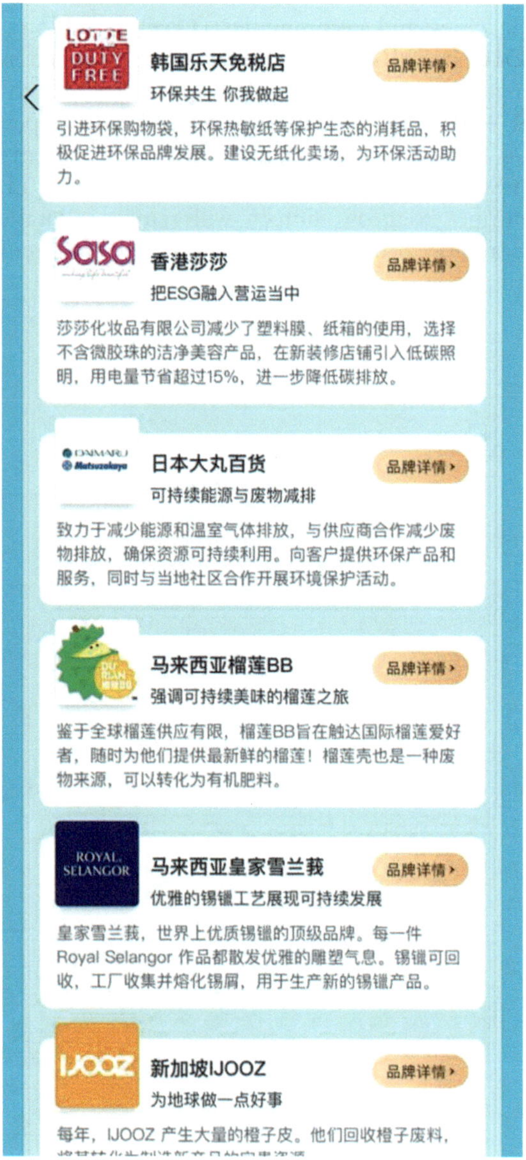
韩国乐天免税店
品牌详情 ›
环保共生 你我做起
引进环保购物袋，环保热敏纸等保护生态的消耗品，积极促进环保品牌发展。建设无纸化卖场，为环保活动助力。
香港莎莎
品牌详情 ›
把ESG融入营运当中
莎莎化妆品有限公司减少了塑料膜、纸箱的使用，选择不含微胶珠的洁净美容产品，在新装修店铺引入低碳照明，用电量节省超过15%，进一步降低碳排放。
日本大丸百货
品牌详情 ›
可持续能源与废物减排
致力于减少能源和温室气体排放，与供应商合作减少废物排放，确保资源可持续利用。向客户提供环保产品和服务，同时与当地社区合作开展环境保护活动。
马来西亚榴莲BB
品牌详情 ›
强调可持续美味的榴莲之旅
鉴于全球榴莲供应有限，榴莲BB旨在触达国际榴莲爱好者，随时为他们提供最新鲜的榴莲！榴莲壳也是一种废物来源，可以转化为有机肥料。
ROYAL SELANGOR
马来西亚皇家雪兰莪
品牌详情 ›
优雅的锡镴工艺展现可持续发展
皇家雪兰莪，世界上优质锡镴的顶级品牌。每一件Royal Selangor 作品都散发优雅的雕塑气息。锡镴可回收，工厂收集并熔化锡屑，用于生产新的锡镴产品。
IJOOZ
新加坡IJOOZ
品牌详情 ›
为地球做一点好事
每年，IJOOZ 产生大量的橙子皮。他们回收橙子废料，

新加坡康福德高
品牌详情 ›
「绿色交通」新加坡环境成就奖
康福德高集团于2022年荣获新加坡环境“绿色交通”类别成就奖。该奖项是新加坡环境理事会对康福德高集团在交通领域去碳化和支持新加坡汽车电气化以及运营节能等方面所作努力的认可。
KING POWER
泰国王权免税
品牌详情 ›
泰国力量
这是一个支持和发展社会的项目，包括体育、音乐、教育，将社区级产品推向全球出口市场等方面。 王权集团热衷于成为泰国社会创造美好和伟大事业的一部分。
跨境游超值好券
香港
新加坡
泰国
马来西亚
4,309人已领
ELEVEN
7-Eleven
维他冷泡无糖...
维他冷泡无糖茶饮...
领取
8,279人已领
sasa
莎莎
立减10港元
满70港元可用
已抢完
2,415人已领
萬寧 mannings
万宁
美贊臣产品满...
适用于万宁香港及...
领取
3,522人已领
萬寧 mannings
万宁
森田药妆产品...
适用于万宁香港及...
1,196人已领
EMPEROR
英皇珠寶
英皇珠宝
8 折优惠
购买定价足金货品...

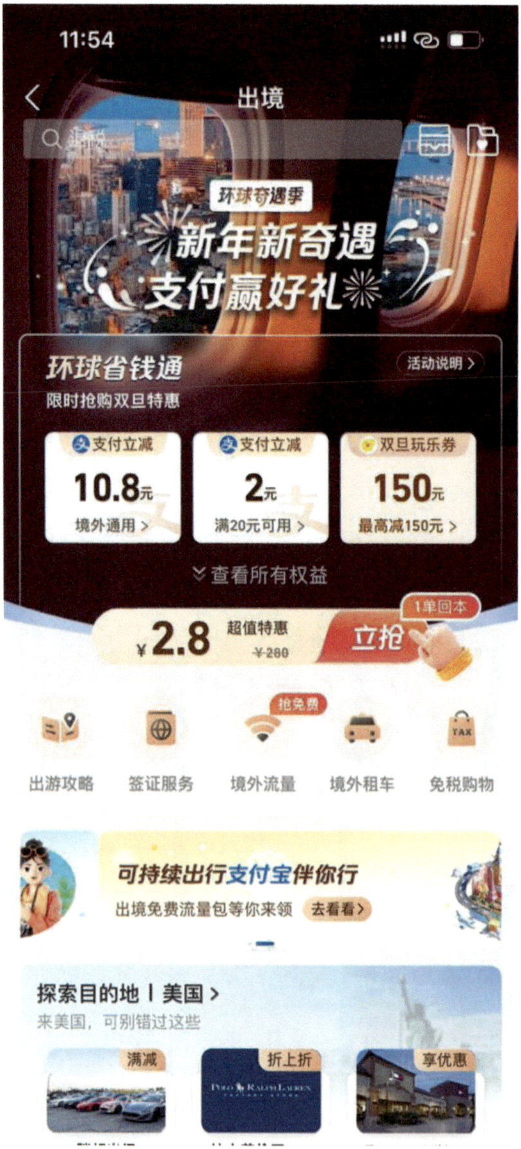
11:54
出境
环球奇遇季
新年新奇遇
支付赢好礼
环球省钱通
活动说明 >
限时抢购双旦特惠
支付立减
10.8元
境外通用 >
支付立减
2元
满20元可用 >
双旦玩乐券
150元
最高减150元 >
查看所有权益
1单回本
¥2.8
超值特惠
¥280
立抢
抢免费
出游攻略
签证服务
境外流量
境外租车
免税购物
可持续出行支付宝伴你行
出境免费流量包等你来领
去看看>
探索目的地 | 美国 >
来美国，可别错过这些
满减
折上折
享优惠

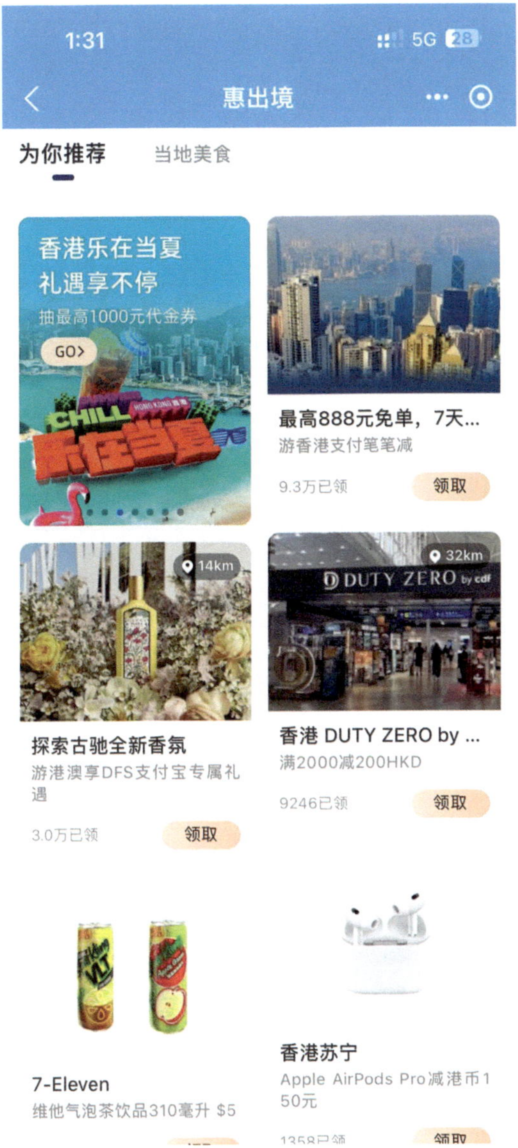
1:31
5G 28
惠出境
为你推荐
当地美食
香港乐在当夏
礼遇享不停
抽最高1000元代金券
GO>
CHILL
乐在当夏
最高888元免单，7天...
游香港支付笔笔减
9.3万已领
领取
14km
探索古驰全新香氛
游港澳享DFS支付宝专属礼遇
3.0万已领
领取
32km
DUTY ZERO by cdf
香港 DUTY ZERO by ...
满2000减200HKD
9246已领
领取
7-Eleven
维他气泡茶饮品310毫升 $5
香港苏宁
Apple AirPods Pro减港币150元

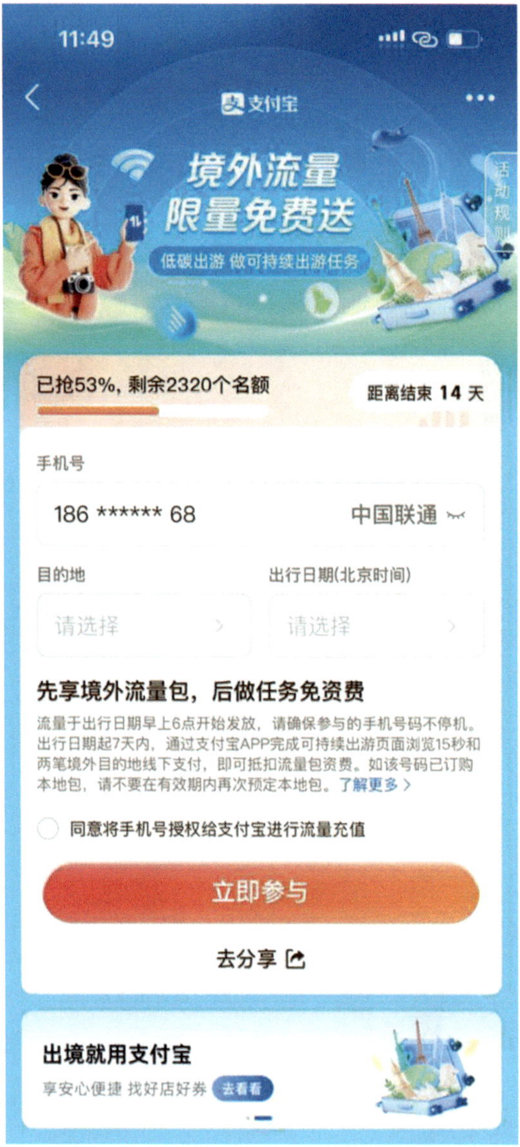
11:49
支付宝
境外流量
限量免费送
低碳出游 做可持续出游任务
活动规则
已抢53%，剩余2320个名额
距离结束 14 天
手机号
186 ****** 68
中国联通
目的地
出行日期(北京时间)
请选择
请选择
先享境外流量包，后做任务免资费
流量于出行日期早上6点开始发放，请确保参与的手机号码不停机。出行日期起7天内，通过支付宝APP完成可持续出游页面浏览15秒和两笔境外目的地线下支付，即可抵扣流量包资费。如该号码已订购本地包，请不要在有效期内再次预定本地包。了解更多 >
同意将手机号授权给支付宝进行流量充值
立即参与
去分享
出境就用支付宝
享安心便捷 找好店好券
去看看

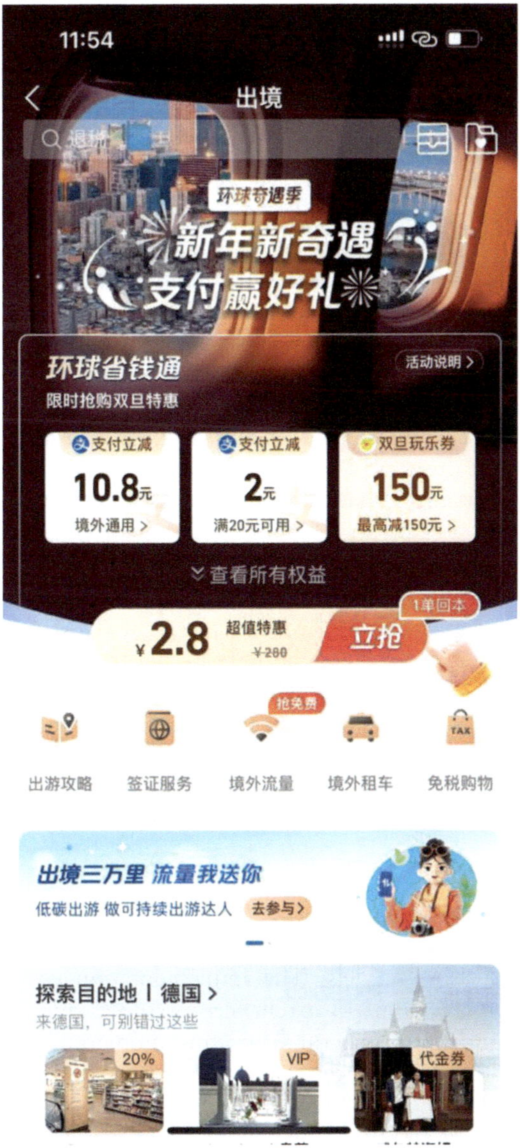
11:54
出境
环球奇遇季
新年新奇遇
支付赢好礼
环球省钱通
活动说明 >
限时抢购双旦特惠
支付立减
10.8元
境外通用 >
支付立减
2元
满20元可用 >
双旦玩乐券
150元
最高减150元 >
查看所有权益
1单回本
¥2.8
超值特惠
¥280
立抢
抢免费
TAX
出游攻略
签证服务
境外流量
境外租车
免税购物
出境三万里 流量我送你
低碳出游 做可持续出游达人
去参与>
探索目的地 | 德国 >
来德国，可别错过这些
20%
VIP
代金券

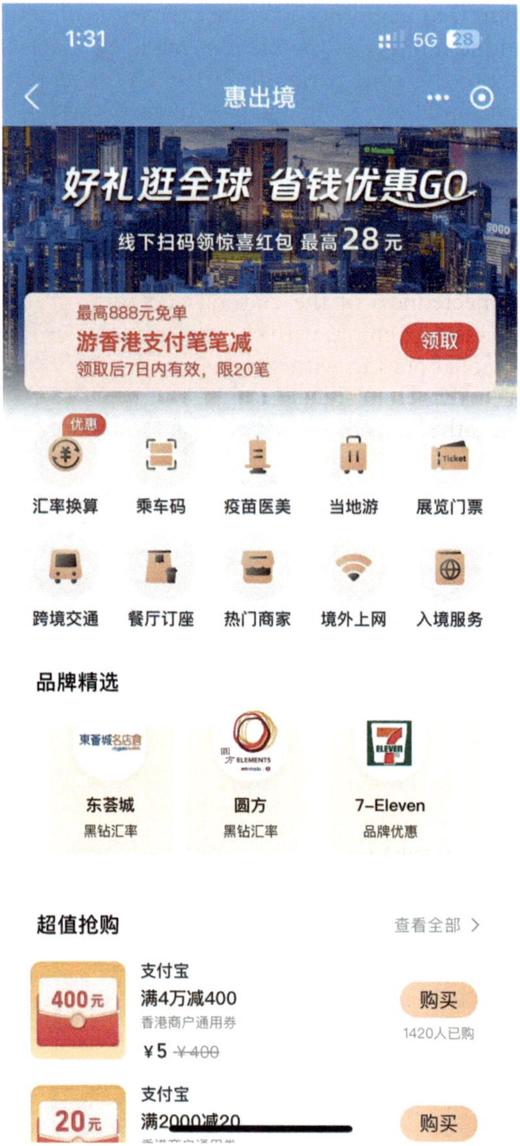
1:31
5G 28
惠出境
好礼逛全球 省钱优惠GO
线下扫码领惊喜红包 最高28元
最高888元免单
游香港支付笔笔减
领取后7日内有效，限20笔
领取
优惠
汇率换算
乘车码
疫苗医美
当地游
展览门票
跨境交通
餐厅订座
热门商家
境外上网
入境服务
品牌精选
東薈城名店倉
东荟城
黑钻汇率
ELEMENTS
圆方
黑钻汇率
7-Eleven
品牌优惠
超值抢购
查看全部 >
400元
支付宝
满4万减400
香港商户通用券
¥5 ¥400
购买
1420人已购
20元
支付宝
满2000减20
购买

低碳出行，美丽脚印
携带非一次性水杯/牙刷等可重复物品，搭乘公共交通，每一次旅行都能减少碳足迹。
支持本地发展：文化保护&经济共荣
深度体验，海外地道风情
打卡当地特色商家、参观艺术馆、传统表演...我们的每一站都有助于保护和传播当地独有的文化遗产。
支持小微，支付宝温暖相伴
选购小微商户和手工艺人精心制造的产品和服务，用支付宝完成每一次意义非凡的支付，点亮小微商户的希望之光。
来看品牌的可持续发展故事
SHISEIDO
资生堂免税
品牌详情
环保之旅 让美资生
资生堂将“美力创新让世界更好”当作企业使命。以成为在百年后仍然持续闪耀、不断发展的公司为目标，一直致力于解决社会问题、减轻环境负荷。
cdf 中免国际
cdfi 中免国际
品牌详情
环保之旅 空瓶换购计划
cdfi中免国际全球cdfBeauty及DUTY ZERO by cdf将设立回收箱，回收美妆产品空瓶和酒瓶，顾客每次回收均可获购物优惠，以鼓励支持环保废料再生。
GALAXY
澳门银河
品牌详情
可持续发展与环保并行
「澳门银河™」作为全球首屈一指的奢华综合度假城，我们致力为社区及环保事业作出贡献。我们把可持续发展纳入经营策略，并透过使用针对节能的专门设备和建立管理

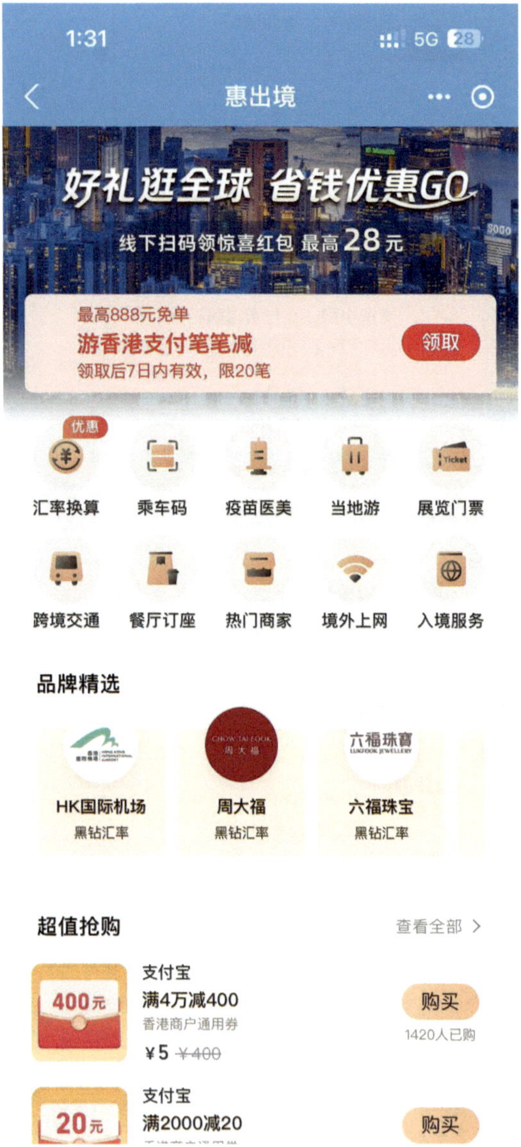
1:31
5G 28
惠出境
好礼逛全球 省钱优惠GO
线下扫码领惊喜红包 最高28元
最高888元免单
游香港支付笔笔减
领取后7日内有效，限20笔
领取
优惠
汇率换算
乘车码
疫苗医美
当地游
展览门票
跨境交通
餐厅订座
热门商家
境外上网
入境服务
品牌精选
HK国际机场
黑钻汇率
周大福
黑钻汇率
六福珠宝
黑钻汇率
超值抢购
查看全部 >
400元
支付宝
满4万减400
香港商户通用券
¥5 ¥400
购买
1420人已购
20元
支付宝
满2000减20
购买

11:40
惠出境
全部
小程序
生活号
理财
市民服务
惠出境
德国
法国
韩国
中国澳门
去遇见一场奇遇
支付宝
反馈与投诉
低碳出行赢免费流量包
你的出境游我来包
去参与
支付宝惠出境
品牌
小程序
出境服务及行中优惠，尽在惠出境!
超优汇率
线下支付笔笔省
支付宝退税
退税快，更便捷
官方账号
小程序
支付宝惠出境
生活号
出境游福利官
支付宝（中国）网络技术有限公司
惠出境 - 视频
大马免签啦！！
出境就用支付宝惠出境
打车点餐超方便!

11:42
可持续出游
惠出境
可持续出游
免费领出境流量包
活动时间：2023.12.14-12.31
境外流量限时免费领
可持续出行才够酷
去领取
可持续出游付款码皮肤 让支付更显态度
规则
付款码皮肤有效期，领取后30天有效
可持续出游
可持续出游付款码皮肤
1积分
去兑换
我们倡导的可持续出游新风尚！
惠行天下：移动支付&低碳出行
海外旅行，轻松一扫
选择移动支付，拥抱海外便捷之旅。划动屏幕，
让小程序点单成为您海外轻松出行的新伙伴。

CHAPTER 16

FinTech for Sustainable Travel, World Heritage, and Economic Transitions for Decent Work and Economic Growth in Macau

16.1 Strong Economic Recovery Driven by the Digital Economy

16.1.1 Localization of the Payment Ecosystem: Alipay+'s Role in Building the Digital Economy in Macau

During the pre-pandemic times, being one of the famous and most visited tourist destinations, Macau underwent serious challenges in terms of visitor influx due to traveling bans and health-related issues. Nevertheless, looking at the post-pandemic era, it is evident that Macau's tourism sector has recovered exceptionally well and has been quite agile making sure that the economy thrives. In this case, the primary contributing factors to the recovery have been the strategic growth and incorporation of Alipay+ in the digital economy of Macau. Due to enabling hassle-free transactions, Alipay+ has improved the experience of tourists and at the same time contributed significantly in assisting the revival of the tourism industry and comprehensively the economy of Macau.

J. Sun, *Sustainable Financialization*, Sustainable Development Goals Series, https://doi.org/10.1007/978-981-96-3580-1_16

To aid in the recovery and support the economic as well as the digital shift for Macau, Alipay+ has resorted to a number of initiatives most notable of which is working with Macau Pass which has recently been authorized to provide acquiring services for Alipay+. The partnership lets local merchants to expand payment options and accept a range of international e-wallets including Gcash from the Philippines Touch 'n Go e-Wallet from Malaysia and TrueMoney from Thailand creating a global digital payments ecosystem. As an added benefit, Alipay+ also partnered with Macau merchants to assist in providing attractive discounts for travelers through AlipayHK, one of the dominant e-wallets in Hong Kong, in order to promote local expenditure.[1]

Acquiring services of Macau pass (*Source* Fintech News Hong Kong)

[1] Alipay+, 2023, https://www.alipayplus.com/news/detail/Macau_AlipayHKoffers.

Furthermore, Alipay+ has expanded its cross-border payment solutions, allowing Macau residents using the Mpay e-wallet developed by the Macau Pass to make payments overseas by switching to the Alipay+ payment code. Busses in Macau have adopted Alipay+ and any visitor on tour is able to pay for the ride with a mobile phone without the fuss of looking for change. Users of Alipay+ including local Macau users, Mpay's residents of the 3A, 3B, and JR tiers have the cross-border payment feature and are able to make payments easily while away from the country solving cross-border payment problems.

Moreover, Mpay e-Wallet shall seek further synergies with Alibaba Group and Ant Group in a bid to extend to the e-commerce and digital media entertainment sectors. As Sun Ho, Chairman and CEO of Macau Pass stated: "Comprehensive and innovative digital solutions is the key to improve the economic growth and the market performance in the transformation of Macau's service industry."

Since January 20th, 2023, AlipayHK users will be able to redeem the highlighted local Macau stores and exclusive offers for tourists like the digital coupons under the Alipay+ rewards function. This is a joint effort involving multiple players which include the Bureau of Economic and Technological Development of Macau, five local business chambers of commerce, Macau Pass, Ant group, and over one hundred small and medium enterprises in Macau.[2]

[2] Ibid.

Macau special offer section in AlipayHK ("澳門優惠專區") (*Source* AlipayHK)

The incorporation of AlipayHK in Macau is central to the role of the digital economy in stimulating the recapture of the tourism market of Macau. Beginning 2023 the movement of mainland and Hong Kong people to Macau began to enter into an active phase. With the latest trends in the post-pandemic economy and the normalization of international flights being slow to be achieved, Macau has emerged as the ideal target for mainland residents once as their appetite for overseas travel requirements had yet to be met.

Also, as noted by the United Nations World Tourism Organization (UNWTO), "accessibility" also stands as one of the significant determinants for the growth of cross-boundary tourism. Since the launch of the Hong Kong-Zhuhai-Macau Bridge, the time and cost involved when traveling to Macau have greatly improved when compared to traveling over sea routes, hence the increase in visitors from Hong Kong apart from when resources were scarce for the three years due to pandemic.

AlipayHK, which has the largest number of active users in Hong Kong, is accepted by nearly 90% of merchants in Macau. Ng Wah Wai, President of the Industry and Commerce Association of Macau, stated: "With the support of Alipay+, this online initiative by AlipayHK, with 3.3 million active users, will help promote local Macau merchants and increase their visibility, positioning Macau as an attractive destination for short- and long-term visits from Hong Kong travelers.[3] This will further explore the potential of online promotion for Macau's unique merchants, enhance the appeal of Macau's tourism consumption, and encourage second and third visits to Macau, creating new momentum for overall economic growth in Macau."

Venetia Lee, General Manager of Ant Group Greater China International Business, stated: "We are pleased to collaborate with the Macau government and five local chambers of commerce to attract more Hong Kong tourists to Macau and stimulate local tourism consumption. This is a recognition of the valuable partnership with AlipayHK."[4]

Not only does this makes the payment and shopping process of the travelers easier, it also signifies the enhancement of cross-border business and tourism opportunities for global merchants. With the onset of Alipay+'s new ecosystem of digital payment solutions, it has played an

[3] Ibid.

[4] Ibid.

integral role in reviving both Macau's tourism and the development of the broader digital economy.

16.1.2 *Partnering with Digital Platforms for Post-Pandemic Tourism Revival*

In 2017, Alibaba signed a strategic cooperation agreement with the Macau SAR government, covering various areas such as tourism, leisure shopping, digital payments, and smart city initiatives, fostering a strong partnership. According to Fliggy, a platform under Alibaba, the annual growth of visitors to Macau over the three years before the pandemic reached an impressive 75%. One of the key attractions for tourists was the emerging trend of leisure travel to urban destinations.

In June 2020, for instance, the MGTO in collaboration with Fliggy and Alipay launched an initiative for the recovery of the tourism sector. They provided tools such as live streaming, pre-sold travel bundles, and discounts in essence hoping to "stock" potential travelers for the time when border restrictions were removed—which was the case in this period of time due to COVID-19. Macau's tourism chief, Maria Helena de Senna Fernandes has also appeared in a live stream on Fliggy trying to promote local tourism. She said, with appreciation for the systematic work of the SAR government and its people, the COVID-19 was reasonably well contained and there was an assurance that these results would be sustained in campaigning for an environment that was healthy, safe, and entirely suitable for a holiday for people around the globe.

Virus and lockdowns presented mountains of difficulties to the tourism sector and the Macau SAR Government was mobilized at all levels to constantly come up with ways to support the sector. One such program that has been running for three years now is known as "Stay, Dine, and See Macau." This initiative brought a lot of relief to the industry, facilitating the businesses of travel agents, hotels, and restaurants while providing sustenance to workers such as travel agents, tour guides, drivers, and so on. For instance, such packages were continuously purchased by the government and offered to the residents to help replace foreign demand with local demand in an attempt to rehabilitate the impacted tourism and related industries.

The program has come a long way since its inception and over 440,000 people have signed up for it, offering 103 local travel routes as well as organizing tour groups that have almost 7000 people in each group.

The hotel experience part of the program was also quite popular, with approximately 75,000 hotel packages purchased—more than 180,000 in hotel stay creating an overall profit of over MOP 190 million. A new type of program focused on students known as "Student Study Tours" was launched this year. More than 29,000 students participated and 525 tours were completed, and these varied tour routes have been greatly appreciated by the people.

As part of Macau's post-pandemic restoration strategy, the focus was on cooperative engagements with social media and digital platforms with a view to increase traffic and stimulate tourism. As explained by the Macau Government Tourism Office (MGTO), a set of marketing plans targeting various regions were employed in the modern art of marketing to boost exposure and experiences. All these strategies involved creating 24 different accounts on social media like WeChat, Weibo, Xiaohongshu, Douyin, YouTube, TikTok, Facebook, Instagram, Twitter, Line and Kakao Talk with a total of 6.13 million followers. Furthermore, around 100 travel influencers from all over the world were invited by MGTO to experience and promote Macau's new ways to experience travel—"tourism+" concept.

In order to attract even more international attention to Macau, the IMTO appealed to users to share clips of their Macau holidays as part of their travel promotion, which has brought initial results. In addition, MGTO has improved its relations with neighboring territories and foreign organizations in order to boost overseas tourist visits to the region as a whole. For instance, Macau and Zhuhai and Hong Kong locations have worked exhibits for "multi-city destination" travel packages, allowing travelers to visit more than one city in a single vacation. Such collaboration did not only enhance the appeal of Macau but also gave tourists more options when it comes to travel planning.

Macau travel guides on the Chinese social media platform Xiaohongshu (*Source* Xiaohongshu)

The Macau Government Tourism Office (MGTO) is reported to have cooperated with mainstream media such as BBC, CNBC, and The Economist as well as major e-commerce sites, celebrities, and KOLs to champion the tourism and cultural resources of Macau. In October 2023, MGTO worked in partnership with popular travel booking websites such as TripAdvisor and Expedia to enhance Macau themed pages on their platforms. These included posting articles that featured innovative ways of exploring Macau, advertising, and promotional travel packages.

Furthermore, Macau's recovery efforts include developing the "Smart Itinerary Planner" and "Smart Tourist Flow" apps, which predict visitor traffic at 117 tourist attractions in Macau. These apps also provide information about specialty shops, cultural and creative stores, and classic brands, allowing users to design personalized itineraries based on their interests.

Such practices assist in showcasing the attractions in Macau and reaching out to potential visitors and providing them with useful real-time information and travel advice. Moreover, they assist in targeting wider audiences and facilitating smooth travels. The combination of digital platforms with classic tourism services has also improved the effectiveness and attractiveness of the tourism emerged in Macau. Social media has become a major driver of travel behavior changes, which is an important area for tourism promotion.

16.2 Sustainable Tourism: Creating the Most Satisfactory Tourist Destination

16.2.1 Leveraging the Local User Base: Building a Deep Travel Brand

16.2.1.1 Comprehensive Layout from Payment Ecosystem to Digital Economy Transformation

During the last Policy Address of the Macau SAR Government in 2023, one of the highlighted goals was to develop an integrated tourism and leisure mechanism at active mode. This industry is of great significance and competitive strength to the economy of Macau SAR in the short run vision as the medium-term perspective would be stable growth of the tourism sector recovering the economies. To the extent possible risks are managed, there are measures being instituted to steadily restore tourism and economic activity including the re-launch of e-visas for mainland residents wishing to visit Macau and recommencing packaged tours of the region.

Macau will now advance strategies to market and enhance its image as a safe tourist destination. Further enhancement of a model that integrates online traffic building up and offline experience, big data technology application, and composite community tourism products development, aimed at broadening domestic and external tourism markets, will be also the priorities. Major cities in the mainland will be targeted through promotional campaigns that would expand the market for the post-covid recovery of the industry.

In addition, it will also enhance the development of the "Tourism + " concept, the integration of the tourism industry with gastronomy and events, health, technology as well as developing tourism for educational purposes.

Alipay+ is widely present in these local tourism consumption activities. In March 2022, AGTech, a member company of Alibaba Group, completed a HK$778 million acquisition of Macau Pass, further integrating the payment ecosystem into Macau's digital economic transformation.

AGTech is a China and Macau-based company that provides electronic payment, lottery, mobile gaming and entertainment, marketing technologies, and non-lottery hardware supply. It is also exclusive in providing lottery-related services and products as it is a member company of Alibaba Group and Ant Group. Macau Pass, a leading electronic payment company in Macau, has a large number of users and a good payment network.[5]

As stated before, the Alipay+ already connected systems of pay in Macau and a wider tourist market. Therefore, with the purchase of Macau Pass in 2022 AGTech already got foothold into the market of Macau, especially in the field of fintech industry. Using Macau Pass' local user base AGTech will try to develop further use of digital payments in retail stores and enhance its presence in the market of Macau. Also, such advantages of its long time operation in Macau market allow Macau Pass to provide AGTech with data about users and the market, which helps to meet customers' needs in Macau more effectively.

After the buyout, AGTech immediately connected Alipay+ to the Macau Pass payment system. With the new app, Macau Pass users can pay using multiple digital wallets like Alipay and WeChat Pay. This improves user payment experience and also provides more payment alternatives to merchants. Other than that, Macau Pass is already in existence and is

[5] AGTech, 2023, https://media-agtech.todayir.com/2024032217240277861110994l_en.pdf.

widely accepted across different industries such as public transport, supermarkets, and restaurants in Macau. There's also QR code-based payment function through Macau Pass' platform where users can conduct transactions without having to carry cash or even cards since it's via the app.

To make cross-border transactions, the Macau Pass app can easily convert and make payments in several currencies which is convenient for foreign tourists shopping in Macau. Foreign Tourists from Mainland China and several other countries can use Macau Pass since the e-wallet is within the app hence users don't have to exchange currencies. Alipay+ at the same time is also promoted globally with AGTech's partnership with Ant Group allowing merchants in Macau to accept payments from all tourists enhancing Macau's tourism business.

Alipay+ supports the local economy of Macau in the diversification process by establishing a digital payment ecosystem. Speaking of the context of focusing on the recovery of tourism as one of the primary economically benefitting, there are great opportunities for future growth for the digital economy and all the industries adjacent to this one. This includes the generation of multitudes of employment opportunities and technological transfer and advancements.

Alipay+ enables the use of e-wallets like Alipay, AlipayHK, MPay, Gcash, Touch 'n Go, and other Alipay+ supported e-wallets in various Asian countries and regions so that international visitors can easily make payments while in Asia. This brings value and satisfaction to all visitors, while increasing the number of visitors to Macau merchants by making it possible for them to reach more customers. D-store digital business solutions are also offered to the merchants by Alipay+ this enables their operations to be carried out digitally in a seamless manner through mini-programs or H5 pages for making payments and carrying out various operations, and marketing to improve brand effects.

Alipay+ boasts strong digital marketing capabilities by collaborating with local Macau brands and merchants as well. The platform offers numerous promotions and coupons, using methods such as live-streaming and online marketing to promote local businesses. These efforts are helping Macau's tourism industry find new growth points and transformation opportunities in the post-pandemic era. Such initiatives not only advance the digital transformation of local merchants but also create a collective impact, establishing an interconnected digital payment ecosystem. This ecosystem helps Macau businesses build a stronger brand presence in the international market and improve customer loyalty.

The "Macau Discount Zone" on Alipay offers a variety of collaborations with local merchants and consumption coupons (*Source* Alipay)

16.2.1.2 *Comprehensive Tourism: Expanding Creative Culture Tourism*

According to the China Tourism Academy's 2023 Mid-Year Outbound Tourism Data Report, Macau ranked as the top outbound travel destination for mainland Chinese tourists in the first half of 2023.[6]Based on the findings of the Macau Government Tourism Office, in the first seven months of 2024, the number of visitors to Macau reached nearly 19.75 million visitors, marking an increase when compared to the previous period by 37% or more. Since the month of August, Macau has averaged about 124,000 visitors each day, hitting about 154,000 visitors on the second Saturday of August's month, and almost back to levels last observed in 2019.[7]

For almost all of the months of year 2024 except these second and third quarter months during June and May, the average occupancy in hotels in Macau reached levels of about 88.5%, increasing from 77.5%-70% in the previous year. This was only 2.6% lower than the same periods in 2019. With summer holiday in the month of July, hotel occupancy levels began recording rates of above 90%. Hotel industry officials in Macau reported that during summer, average room rates in three-to-four-star hotels are in the region of 1,000 MOP per night while five-star hotel rooms are also cheap as compared to the current prices in the respective neighboring areas. Predictably, occupation rates in the summer months are forecast to be at least 90%.[8]

[6] Ministry of Culture and Tourism of the PRC, 2023, https://www.ctnews.com.cn/baogao/m/content/2023-08/25/content_148560.html.

[7] Macau Government Tourism Office, 2024, https://www.Macautourism.gov.mo/zh-hans/.

[8] Macau Government Tourism Office, 2024, https://dataplus.Macautourism.gov.mo/Publication/StatisticalTables?lang=T.

酒店入住率及平均房價
Hotel's Occupancy Rate and Average Room Rate

二零二四年七月 July 2024

類別 Type	入住率 Occupancy Rate (%)			房價 Room Rate (澳門元MOP)		
	2024	2023r	相差 Diff.	2024	2023r	變動率 Change %
所有酒店 All type of Hotels	94.3	90.8	+3.5	1,382.1	1,433.0	-3.6%
五星級 5 star	94.9	89.8	+5.1	1,552.4	1,623.1	-4.4%
四星級 4 star	89.0	90.2	-1.2	959.6	909.6	+5.5%
三星級 3 star	97.2	97.4	-0.2	949.3	1,046.2	-9.3%

二零二四年一至七月 January - July 2024

類別 Type	入住率 Occupancy Rate (%)			房價 Room Rate (澳門元MOP)		
	2024	2023r	相差 Diff.	2024	2023r	變動率 Change %
所有酒店 All type of Hotels	90.3	82.4	+7.9	1,399.8	1,287.4	+8.7%
五星級 5 star	91.4	81.9	+9.5	1,578.1	1,438.3	+9.7%
四星級 4 star	83.0	78.0	+5.0	925.8	920.5	+0.6%
三星級 3 star	93.9	89.9	+4.0	942.0	952.0	-1.1%

r:修正數據 Revised figures.

Hotel's occupancy rate and average room rate in Macau in 2024 (*Source* Macau Government Tourism Office)

Macau's cultural and tourism activities are increasingly diverse and abundant. The summer 2024 is expected to be busy for Macau with the Government Tourism Board already conducting events and advertisement ahead of the tourist season. "Cool Play in Macau" was a summer event that involved placing multiple check-in points throughout the community. Travelers were asked to upload images of their social media with receipts of minimum 100 MOP along with the post where they could win free McDonald's meals or specially made ice cream items.

The tourism board-operated Macau Grand Prix Museum also continued the popular free bus tour of the Guia Circuit for the second time in 2024, for five selected public holidays in the second half of the year. The program was initiated in 2023 and other bus tours were held during summer holidays.

Various museums in Macau have launched diverse exhibitions, such as the "Historical Photo Exhibition Celebrating the 25th Anniversary of Macau's Return to the Motherland" at the Macau Museum, the "Integration of Art between China and the West in the eighteenth-nineteenthth Centuries" at the Macau Museum of Art, and the "The Way of Huang Tingjian" at the Taipa Houses Museum.

Concerts have also become more popular in Macau, attracting extra tourists who come to view exhibits and spend the night in the hotels. Ms. Liu who is an incoming tourist from Guangdong observed how new the indoor places are, the good infrastructural and acoustical qualities of these places, the provision of restrooms and several food courts. Mr. Zheng, who traveled with his family to Macau for four days and three nights, mentioned the "Supercar Exhibition," featuring 30 legendary supercars, and a newly created Hanfu experience area that showcased Macau's diverse charm.[9]

The tourism board has been implementing its activities by launching promotional campaigns aggressively both offline and online. Each month's targeted promotions have managed to embed Macau's wealth of tourism advertisement on popular social networks. Another initiative involves the government partnering with online travel agencies and e-commerce companies to offer campaign specific products. The tourism board is also committed to enhancing Macau's reputation as a "Creative City of Gastronomy," continuing to deepen the integration of "tourism + gastronomy" and attracting major events and television programs to Macau. Notably, the food and cultural tourism program "Macau Dual Track—Culinary and Cultural Tourism Check-in," created by the Macau SAR government and the China Media Group, has received accolades from the National Radio and Television Administration for its innovative broadcasting. Within a short period after it began, it has gained huge viewership of over 55.84 million total video views and has reached 100

[9] Radio the Greater Bay, 2024, https://mp.weixin.qq.com/s/zOGiitvnU_mZB-YP8ddjxQ.

million media audiences, while advertisement coverage across social media platforms has surpassed 103 million.[10]

Moreover, in order to attract tourists from different regions to Macau, the tourism board provides shared bargains for travel. Travelers may take particular phases discounts on round trip flights and accommodations in hotels of their choice. More and more domestic and Taiwanese tourist groups and even international tourism groups are starting to come to Macau to experience what it has to offer. The Macau Tourism Board has also partnered with airlines to enhance ticket sales and offer temporary campaigns in addition to the "Buy One Get One Free" offer on flight tickets standing in favor of the Macau Airlines. The same offers are also available for ferry tickets and train tickets.

According to data from the Statistics and Census Service of the Macau Government, in the first quarter of 2024, Macau's economy experienced a substantial growth of 25.7%, with inbound tourist numbers rising by 79.4% to 8.8 million, recovering to 85.7% of the figures from the same period in 2019. The total number of visitors to Macau in 2024 is expected to reach 33 million. In 2023, the total number of inbound tourists exceeded 28 million, a year-on-year increase of 3.9 times, with 19 million visitors from mainland China, accounting for 67.5% of all inbound tourists and representing a year-on-year increase of 2.7 times.[11] It can be said that the travel experiences for tourists in Macau are comprehensively covered by discounts and creative activities. Due to these friendly and sincere promotional measures, Macau was awarded the title of the Most Satisfactory Destination for Chinese Mainland Travelers in Q1 2024.

16.2.2 Expanding Beyond Macau: Targeting Broader International Markets

To further stimulate the vast consumer market among visitors from Greater China, Macau is actively leveraging the central government's supportive policies to strengthen its appeal to mainland tourists.

The experience group of promoters from Macau traveled to nine cities in the Greater Bay by road under the caption "Experience Macau—it's more than just gambling" from 2022 to April 2023, covering over 2.828 million peoples' views. There were around 850 million views for the

[10] Bauhinia Magazine, 2023, https://zijing.com.cn/article/2023-09/04/content_1148186265224183808.html.

[11] Government of Macau Special Administrative Region Statistics and Census Service, 2024, https://www.dsec.gov.mo/en-US/.

marketing channels of the tourism bureau and 170 million views for the views of industry participants. The tourism board teamed up with local businesses to promote “Macau Week” held in Qingdao which lasted for 5 days and had over 200,000 on-site attendees and over 130 million viewers online. Similar events termed “Macau Week” were also hosted in Tianjin and Xiamen in the last part of the year, majority showcasing Macau’s offers in their culture and tourism, as well as promotional opportunities. Besides, the tourism authorities within the Greater Bay Area are pushing multi-destination travel products and make visitors from the region come for different attractions.

In addition, effective March 6, 2024, Shaanxi’s Xi’an and Shandong’s Qingdao have been permitted to apply for individual travel to Hong Kong and Macau. In addition, starting May 27, 2024, a further eight cities, Taiyuan and Xining included, will issue individual travel permits for qualified residents to Hong Kong and Macau.

In August 2024, the MACAU Week in Xi’an, Shaanxi was grandly inaugurated. The roadshow featured three main thematic zones showcasing tourism, business, culture, cuisine, sports, and entertainment in Macau, as well as outstanding performers in these sectors. It was also possible to purchase tickets for the event with travel packages that included airfare, hotel stays, meals, and other perks worth more than 38 million RMB, both online and on-site.[12]

Tourist Queueing for a Lucky Draw at the Macau week event (*Source* Macau Government Tourism Office)

[12] Macau Government Tourism Office, 2024, https://www.Macautourism.gov.mo/zh-hans/article/promotion/Macau-week-in-shaanxi-xian-2024.

In addition to focusing on the Greater China region, Macau's tourism recovery strategy has expanded to encompass a broader international market. In 2023, the Macau Tourism Board took its promotional campaigns on Portuguese soil and remained active in the regional and international tourism fairs all through the year 2024. Interestingly, Macau was involved in the Lisbon Tourism Exhibition in Portugal and worked with Portuguese Travel Agents Association to carry out promotional activities during the "2024 Spain Tourism Fair." Even the Portuguese Travel Agents Association has named Macau "the International Recommended Tourism during the year 2024."

Through close collaboration with the Portuguese Travel Agents Association, which is a member of the European Travel Agents and Tour Operators Association, Macau was chosen as the host city for the "2025 Macau Summit" in April 2024. In this forum, members from 30 countries across Europe will come together making it a very unique marketing occasion considering the top travel agencies and tour operators based in Europe and Macau's interesting various "tourism plus" offerings at the market.

Since 2023, the Macau Tourism Board has hosted hundreds of industry experts, media, and travel influencers from mainland China, Hong Kong, Taiwan, Thailand, Singapore, Indonesia, Malaysia, Japan, Korea, and Portugal to go to Macau. Such visits are usually characterized by tourism promotion activities in which the aim is to sell Macau through a range of means so as to draw people from all parts of the world.

Macau Tourism Board's Roadshow in Thailand (*Source* Bauhinia Magazine)

To further expand into multiple overseas markets, the Macau Tourism Board has introduced the "My Treat for You, See You in Macau" initiative, which provides international travelers who spend two nights in Macau or more with a hotel voucher worth half the cost for international travelers through Trip.com. Other offers include discounts on international airfare, bus, and ferry tickets bought one and free for travelers from HK and Taiwan and other overseas places as well as subsidy programs for tour groups from Taiwan and other international markets.

The tourism board has also sponsored tourism representatives from Osaka and Tokyo in Japan to go and seek business in Macau with a view of attracting foreign tourists. While on those visits, the delegation engages in local industry meetings to interact with residents and seek joint venture prospects with them.

Furthermore, the Macau Tourism Board collaborates with major e-commerce platforms like Tripadvisor and Expedia to promote Macau tourism in various countries, including Singapore, Malaysia, Indonesia, the Philippines, Japan, South Korea, Australia, Europe, the United States, and the Middle East. The board has properly revised the thematic pages and such resources, as informative articles promoting original recreation in Macau, as well as special travel offers were advertised. In order to satisfy the language and the social media needs of the different groups of tourists, the Macau Tourism Board created the official account of KakaoTalk for the Korean market and revived the Instagram, Facebook, and Twitter accounts for the Korean and Japanese markets.

16.3 A Broader Vision: "1 + 4" Industrial Transformation of World Heritage Tourism and Leisure Destinations

In the report titled "2023 Fiscal Year Policy Report World Heritage Tourism and Leisure Destinations" the government unequivocally undertakes to enhance the prospects of smooth economic recovery through steadfast application of relatively moderate "1 + 4" diversification strategy which aims at industrial structure optimization. The "1" is about pursuing the goals of making the world a point for tourism and leisure, while promoting the diversification of economy based on the "tourism and leisure" industry; the "4" is about having continuous emphasis on health, modern finance, high technology and exhibition business and cultural sports industries. The objective of this strategy is to raise the volume of the contribution of these four industries so that they contribute towards the growth of the economy and improve competitive edge and likewise ensure that a balanced industrial structure which is appropriate to the conditions prevailing in Macau is formed.

In the policy report, the government sets a development target where the gaming industry constitutes 40% of the local GDP, and the four key industries each account for 15%. In 2023, the government actively promotes cross-industry integration with tourism, enhancing collaborative development between the tourism sector and these industries. Specifically, it is highlighted that promoting additional tourism features and stimulating local community economies are the two important actions that are attainable in facilitating the revival of total tourism and leisure activity. Several rejuvenation projects in historical and cultural areas have been carried out to entice residents and tourists. Provisional statistics so far reveal that the figure for inbound tourists to Macau in the first three-quarters sky rocketed by 3.6 times as compared to the corresponding quarters the previous year.[13]

This kind of transformation is not only useful in lessening dependence over a certain industry however gives a chance to expand the available job opportunities for the people, thus improving the well-being of the society as a whole. The strategy of "1 + 4" shows the capability of the Macau's economy to adjust to global changes and shows the capabilities of its future economic plans. Under the current strategies, it is anticipated that Macau will be developed into a very attractive worldwide tourist and leisure venue as well as a major commercial center within the region and in the international sector.

16.3.1 Economic Integration Centered on Tourism to Attract a Broad Flow of Visitors

Regarding the "1," which refers to the development of a "World Tourism and Leisure Center," Macau is focusing on expanding its comprehensive tourism and leisure industry by fostering non-gaming elements within

[13] Xinhua News, 2023, https://www.gov.cn/yaowen/liebiao/202312/content_6922345.htm#:~:text=%E6%BE%B3%E9%97%A8%E7%89%B9%E5%8C%BA%E6%94%BF%E5%BA%9C20,%E5%A4%A7%E9%87%8D%E7%82%B9%E4%BA%A7%E4%B8%9A%E5%8F%91%E5%B1%95%E3%80%82.

its tourism sector. Specifically, Macau is leveraging its status as a World Cultural Heritage Site to strengthen the development of cultural tourism. Efforts are being made to improve all aspects, including guest reception, tourism service facilities, and service quality.

Community tourism is also empowering the local economy in Macau. The Macau Government Tourism Office (MGTO) has set up three funding programs: "Community Tourism Economic Expansion," "Gastronomy Culture Promotion," and "Coastal Tourism." These programs enable local communities to conduct activities that would make use of particular community attributes to attract tourists, thus helping the local economies recover. Among them, the "Community Tourism Economic Expansion" project introduced a new "Tourism+ " category, promoting cross-industry integration such as "Tourism + Sports," "Tourism + Cultural and Creative Industries," and "Tourism + Technology."

The Macau Economic and Technological Development Bureau (DSEDT) is also focused on fostering sustainable growth in community business areas through innovative technology. In 2023, they launched the new phase of the "Kwan Chai Wui 2023" event, creating innovative elements for special community areas and providing a platform for local tech companies. This event uses visual technology to attract visitors to explore the local area, injecting new vitality into the Kwan Chai District. Through the joint efforts of DSEDT, MGTO, and the Macau Trade and Investment Promotion Institute (IPIM), different areas such as tourism, gastronomy, technology, and conventions were brought together, which directed the visitors into the community and contributed to the local economy development.

The "Kwan Chai Wui 2023" event (*Source* Government of the Macau SAR—Economic and Technological Development Bureau)

Expanding the High-End Tourism Market

Expanding the high-end tourism market is a key priority, particularly by developing MICE tourism (Meetings, Incentives, Conferences, and Exhibitions). This kind of tourism which is classified as of high value not only brings in monetary gains to Macau but also increase its standing in the advances of business and conferencing globally. To entice more MICE activities, the government of Macau has undertaken a number of measures that include upgrading of the conference and exhibition centers and dedicating supporting services, policies, and marketing strategies.

For instance, large integrated resorts such as The Venetian Macau and Galaxy Macau offer a comprehensive range of technologically supported meeting venues, suitable for international large-scale conferences and exhibitions. Moreover, the government works in partnership with internationally recognized exhibition companies in order to host international Md conferences and exhibitions in Macau. Events like the Macau International Trade and Investment Fair (MIF) demonstrate Macau's strength as a regional MICE hub and attract many business travelers and exhibitors.

Promoting Non-Gaming Industries Through Gaming

The government calls on gaming concessionaires to not only run casinos but also invest their resources in off-gaming development. For example, concessionaires are expected to develop attractive themed retailing and entertainment complexes and to promote large-scale and important events and festivals. This policy seeks to increase the entertainment quotient and diversify offerings in order to broaden the appeal to families and other categories of tourists.

While seeking wholesale linkages between the gaming sector and other sectors, Macau has also offered active policies to control gambling activities and promote responsible gaming. In recent years, the government has been taking steps toward sustainable growth of the gaming industry, such as improved regulation of casino operators, exclusion of minors from gambling establishments, and mass media educational campaigns. The public is regularly assisted by the Macau Gaming Inspection and Coordination Bureau, which conducts responsible gambling campaigns and provides appropriate consultations.

Furthermore, the Macanese authorities are making use of advanced technologies such as facial recognition systems that help regulators in masking illegal practices while also aiding in promoting transparency and fairness in casino business activities. Such endeavors do not only help in promoting a good image of the industry but also help in providing a better and safe entertainment environment for the guests.

Also, the model of public–private partnership continues to be advocated for the recouping of the economy. In collaboration with new gaming licenses recently issued, the government encourages concessionaires to seek prospects of foreign tourists and promote non-gaming ventures like conferences, exhibitions, and entertainment performances. As the Macau resorts' advertising campaigns throughout international media, social networks, and travel agencies are still running, the diversity of the tourism offer is emphasized as well. Concessionaires are specifically advised to provide large-scale and attractive themed attractions and love to invest in major brand events and festivals which compel tourism and non-gaming elements further.

Last but not least, by facilitating to development of the non-gaming tourism sector, promoting integration of the gaming-related activities with the non-gaming-related activities and strengthening the regulatory framework together with responsible gambling policies, the Macau government is making progress towards its objective of economic diversification. These solutions not only increase the position of Macau as a high-quality tourism and leisure center, but it also creates a very strong platform for sustainable development of the city in the future.

Tourism + X

Macau is actively promoting the development of "Tourism + Gastronomy," bringing together representatives and chefs from 23 "Creative Cities of Gastronomy," including Macau, for an impressive series of culinary demonstrations—24 in total. Furthermore, the large-scale culinary tourism program, Macau Double Line: Culinary Tourism Check-In, co-produced with China Central Television, has won recognition as an "Innovative and Outstanding Program" from the National Radio and Television Administration. Macau also supports various integrated resort businesses in hosting world-renowned culinary award ceremonies to enhance its attractiveness in the international food scene.

Likewise, different relevant government departments join forces with the operators of integrated resorts in the promotion of tourism, cultural, and sporting events through established communication channels, through the government and local business organizations. This teamwork of cooperation optimally utilizes resources and strengthens synergy in boosting the advancement of "Tourism + Events" and "Tourism + Exhibitions." Such initiatives are very beneficial to the local small and medium enterprises (SMEs) and are good in boosting growth in the provision of services such as eating out, travel, building, advertising, and lodging. In addition, there are rich historical and cultural resources in Macau which are pillars of the cultural and creative industries. Many different creative projects were successful in marketing and making products with Macau elements, which helps the development of the "Tourism + Cultural Creativity" brand. In 2023, Macau also hosted the inaugural "Macau Study Tour Development Forum," aimed at advancing "Tourism + Education."

Apart from the tourism centered economic integration, Macau has an ambition of economic evolution through the use of digital finance enabling the enhancement of smart technology and its utilization on more scenarios. The launching of the "2023 Digital Support Services for SMEs" program enables eligible enterprises to access different digital options and strategies for the marketing of their businesses. Multiple foreign payment methods and various regional QR code payment services which merchant banks provide help business establishments to receive clients visiting Macau. Utilizing big data technology, Macau monitors visitor traffic at over 110 attractions, helping travelers adjust their itineraries for a smoother experience.

Macau also seeks to enhance inter-regional synergy which is tourism-based. A tourism developing organization comprised of Guangdong, Hong Kong and Macau has officially announced 10 more-and this time-one-stop trendy touristic routes within the Greater Bay Area. The ferry line between Taipa Passenger Terminal and Guishan Island in Zhuhai commenced on the 15th July 2023 enhancing the development of maritime tourism. Different government institutions are working hard to enhance collaboration with the Hengqin Deep Cooperation Zone in areas such as tourism, exhibitions, or culture, for example, bringing Hengqin performers to major annual events in Macau, organizing brand several simultaneous conceptual marketing campaigns and promoting brand promotions related to exhibition development together.

16.3.2 Sustainable Development Momentum Through Economic Diversification

In the second phase review of the Macau Tourism Industry Development Master Plan released by the Macau Special Administrative Region (SAR), actions for 2023 were summarized. Overall, Macau is dedicated to building a cross-sector integrated and diversified economy centered around tourism. As pointed out by Chief Executive Ho Iat Seng in the 2024 Policy Address, titled "Consolidation, Enhancement, and Diversified Development," the general direction of the government's policy for 2024 is to "consolidate recovery, focus on diversification, improve people's livelihood, and enhance development." This portrays building on the current pace of recovery of the economy, undertaking the consolidation of Diversification framework dubbed as "1 + 4" and accelerating the gradual shift of the Macau's economy to more diversification of the economy faster, better, and in a practical manner.

Implementing the "1 + 4" policy will mean continuing the traditional pillar industry of gaming and tourism while also growing the TCM(Traditional Chinese Medicine) and health industry, finance, high-tech industries, the MICE(Meetings, Incentives, Conferences, and Exhibitions) and cultural sports industries. Year 2024 is hailed as the formation year for the commencement of the "1 + 4" in Macau.

Health and Wellness

Health is quite evidently a universal concern. The term "big health" expands the health concept in accordance with socioeconomic progress, elevation of consumer needs, and the dynamism of disease patterns. It includes the entire lifecycle, from avoidance and cure, to life utilization and health in all its dimensions.

The administration of Macau SAR treats the industry of health care as an industry development path with emphasis on R&D, production and medical services in TCM while making use of the unique geographical setting, abundant TCM resources, and policy supports. The government advocates the building of TCM industrial parks and invite prominent local and foreign firms and research institutions to help in the modernization and global expansion of TCM. Further, Macau also actively markets TCM-orientated medical tourism blending the local tourism natural and culture resources to provide these services to more tourists.

To further institutionalize, professionalize, and internationalize the TCM industry, the Macau SAR government formally implemented the Pharmaceutical Activities and Registration of Chinese Proprietary Medicines law in 2022. This law has brought forth new registration and drug regulation systems which promoted standardized management of the TCM industry which in turn instilled the confidence of the development of the industry in the public. Since the law's implementation, it has attracted several major pharmaceutical companies from mainland China to set up operations in Macau, with ongoing improvements in the regulation of TCM activities, presenting new opportunities for the TCM industry.

In the TCM R&D scenario, Macau gives importance to technological improvements, providing incentives, and participating in the relevant research endeavors. Institutions of higher learning like the University of Macau and Macau University of Science and Technology possess specialized TCM research faculties and have created alliances with well-known local and foreign universities and research institutions. Not only do such cooperations foster academic interaction, but they also hasten the transfer of research output. For instance, it has made significant strides in developing new and more effective TCM medications, especially for patients suffering from chronic ailments as well as elderly people who have a higher demand for effective TCM treatments.

Certainly, the SAR Government of Macau is also keen to encourage the establishment of TCM industrial parks with a view of developing an all-rounded platform for research, production, and sales. These parks include modern laboratories and factories and provide an optimal atmosphere for collaboration between businesses and scientific and research organizations. Wishing to bring TCM in line with global standards, Macau intends to modernize it by combining advanced research practices and management know-how with the more traditional principles of TCM.

Modern Finance

Modern finance is an important component in Macau's economic diversification. Having entered a saturation in the gaming sector and with global trends leaning towards sustainable finance, the government of Macau SAR recognizes that there's a necessity to optimize and upgrade the economic structure through the embracement of modern finance. The Macau SAR is becoming a regional wealth management center where the people offer classical financial services such as private banking and asset management.

Macau possesses distinctive benefits in the realization of becoming a regional wealth management center. There is a stable financial system, a sound legal environment, and being in the Guangdong-Hong Kong-Macau Greater Bay Area, puts Macau within reach of major financial centers like Hong Kong and Shenzhen enabling it to harness its geographical advantage to attract both domestic and foreign investments. To enhance the competitiveness of the Region, the Macau government has also implemented a number of policies to attract financial institutions including some tax relief, financial and technical assistance policies. In addition, Macau has increased cooperation with international financial institutions and integrated advanced management experience and skills into the development of local financial services.

FinTech is a major key focus in terms of financial innovation in Macau as well. Digital services offered by the Macau government also include blockchain technology and digital currency projects that will assist in transitioning the classical finance industry into modified versions. It is also expected that further progress will be made in Macau's efforts to create financial transactions that will incorporate blockchain compliance. However, efforts are also being made in the field of digital currency among other cross-border payment systems that will be cost-effective and secure the current day financial systems.

Green finance is also a crucial area which Macau aims to develop further and capitalize on. With the world progressively focusing on promoting sustainable development initiatives, there is yet another relief as Macau is working on providing sustainable financial solutions. The government has put in place a number of measures to stimulate the financial institutions in the innovation of green linked credit, green bond, and other financial products which aid the low-carbon economic and environmental protection initiatives. For example, Macau has provided for the issuance of several green bonds directing them towards renewable energy or clean transportation projects. Additionally, Macau set up a green finance fund for backing businesses in need.

Private banking and asset management in Macau have expanded to become the cornerstones of new financial industry. Macau managed to convince a number of prestigious private banking and asset management houses to open offices providing high net worth individuals with tailor-created wealth management strategies. Such services include the provision of investment advisory, assisting with asset allocations, and formulation of tax strategies with an aim of wealth creation and preservation among many other goals. Furthermore, it is evident that Macau is also keen on encouraging the growth of the insurance sector with an emphasis on high-value health and life insurance products to serve its evolving clientele.

In further diversifying and modernizing its economy, Macau is optimizing its economic structure through the establishment of a regional commercial wealth management hub, growth of FinTech, provision of certain financial services, development of green finance among others. Going forward, it is envisaged that Macau will persist in enhancing the activities associated with the contemporary financial services encouraging more financial institutions and high net worth individuals to help enhance Macau's position on the global financial services platforms.

High-Tech Industries

High-tech industries make an important contribution to the deep transformation and the upgrading of the Macau economy. The activities of R&D are appreciated and backed by the Macau government especially in regards to the biotechnology, Information Communication technology (ICT), and Artificial Intelligence (AI) fields. Related to this, Macau has set up a number of R&D centers and Laboratories, while establishing relations with various local and foreign universities, and research institutions.

Eleven years ago, the University of Macau Hengqin Campus was put into use, marking a successful model of collaboration between Hengqin and Macau. Besides, the University of Macau intends to create Higher Research Institute in Hengqin to advance further the collaboration of the industry, educational, and research sectors. The economy of Macau is encouraging the moderate diversification of the economy with the moderate focus on the fintech, advanced manufacturing, cultural and creative industries, and the TCM industry.

Considering these strategic layouts and commitments, one can conclude that Macau in the near future will execute more research and development (R&D) and commercialization activities, enhancing the distinctive nature of the scientific and technological accomplishments.

The education footprint of the University of Macau grows larger with the Deep Cooperation Zone in Zhuhai Hengqin. During the UM Media Open Day on January 14, 2024, UM President Song Yonghua told that there are future plans to construct a new campus in the cooperation zone and it will be approximately half the size of UM. The new campus will provide education in medicine, emerging engineering, microelectronics, Internet of Things (IoT), AI, robotics, and quantitative business analysis.

In addition, talent cultivation is a key component which is also related to the development of high-tech industries. Addressing educational cooperation programs, Macau will also be focusing on imaginative facets and will be working towards a skilled base which would be prepared for the needs of the present industrial era. There has also been a set of proactive measures undertaken by the administration like tax waivers and subsidies, to entice more tech companies to establish their presence in Macau.

Notably, Macau is a key research base for microelectronics. Although Macau may not be widely known for its high-tech industry, it is one of the leading regions in China for semiconductor R&D. For instance, in the 2023 International Solid-State Circuits Conference (ISSCC), the University of Macau contributed 15 papers, ranking first among global universities.[14] Yu Chengbin, the founding chairman of the Macau Microelectronics Association, noted that the coordination of enterprises in the industry is vital, as semiconductors are said to be an industry with a world presence, "decoupling" is not a universal trend but rather a situation that arises because these two nations, China and the United States, have differing priorities and system requirements. This underscores the resolve and self-assured feeling that residents of Macau have in respect to the quest for knowledge and any industrial-related developments, depicting the significance of this growth strategy to the economic diversification and sustainability of Macau.

MICE and Cultural Sports Industries

The MICE (Meetings, Incentives, Conferences, and Exhibitions) and cultural sports industries are another crucial component of Macau's diversified economic development. They play a significant role in enhancing Macau's international influence, attracting high-end business travelers and exhibitors, enriching the cultural life of local residents, and promoting community economic growth.

Macau has accomplished significant achievements in the MICE segment. For example, the annual Macau International Trade and Investment Fair (MIF) draws exhibitors and buyers worldwide, demonstrating

[14] Caixin, 2024, https://economy.caixin.com/2023-03-29/102013308.html?originReferrer=caixinsearch_pc.

Macau's capacity to hold exhibitions of a national scale as well as global scope need not be overemphasized. Such events have both direct economic return and create demand for related activities like hospitality, catering, and transport services. In order to further boost the competitiveness of the MICE market, the Hong Kong SAR government keeps on expanding the range of MICE facilities, enhancing the quality of service provided, and utilizing several policy tools including tax waivers and financial assistance to draw more international meetings and exhibitions to Hong Kong. On a global scale, the authorities are willing to routinely attend events hosted by cities that are members of the UNESCO Creative Cities Network and participate in meetings of international organizations in order to further develop its position.

At the same time, Macau is also focusing on other overseas cities in seeking trade and commercial cooperation, aside from just holding important exhibitions. Working with other cities in the Greater Bay Area will enable Macau to make impressive use of its role as an economic and trade cooperation platform between China and Portuguese-speaking countries, and promote further international trade contact.

Culture is another significant advantage for Macau, a UNESCO World Heritage Site. With its rich historical and cultural heritage, Macau serves as a witness to colonial history and a hub for East–West cultural exchange. The government promotes cross-cultural understanding and respect through various cultural festivals, such as the Macau Arts Festival and the Macau International Music Festival.

The concerts held at the Cotai Arena in Macau have been met with enthusiastic responses (*Source* Eason Chan's Fear & Dreams Concert)

Macau has also invested heavily in the development of the sports industry. To attract more major sporting events, the government has increased investment in sports infrastructure, planning to build or upgrade existing venues to meet international standards. For instance, Macau is constructing new competition venues that will be capable of hosting international events. Additionally, Macau is actively bidding to host various international sports events, such as the 71st Macau Grand Prix, the 6th Asian Squash Masters Championships, and the 2025

National Games (co-hosted by Guangdong, Hong Kong, and Macau), to enhance the city's international profile.

The audience comprises definite international and domestic guests hence boosting the revenues from accommodation, catering, and retail activities and overall impact of the tourism industry. Construction of new sports facilities and improvement of the existing ones will also enable Macau to host more conferences and exhibitions worldwide, thereby consolidating the region's status as a MICE hub. Besides, organizing and conducting these events will be accompanied by a lot of employment opportunities in a number of branches such as construction, services, and security. The growth of the sports industry not only enhances the development of local sports but also enhances the economy of the local community both socially and economically which enhances the growth and development of a multifaceted economy of Macau.

In brief, the MICE facilities and the cultural sports industries are new areas in which the economy of Macau can grow. If MICE facilities and standards of service are perfected, international trade contacts are established and enriched with cultural contacts, as well as the development of the sports industry, it is possible to increase the momentum of business trips and exhibitions in Macau and create more jobs for local people as well as expose them to wider culture. This will complement towards the sustainable development of economy and society as well as general development.

Macau's efforts demonstrate that the "1 + 4" vision plan has a comprehensive impact on economic transformation, encompassing not only short-term economic recovery but also structural adjustments and shifts in development models. The implementation of this plan helps Macau transition from an over-reliance on the gaming industry to a more balanced industrial structure, injecting momentum into sustainable economic development. Furthermore, the plan takes into account the coordinated development of the Guangdong-Hong Kong-Macau Greater Bay Area, redefining and directing Macau's role and contributions within the regional economy. Through this plan, the Macau SAR government has demonstrated a firm commitment to moderate economic diversification and urban transformation, laying a solid foundation for the long-term prosperity and stability of Macau.

16.4 Conclusion

To summarize, the initiatives of the digital economy, especially the introduction of Alipay+, have greatly assisted in the recovery of Macau's economy and boosted its tourism industry. The strategically developed payment ecosystem not only facilitated a quick recovery from the pandemic-induced economic downturn but also opened avenues for sustained and diversified economic growth. The partnership between Alipay+, Macau Pass, and other organizations has made payment processes easier and more global. Such a balance is necessary for fostering local economic activities while reaching out to customers worldwide.

These practices highlight the new direction Macau's economy is taking. By positioning visitors at the core of its economic strategy, these measures have the potential to enhance the tourism experience, deepening visitors' connections with Macau. Most significantly, Macau is transforming into a hub for tourism and culture through the strategic development of an integrated digital economy system, spanning payment solutions to cross-border services.

These partnerships also emphasize the broader vision of Macau's "1 + 4" industrial restructuring, which seeks to diversify its economy beyond the gaming industry. By focusing on its World Heritage sites while expanding into areas like technology, finance, and creative sectors, Macau aims to achieve a more stable economic future.

Sustainable FinTech advances Macau's economic transformation by supporting the achievement of Environmental, Social, and Governance (ESG) goals as well as the United Nations Sustainable Development Goals (SDGs). The introduction of green financial products, such as green bonds and green loans, will help finance environmentally friendly projects and establish a low-carbon economy. This effort is already underway, with financial institutions in Macau offering green bonds to fund renewable energy ventures and clean transportation infrastructure development.

Additionally, Sustainable FinTech aims to support Macau's residents and attract new businesses by promoting social inclusion and equity. For example, blockchain technology can create a more efficient and transparent supply chain financing framework, enabling access to funding for SMEs and other vulnerable groups. Furthermore, the evolution and adoption of digital payment systems like Alipay+ allow more people to access seamless and secure financial services, helping to narrow the financial divide.

In conclusion, Sustainable FinTech is not only essential for enhancing Macau's economy but also critical for ensuring long-term sustainable development. It offers valuable practical experience and a developmental model that can inspire the world. The sustainable growth of Macau should be coupled with its economic advancement through digital means within a new framework of diversification and integration into the global economy. Clearly, Macau's active and strategic positioning in digital financial technology, combined with its commitment to sustainable tourism development, will propel the region toward becoming an advanced, multifaceted economy, setting an example for other regions to follow.

CHAPTER 17

Sustainable FinTech for Sustainable Tourism, Inclusive Partnership, and Smart City in Low-Carbon Transition: Alipay+ Ecosystem in Singapore (Changi Pay, EZ-Link Wallet, ComfortDelgro, GrabPay, OCBC Digital), South Korea (Kakao Pay, NAVER Pay, Toss) and Japan (Paypay)

17.1 Inclusive Fintech, Sustainable Tourism, and Smart City Construction in Singapore

17.1.1 Late Development, Rapid Advancement: A Competitive Mobile Payment Market in Singapore

As a representative developed economy in Southeast Asia, Singapore has established a robust economic infrastructure. But when the former Prime Minister of Singapore Mr. Lee Hsien Loong traveled to China in 2017, he pointed out that while China had already largely moved to online payment methods such as Alipay and WeChat Pay, Singapore in his words seems to be a "country bumpkin" in comparison to China.

Perhaps China's development acted as a catalyst for the development of Singapore's online payment world. In July 2017, the Association of Banks in Singapore ABS together with the major banks implemented PayNow, a China UnionPay look like system. The same year, NETS which was

J. Sun, *Sustainable Financialization*, Sustainable Development Goals Series, https://doi.org/10.1007/978-981-96-3580-1_17

founded by Singaporean banks introduced its contactless mobile wallet service NETSPay which also provides QR code payments in addition to contactless payments. In addition, the Malaysian ride-hailing company Grab launched GrabPay in late 2016.

In 2017, the Monetary Authority of Singapore (MAS) put forward the new consolidated payment enhancement called the Payment services act PSA whose aim was to expand the regulatory boundaries of payment systems services and streamline rules concerning payment activities risk. The act also created a licensing regime for retail payment services, which standardized the payment system as well.

E-wallets began to rise in popularity in Singapore, with mobile payments based on e-wallets generally divided into three main categories. The first category includes e-wallet payments linked to traditional bank cards or accounts, such as DBS's PayLah!, OCBC's Pay Anyone, UOB's Mighty, and China UnionPay's QuickPass. These networks typically use open-loop acceptance networks and standardized QR codes, such as NETS and PayNow for PayLah! and UnionPay standards for QuickPass. The second category consists of e-wallets based on prepaid accounts, such as Singtel's Dash and ShopeePay from Sea Group. Merchants in this category typically establish independent acceptance networks based on proprietary standards. Alipay and WeChat Pay also fall under this category but with added functionality that supports linking to bank cards while using their proprietary standards and acceptance networks in China, blending characteristics of the first and third categories. The third category involves e-wallets that enable payments through linked traditional bank cards, such as FavePay from Fave, which was acquired by the Indian POS company Pine Labs. Acceptance networks here are typically developed independently based on proprietary standards.

With the emergence of these new electronic payment methods, various organizations sought to "get a piece of the pie," intensifying competition among market players and between different clearing organizations. This competition fast tracked diffusion and iteration in Singapore's payment market moving it towards saturation very rapidly, while also promoting fragmentation in the market.

17.1.2 The Trust Crisis of Foreign Financial Systems in Privacy Protection: A Privacy-Conscious Singapore Market

According to the National Privacy Test (NPT), Singapore ranks among the top global performers in cybersecurity and privacy awareness, along-side Poland (NPT score = 64), Germany, and the United States (NPT = 63), as well as the UK, Austria, and Portugal (NPT = 62).

Singaporeans excel at creating strong passwords (98%) and understand how devices can be infected with malware (94%). They are also aware of which types of sensitive data should not be shared on social media (93%) and how to respond to suspicious streaming service offers (91%). However, only 8% of Singaporeans are familiar with online tools that protect digital privacy, and just 10% are aware that the data collected by internet service providers (ISPs) is part of the metadata.

In Singapore, 1% can be considered "digital drifters" who are almost oblivious of privacy and cybersecurity issues; on the other hand, 20% fall into the range between 75–100 on the cybersecurity scale being called well-informed "cyberstars."

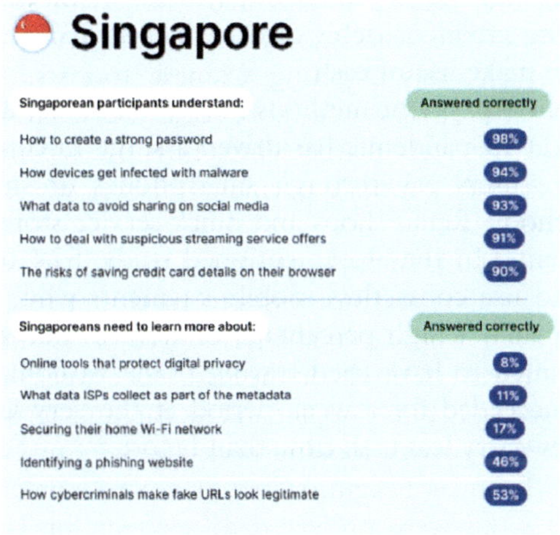

Cybersecurity awareness of Singaporean Netizens: Strengths and weaknesses (*Source* Internet)

Since 2021, there has been a steady increase in the amount of privacy awareness in the Singaporeans. Globally, more participants have recognized the importance of reading terms of service agreements, and in Singapore, the percentage of people who perceive this as relevant increasing from 24% in 2021 to 56% in 2023.

This trend is also reflected in the perception of phishing fraud risks. For example, in 2021, only 48% of respondents could say how they would react to such fraud, while this percentage rose to 65% by 2023. In 2021, when the respondents were asked what actions they would take in case their information was leaked as a result of a data breach, 45% gave an appropriate reply, this number was 57% in 2023. This involves looking for any suspicious activity on the accounts and changing the passwords in question.

This recall concerning the sensitivity of information explains the widespread skepticism towards foreign payment systems among Singaporeans. Furthermore, a robust local banking structure has resulted in the local market being close to over saturation of payment solutions. According to Visa's 2023 Consumer Payment Attitude Study, Singaporeans strongly favor cashless payment methods, with 97% of respondents using credit or debit cards, which is 15% more than those who continue to make use of cash.

Besides common payment methods like credit cards, debit cards, and cash, the COVID-19 pandemic has driven a surge in contactless mobile payments, where users pay through smartphones or smartwatches via tap or scan methods. Retail shops and quick-service stores also embrace QR code payments. In the aforementioned study, 52% of Singaporeans were reported to use contactless mobile payments while 48% QR code payments. With such a high percentage of cashless transactions, 66% of respondents claimed to have tried leaving home without cash, and 42% of them have succeeded for a week period at the very least, while 39% claimed that they carry less cash now than before.[1]

[1] Visa, 2023, https://www.visa.com.sg/content/dam/VCOM/regional/ap/singapore/global-elements/documents/visa-cpa-2023-report-smt-final.pdf.

17.1.3 Online and Offline Multi-Scenario: Alipay+ Digital Marketing Solutions Empower Sustainable Tourism in Singapore

In October 2022, Ant Group developed its partnership with the Singapore Tourism Board namely the STB which was established back in 2019 and seeks to propel the growth of tourism in Singapore over the next 2 years through its marketing and digital efforts. To promote Singapore to Asian tourists from key markets, Ant Group has begun, in conjunction with the Singapore Tourism Board subsidiary STB, the SingapoReimagine with Alipay+ campaign targeted at tourists from Malaysia, the Philippines, South Korea, and Thailand. This campaign is part of STB's wider reconstruction of international outreach and is aimed at getting new and exciting experiences of visitors from those regions. Concurrently, it facilitates the use of their familiar mobile wallets for transactions within Singapore.

With this kind of cooperation, the following mobile wallet GCash from the Philippines, TrueMoney from Thailand and AlipayHK from the Hong Kong SAR, China, were added to the already existing partners Kakao Pay from South Korea, Touch n Go e-Wallet from Malaysia and Alipay from the mainland China.[2] This expansion raises the total count of mobile wallets supported by Alipay+ in Singapore to six. Tourists utilizing these digital wallets can now carry out transactions with ease and securely with local merchants in the tourism and lifestyle industries that have adopted Alipay Plus. This includes several retail shops in Changi Airport, ComfortDelgro taxi services, hotels, and such cruise attractions as Gardens by the Bay.

At the 2022 Singapore FinTech Festival (SFF), Ant Group launched the Alipay+ D-store digital business solution.[3] Alipay+ D-store utilizes an innovative concept where merchants integrate all their shops into online stores or does not integrate the shops presence within multiple wallets or apps together to connect with the consumer effectively. It makes business more efficient by utilizing a one store multiple platforms model.

[2] The Digital Banker, 2024, https://thedigitalbanker.com/alipay-connects-2-million-merchants-in-japan-as-global-tourists-travel-to-the-country-for-cherry-blossom-season/.

[3] Marketech Apec, 2024, https://marketech-apac.com/tag/alipay/.

Huggs Coffee, the biggest local chain of coffee in Singapore was one of the most highlighted brands during the SFF as well where the participants were shown how to order a "digital coffee" from their phones. Even the Prime Minister Lawrence Wong of Singapore had the curiosity to scan in and give it a try. With the launch of Alipay+ D-store, Huggs Coffee's marketing strategies included the establishment of multiple online stores in different foreign digital wallets. With this shift towards digital, the business saw a rise in new memberships by 65% and 87% new online traffic.

D-store also showcased its ability to help merchants establish a digital store in just 10 minutes, enabling 5,000 small business owners across Southeast Asia to launch online operations with just a smartphone, lowering the barriers to entry.

Ant Group's booth was visited by Deputy Prime Minister Heng Swee Keat, and he saw how one of Singapore's local tea brands, LiHo TEA, functions without cashiers. All transactions were done and processed online which reduced the cost for the merchants and enabled the customer to order from the multiple e-wallets available.

Alipay+ has penetrated the global trade market and currently connects millions of merchants with their services both physically and virtually. Internally, it enables American brands that include Apple, Google, Netflix, Agoda, and Uber alongside regional counterparts such as Lazada, Foodpanda, TikTok, and others. Also, SHEIN and iQIYI, the Chinese brands have received some international exposure through Alipay+.

Offline, Alipay+ is continuing expanding its network and as of April 2022 was reported to connect up to 70,000 merchants in Germany, Austria, Malaysia, and South Korea. These include drugstores, convenience stores, grocery stores, online travel, ride-hailing services, and other merchants that further supplement both offline and online sales.

17.1.3.1 Unified Digital Payment Ecosystem

American scholar Francis Fukuyama, in his analysis of why post-colonial non-Western countries often struggle with weak state capacities, identifies three essential elements for a well-ordered society: strong government, rule of law, and democratic accountability.

Singapore offers a good example as it is characterized by strong state governance since there has been a single political party that has reigned rather undisturbed enabling the establishment of political order based on modern legal institutions. These and other features of the government allow a high level of efficiency and integrity, which is important in the economic and social development of the country. Despite that, the size of the government is rather small and has established sort of a democratic rule suitable for a complex and dynamic economy.

The active cooperation with civil society in all the processes and projects certainly contributes to the relatively high competence of the Singaporean state institutions within the international context and the focus on "economic constraint" to achieve an interdependent governance model. This interdependence is evident in the governance in Singapore, where the government engages diverse actors in the implementation of projects and policies, rather having a unilateral approach. This was not always the case as historically, most countries operated on a singular governance model. Over time, and through promoting a civil society, Singapore adopted an interdependent model of government, providing better relations between the state and civil actors.

Alipay+ has become a valuable addition to Singapore's efforts to unify the fragmented digital payment market and enhance its digital economy capabilities.

Before the pandemic, tens of thousands of Malaysian workers commuted daily across the Johor-Singapore Causeway to work in Singapore, returning home to Johor Bahru each evening. Similarly, on long weekends, Singaporeans frequently drove to Johor Bahru, took ferries to Indonesia's Bintan and Bali, or flew to Thailand's Phuket for vacations. Indeed, even just prior to the pandemic, intra-regional tourists contributed for 51.7% of total tourist arrivals in the Southeast Asian region. As of 2022, the leading 10 markets for tourists in Singapore were Indonesia, Australia, India, the USA, Malaysia, China, the United Kingdom, Philippines, Vietnam, and Thailand.

Prior to COVID-19, approximately one-third of Singapore's tourists came from Southeast Asia, another third from Mainland China and other parts of Northeast Asia, and the remaining third from other regions of the world.[4] As cross-border tourism gradually recovers, consumers who have become accustomed to mobile payments during the pandemic naturally expect to use the same payment methods across different countries,

[4] Singapore Tourism Analytics Network, 2023, https://stan.stb.gov.sg/public/sense/app/289a7ba7-52da-45af-bcc7-b8821a76c87f/sheet/5a2be343-2281-45bb-9ef0-7b421c69d88d/state/analysis.

aiming for a seamless "one wallet for global purchases" experience to enhance transaction efficiency.

Overview of overseas tourists visiting Singapore (*Source* Singapore Tourism Analytics Network)

17.1.3.2 Alipay+ Payment Partnership: Seamless "One Mobile, Travel the World" Experience

However, after the pandemic has calmed, China is estimated to be the country of origin for one-third of Singapore's tourists. The visa-free travel between China and Singapore was put into effect as of February 9, 2024. Based on the data from the Singapore Tourism Board (STB), the attractive features of Singapore such as its family friendly environment and safety is likely to increase with the variety of tourism products. In the year 2024, air traffic from China to Singapore and vice versa has returned to the levels that were in place before the pandemic. With the sustained recovery of outbound tourism and flights in addition to those who are leaving China, an increase in the tourism and expenditure by the Chinese in Singapore is to be anticipated as well.

As of 2024, statistics from the Singapore Tourism Board show that China remains one of Singapore's key source markets.

Key source market: China (*Source* Singapore Tourism Analytics Network)

In light of this tendency, an agreement was signed between Alipay and STB on September, 2017, which is designed to enhance the general experience of Chinese visitors in Singapore. The concession included common responsibility from both parties to promote markets so that respective tourists would be encouraged to use Alipay in Singapore.

After the signing of the MoU, Alipay's usage by Chinese nationals visiting Singapore increased by leap and bounce. In 2017, China became the top source market for arrivals and tourism receipts with over 3.2 million visitors and above SGD 3 billion dollars in tourism spending on Singapore.

Alipay+ has broadened its scope of partnership to include more local digital payment players in Singapore. Such collaborations also bear the benefit of technology and infrastructure provided by Ant Group who make it possible to develop tools for Singapore's local payment ecosystem in an efficient and viable manner. Alipay+ has cooperated with most of the local digital wallets as well as banks and public transport platforms which are relevant to the easy and ecological travel betterment.

Changi Pay

Changi Pay is a digital wallet that has been started by Changi Airport Group (CAG) with the aim of facilitating payments and shopping for customers at Changi Airport as well as Jewel Changi Airport. It allows users to make payments using Liquid accounts, linked credit/debit cards, or electronic vouchers in a swift manner.

In 2023, Changi Pay engaged in a collaboration with Ant Group, joining the Alipay+ payment network. Located at a key vantage point along the busiest international air corridors, Singapore has been served through its flagship Changi Airport, recognized as the best airport in the globes. Changi receives over 62 million passengers per annum with over 100 foreign airlines connecting to over 100 countries and regions.[5]

This collaboration enables Singapore residents to use Changi Pay for payments at millions of merchants in Mainland China and other Alipay+ acceptance points, including in Australia, the European Economic Area (EEA), Japan, Malaysia, Hong Kong, Macau, Qatar, San Marino, South Korea, Switzerland, Thailand, the UK, and the USA. Furthermore, the collaboration eases payment problems for Singaporeans when they are in China as users pay attractive rates without foreign fees, and users earn electronic vouchers to be used for shopping at Changi Airport when they arrive back home.

Changi Pay, for example, is beginning to gain traction among merchants at Changi Airport, Jewel Changi, and on the iShopChangi platform. There are plans to expand this service to iShopChangi, which would enable users to buy duty-free items online. Users can also trigger and redeem vouchers from interested merchants at Changi Airport and Jewel Changi effectively making Changi Pay a multifaceted and all-encompassing payment support for the residents both within the region and worldwide.

Through its partnership with Alipay+, Changi Pay has enhanced its international payment capabilities, reinforcing Changi Airport's position as a premier global aviation hub. This collaboration benefits both Changi Pay, by expanding its service reach, and Alipay+, by attracting a broader international user base and diversifying its global payment network. This partnership contributes to more inclusive digital financial services, supporting both companies' growth in the digital economy.

EZ-link Wallet

EZ-Link Wallet, enabling users to conduct retail merchants' QR Code scan payments, is among the services offered in the EZ-Link mobile application, which is, in turn, held by an electronic entity such as a transit card wallet. In Singapore, EZ-Link is credited with having issued the most transit cards since 2002, having already distributed over 40 million EZ-Link cards. As of 2022, EZ-Link has been incorporated into the Alipay+ wallet.

[5] Changi Airport Group, 2024, https://www.changiairport.com/corporate.html.

EZ-Link Wallet enjoys great popularity not only in Singapore but also internationally through Alipay+ in places like Japan, allowing users to make payments effortlessly. There is widespread acceptance of the EZ-Link Wallet within Singapore, and includes offerings in a number of segments like public transport, food, and shops. Payments can be made with participating merchants either through SGQR codes or EZ-Link Wallet. Blockchain concepts are also coupled with the EZ-Link Wallet such as SimplyGo which enables the easy charging of transportation fares facilitating its further application to small-scale commercial transactions. The list of joint ventures of EZ-Link Wallet includes such famous merchants as Old Chang Kee, Gong Cha, Skechers, A-One Claypot, and Dunkin' Donuts.

EZ-Link Wallet has played an essential role in promoting cashless transactions in Singapore which is beneficial for residents and visitors alike. It is accepted across 30,000 outlets throughout the country including shopping, eating, private transportation, and government and community services.[6]

The wallet's ecosystem is enhanced by partnerships with the National University of Singapore (NUS) and Alibaba Cloud. Since 2017, these collaborations have focused on boosting Singapore's data intelligence and computing capacity, improving EZ-Link's data analytics and overall user experience.[7]

Through its integration with Alipay+, EZ-Link Wallet users can access competitive exchange rates and promotional benefits through Alipay+ Rewards when paying at point of sale globally accepted by Alipay+. This collaboration further enables payment support in Mainland China with Alipay+ as the underlying payment network.

This cooperation is beneficial to EZ-Link and Alipay+ as well. For users of EZ-Link Wallet, it facilitates carrying out cashless transaction while traveling in the Mainland China. For Alipay+, it widens its footprint globally, enhances the trust of overseas users and makes the platform attractive to more global merchants and sightseeing tourists.

[6] Thoughtworks, 2024, https://www.thoughtworks.com/zh-cn/clients/ezlink.

[7] Alibaba Cloud, 2017, https://www.alibabacloud.com/zh/press-room/alibaba-cloud-national-university-of-singapore-and-ez-link?_p_lc=1.

ComfortDelGro

ComfortDelGro is Singapore's largest land transport company, operating around 35,000 busses, taxis, and rental vehicles.[8] It also manages a 177-kilometer light rail and heavy rail network in Singapore and New Zealand.[9] With global operations across seven countries, including Singapore, Australia, the UK, New Zealand, China, Ireland, and Malaysia, ComfortDelGro plays a vital role in international transport. In Singapore, the company operates approximately 9,000 Comfort and CityCab taxis, supporting around 20 different cashless payment methods.[10]

In 2022, ComfortDelGro has integrated Alypay+ , this integration allowed visitors from Malaysia and South Korea to utilize Touch 'n Go e-Wallet and Kakao Pay for cashless transactions in all Comfort and CityCap taxis in Singapore. With respect to countries, this collaboration works better for travelers from these countries as they can easily avail taxi ride without converting their money to local currency.

Touch 'n Go e-Wallet and Kakao Pay were integrated into ComfortDelGro's payment system through Alipay+ (*Source* ComfortDelGro)

The collaboration brought additional marketing opportunities, including promotional periods with immediate discounts for Touch 'n Go e-Wallet users, designed to strengthen customer loyalty and promote the brand.

Amidst the growing business competition, this collaboration brought about superior operational efficiency and brilliant customer experience to ComfortDelGro while expanding her international footprint as the

[8] Alipay+, 2022, https://www.alipayplus.com/news/detail/comfortdelgro_alipayplus.

[9] Ibid.

[10] ComfortDelGro, 2024, https://www.comfortdelgro.com/about/corporate-profile/.

biggest taxi operator in Singapore through the global network of Alipay+ . As the pandemic improved and the borders of Singapore opened in 2022, the resumption of travel was linked to the economic exterior of the country.

The Chief Executive Officer of ComfortDelGro Taxi, Jackson Chia said "With the travel bans, Singapore welcomed tourists, and safety measures are in place, we expect the increase of travelers once again. In this regard, our partnership with Alipay+ is extremely important as it allows ComfortDelGro taxis to accept cashless payments via Malaysia's Touch 'n Go e-Wallet and soon Kakao Pay. This makes it much easier for visitors from Malaysia and South Korea to pay for rides on taxis and greatly improves their comfort levels while in the country74."

As a result of integrating Alipay+, ComfortDelGro was able to fully integrate the services of such mobile payment systems popular in different countries and directly interact with the growing audience of digital payments users.

"By working in conjunction with ComfortDelGro Taxi, our goal is to provide a continuous and intelligent digital travel experience for all Alipay+ mobile payment users in Singapore, enabling them to use their familiar mobile wallets for transactions. This alliance is timely, coinciding with Singapore's proactive border reopening and the return of tourists to the city-state. We see this as an optimal opportunity for tourists to reconceptualize their travel experiences in the post-pandemic era," said Cherry Huang, General Manager, Global Merchant Partnership, South and Southeast Asia at Ant Group.[11]

Grabpay

Grab is a super app developer that offers mobile services such as ride-hailing, food delivery, and digital payments. Founded in 2012 in Malaysia, the company quickly gained prominence through its ride-hailing service.

Grab made a deal in 2016 with Ant Financial Services Group in order to support the payments of Chinese tourists visiting or living in Singapore or Thailand. Due to this partnership, Alipay users were able to pay for the rides booked through Grab app using their Alipay accounts.

[11] Fintech News Singapore 2022, https://fintechnews.sg/61817/payments/comfortdelgro-partners-alipay-to-accept-mobile-payments-from-malaysia-s-korea/.

In the past, Chinese tourists who visited several countries were advised to use either dual-currency credit cards or cash to settle the rides. By integrating Alipay as the payment option, Chinese foreigners paid in RMB without having to deal with exchange problems. In order to make a payment, the passengers had to open the Grab app, go to the menu and select GrabPay and link it to the Alipay account.

Grab Thailand also conducted a survey of 3,000 Chinese residents from major cities such as Beijing, Shanghai, Guangzhou, and Shenzhen to understand their short-term travel plans to Southeast Asian countries. The survey revealed that Thailand (71%) was the most popular destination for Chinese tourists, followed by Singapore (66%) and Malaysia (53%). Most Chinese travelers preferred to travel in small groups of 2–4 people (81%), while 10% opted for larger groups, and 9% traveled solo.[12]

Worachat explained, "Chinese tourists are one of Grab's key focus areas and strategic target markets this year. Our platform is very well-known among Chinese travelers because it offers a wide range of services, from ride-hailing to delivery, with high safety standards, transparent pricing, and an easily recognizable brand."[13]

Grab's collaboration with Alipay+ highlights the company's continued commitment to providing a more seamless experience for both local and international users of the Grab app. Passengers can choose their preferred payment methods, whether it's cash, credit cards, or Alipay. Grab has also partnered with local mobile operators to offer mobile wallet solutions.

OCBC Digital

OCBC Digital is the digital banking service of Oversea-Chinese Banking Corporation Limited (OCBC), offering a range of digital solutions, including mobile banking apps, online banking, and corporate banking solutions. The primary markets for OCBC Digital are Malaysia, Singapore, Indonesia, and Greater China. OCBC has been consistently ranked among the world's 50 safest banks by Global Finance magazine.[14]

OCBC became the first bank in Singapore to integrate Alipay+ into its OCBC Digital app. Since September 22, 2023, Singaporean users have

[12] Grab TH, 2023, https://www.grab.com/th/en/press/others/grabchinesetrend/.

[13] Ibid.

[14] Global Finance, 2023, https://gfmag.com/banking/worlds-safest-banks-2023-global-100/.

been able to use OCBC Digital for cashless payments in China. With this partnership, OCBC customers can pay at millions of merchants within the Alipay+ ecosystem, with payments directly debited from their OCBC Singapore bank accounts. This allows users to avoid the need for downloading third-party payment apps, setting up accounts, or exchanging foreign currency.[15] Alipay+ provides competitive, real-time exchange rates without extra fees, which are displayed in the app before transactions. Users can either present a QR code to the merchant (customer-presented mode) or scan the merchant's QR code (merchant-presented mode) to complete payments.

The partnership between OCBC Bank and Alipay+ enables OCBC Singapore customers traveling to mainland China, Malaysia, and South Korea to be the first to benefit from this service. OCBC China is committed to supporting the bidirectional flow of trade and investment between the Greater China region and ASEAN, aiming to achieve RMB 1 billion in cross-border collaboration revenue by 2026. By September 8, 2023, OCBC customers were able to make payments at merchants in Malaysia and South Korea that accept Alipay+ QR codes, and from September 22, 2023, they gained access to cashless payments at millions of merchants across mainland China. These merchants span key travel sectors like retail, dining, attractions, and transportation, enhancing convenience for travelers during their daily activities.[16]

This facility is obviously going to be in high demand since traveling OCBC customers based in Singapore visit mainland China where cashless payments are the norm and Alipay can be used for just about anything from taxi rides and restaurant meals to shopping online.

The CEO of OCBC Group, Helen Wong, has pointed out that Greater China is a core region for the bank, especially with its current and even future technologies being growth engines of the region which also create positive spilling effect across Greater China. By the year 2026, OCBC interests China's growth center and will earmark RMB 1.5 bilion to invest in data and AI development as well as standardizing platforms to serve clients and employees better. The group will develop its engineering center with plans to recruit 300 engineers in mainland China in the next

[15] OCBC, 2023, https://www.ocbc.com/group/media/release/2023/ocbc-is-first-bank-in-singapore-to-offer-global-cross-border-payments-via-alipay-plus.

[16] OCBC China, 2023, https://www.ocbc.com.cn/CN/group/medium/202312/P020231207432679171655.pdf.

three years. To improve its technology infrastructure, OCBC Hong Kong will spend HKD 1 billion to improve all its systems within three years and expects to have double digit increase in income in five years.

The broad market reach brought by Alipay+ is helping the Singapore government improve the experience of foreign travelers, while fostering greater connectivity among businesses. This advances the building of an integrated, open, and friendly digital economy. Partnering with local companies and institutions, Alipay+ helps expand a more accessible financial ecosystem by deploying capital, technological, and know-how resources.

17.1.4 Digital Finance and Low-Carbon Transition under Singapore's Smart City Construction

17.1.4.1 Singapore's Pioneering Smart City Layout and Practice

Singapore is one of the first countries to introduce "government informatization" globally. The journey of Singapore's smart city development dates back to the 1980s when the country began implementing the "National Computerization Plan," gradually promoting government digitalization. In 1992, Singapore launched the "IT2000 Intelligent Island Plan," followed by the "Infocomm 21" plan in 2000, and later introduced the "Intelligent Nation 2015" initiative in 2006. In 2014, Singapore unveiled the "Smart Nation 2025" plan, aiming to become the world's first Smart Nation.

Under the "Smart Nation 2025" framework, Singapore adopted the "3C" concept: Connect, Collect, and Comprehend. This approach focuses on using advanced digital technologies to establish secure, high-speed, and scalable connectivity, collecting real-time data through a nationwide sensor network, and leveraging data analytics to accurately sense and predict citizens' needs, thereby enhancing public service delivery.

IT penetration and development has also offered Singapore a chance to take its competitive advantage a notch higher while coping with the constraints of having limited land, resources, and a large aging demographic. The Smart Nation initiative has been grafted well to the next phase of national development of Singapore where the opportunities of the digital age are utilized to gain growth and remain globally competitive.

Smart Singapore, the city-state's smart city initiatives have spanned transportation, healthcare, education, and finance, to cite a few examples. For instance, urban traffic in the city-state has been optimized using an intelligent transport system to reduce congestion. As for healthcare, e-health records and telemedicine technologies have increased efficiency and improved the quality of medical services. In the field of education, the availability of digital platforms and tools has also improved the delivery and personalization of learning.

In the Sustainable Trade Index 2023 launched by the Hinrich Foundation, Singapore has risen to third place globally. The index assesses 30 economies based on ESG (Economic, Social, Environmental) factors, evaluating how well these economies balance trade with sustainability. Singapore achieved a score of 94.1, ranking just behind New Zealand and the UK. Economically, the high score is a reflection of Singapore's strong infrastructure, technological advancement, and ample funding opportunities for the private sector. Socially, Singapore enjoys a sound political and economic atmosphere, advocates for education and allows for upward social mobility. Environmentally, Singapore is at the front in Asia in terms of pollution control, carbon cutting, energy savings, and wastewater treatment based on the high level of policies applied in these areas.

Sustainable trade index 2023 (*Source* Hinrich Foundation)

The Framework of smart cities' presents new solutions for building and developing urban space. Cities are enabled through the ubiquitous data transmission, the Internet of Things, and big data integration to optimally use their strategic resources to create new means of development. Smart city means not only enhancing infrastructure with smart devices but in addition integrating technologies and innovations that will make artificial intelligence to enhance public services, make economies move, and develop people-focused services.

17.1.4.2 Building Inclusive Digital Finance

There exists a strong relationship between smart cities and their digital finance ecosystems, where the latter acts as one of the building blocks that enables the growth and functionality of the former. A broad definition of digital finance is the delivery of financial services through technology, and given the sort of and the nature of urban growth taking place, it is imperative to promote and expand the outreach of financial services in cities. It advances the benefits of the smart city by facilitating transactions, enhancing connectivity to wide networks of financial services, and stimulating growth in the financial industry.

Smart cities and their technology infrastructure incorporated digital finance within its practices like contactless payments, mobile banking, or incorporation of blockchain to facilitate secured transactions. These activities do not only facilitate the ease of conducting financial transactions but also enhance the capacities of smart cities by providing data for urban and policy development.

The development of digital finance practices in Singapore can be described as fast and significant. The Monetary Authority of Singapore (MAS) in 2018 launched the Unified Point of Sales System (UPOS) to ease the limited number of terminals, as well as enhance the capabilities of cashiers to distinguish between a variety of payment instruments quickly.

Emerging e-wallets received a boost in 2018 when MAS and the Infocomm Media Development Authority introduced the SGQR unified QR code standard. What began as an additional incentive for e-wallet adoption is now an industry norm.

Ant Group, which had already collaborated with the Singapore Tourism Board, recognized Singapore's forward-looking development vision and immense growth potential, extending the Alipay+ payment partnership to the country. As Southeast Asia becomes the fastest-growing region for mobile payments globally, with users gradually adapting to their local e-wallets, Alipay+ offers an open digital payment solution with limitless connectivity. It can integrate millions of merchants worldwide into a single technology service, connecting directly to over a billion Asian consumers through numerous overseas digital wallets. Alipay+ 's marketing tools also enable local merchants undergoing digital transformation to perform precise marketing to overseas customers, achieving reach on both the B2B and B2C fronts.

In 2021, NETS implemented the Alipay+ solution to support foreign travelers spending in Singapore. Developed by NETS, SGQR is the world's first integrated payment QR code. Dr. Huang Yin, the General Manager of Ant Group Alipay+ Merchant Services, shared that they in collaboration with NETS will focus on advancing digital payment technologies that connect thousands of merchants across the island to consumers from Asia.

In November 2023, the MAS published a concept validation test for SGQR + . The improved version of SGQR + offers merchants the facility to accept 23 local and cross-border payment types, such as Changi Pay, EZ-Link, GrabPay, Google Pay, Visa, Alipay, WeChat, DuitNow, and PromptPay. Merchants are allowed to sign a contract with one financial institution, allowing them to accept a variety of local and international payment methods, thereby increasing sales potential.

Previously, with the SGQR code, merchants had to maintain separate commercial relationships with individual financial institutions to accept a range of payment options such as Alipay, Grabpay, and NETS. Consumers could only pay with their preferred options at specific points of sale that accepted those options. The upgraded QR code, which combines the strengths to facilitate Alipay+ and NETS to bring together the essential functions, aims to promote and standardize electronic consumption in the region, integrating major electronic wallet companies into Singapore and enhancing linkages with neighboring countries such as Malaysia and Indonesia. It also allows Alipay+ to connect more small and micro enterprises in Singapore and make the market more integrated, encourage sustainable tourism, and make payment experience smoother and more effortless.

It is very clear that the growth of digital finance in the smart city of Singapore is a multifaceted and multi-layered undertaking. Apart from the aforementioned digital payment systems, other factors such as financial services, technological infrastructure, and policies also need to be in place.

The new financial service framework developed by Singapore's urban center uses its major urban resources to chart new areas for city growth by including activities like business cloud migration, expanding access to financial services, increasing the digital competency of the citizens of Singapore, and helping the local communities and businesses to embrace technology in a big way. Besides, the MAS has begun the implementation of the Financial Sector Technology and Innovation (FSTI) program, which included the development of quantum computing and AI capabilities in finance as well as other innovative ventures. Stakeholder collaboration is crucial, and the MAS also actively supports financial network security and offers funding for innovative approaches such as post-quantum cryptography and quantum key distribution to safeguard essential enterprise information and minimize the impact of cyber threats.

In the case of technological infrastructure, the Smart Nation Sensor Platform (SNSP) in Singaporean context is a country-wide integrated platform using sensors to gather important data for further analysis in order to develop appropriate solutions to problems. For example, wireless sensors embedded in smart meters enable households to remotely manage their water usage data and leakages using a smartphone.

Moreover, the government of Singapore undertook implementation of a number of straightforward projects supported by the Smart Nation policy initiative so as to enhance the interactions of Singaporeans, businesses and the government through the application of modern technologies. These projects are built on three key pillars: fostering technological reach out to all Singaporeans, improving the digital skills of Singaporeans, and encouraging local communities and businesses to support widespread adoption of technologies.

Digital finance has many layers of significance on the development of smart cities. It enables easy and fast to access financial services to residents which enhance their living standards, it enables easy access to local businesses financing thereby providing a source of growth and development to the city's economy as well as making it attractive to both investors and tourists. Digital finance also enhances the sustainability of cities by providing a robust financial ecosystem that is capable of withstanding and recovering from economic disruptions.

In terms of scale, the adoption of digital finance in smart cities is often all-encompassing incorporating many aspects, including public transport and utility bills, government services, and business transactions. A huge impact integration of new finance technologies has also as it transforms economic processes, promotes tech progress, and expands possibilities of green growth of the cities.

Smart cities and digital finance are created for each other: the first one develops due to advances in technologies powered by digital finance, the second one gets stimulated through the wide implementation of innovative digital financial tools. In this regard emergence and advancement of digital finance tools is crucial to the establishment of the smart city concept, which is oriented towards the development of advanced, functional, and environmentally friendly cities.

17.1.4.3 Low-Carbon Transition of the Garden City

In 1965, as Singapore dealt with the problems of urban expansion and loss of natural ecosystems, its then-leader Lee Kuan Yew advocated for the beautification of the city with the purpose of propelling the economic growth of Singapore through clean and green spaces. The initiative was so successful that by 1980s, Singapore earned the title "The Garden City." Since the late 1990s, the government has embraced a "city in a garden" concept which aimed at incorporating the natural ecosystem in urban planning with an aim of making the city more hospitable for international business, investment, and its inhabitants, thereby committing to sustainable development from the onset.

Environmental and sustainable development have been central narratives since the birth of Singapore. The country has taken several measures to assist in balancing its low-carbon economy transition and smart cities building, with an intention of attaining both sustainability and digital transformation within the nation.

As a model of smart city development, integration of transport in Singapore is one of its key elements demonstrating the ability to think ahead in regard to smart mobility solutions. Singapore has established an integrated transport system through the use of advanced sensor networks, real-time traffic management systems, and a multitude of public transport options. These days, commuters have it easier as they access real-time information about travel as well as book their trips via their smart cards or mobile devices. Such an intelligent transportation system is also helpful in monitoring and controlling the traffic hence efficiency on the roads is achieved while wastage of energy and pollution to the environment as a result of traffic jams is also kept in check. Besides those advancements, these strategical innovations can also reduce traffic congestion and travel time and will help to set up more environmentally friendly links in the transport network.

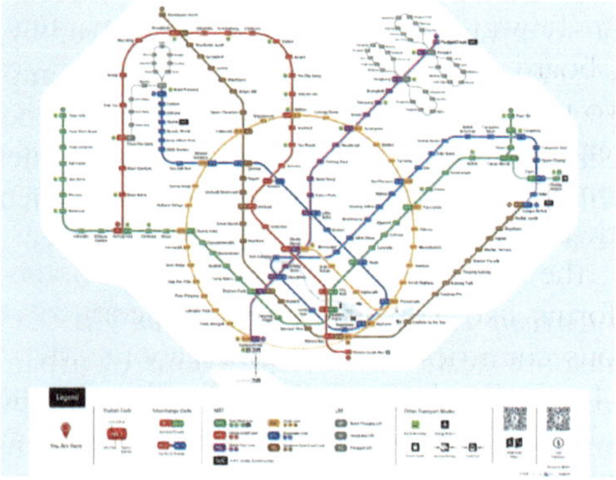

Singapore MRT (Mass Rapid Transit) Map (*Source* Land Transport Authority)

Singapore's smart city projects have greatly advanced in fostering the environment and system sustainability. When it comes to waste management, for instance, Singapore has adopted an automatic pneumatic waste collection system that is novel. This technology allows for the collection of waste via a pipe system embedded within the ground which minimizes the need to use garbage trucks making them more efficient and also minimizes carbon emissions. Furthermore, Singapore has a holistic approach to the management of water resources and energy efficiency through substantial investments in smart grids and solar energy infrastructure to maximize resource utilization and environmentally friendly in the long run.

As far as green characteristics of the buildings are concerned, Building and Construction Authority (BCA) took the initiative by launching the Green Mark Certification Scheme, which appeals to building owners and operators to construct greener buildings in an energy and resource-efficient manner. By the end of year 2020, more than 4000 projects within Singapore have managed to achieve the BCA's Green Mark certification standards.[17]

[17] Singapore Economic Development Board, 2023, https://www.edb.gov.sg/cn/insights/towards-a-green-future-unveiling-singapore-sustainable-development-dna.html.

The efficient use of solar energy is also a key focus of the Singaporean government. The Singapore Economic Development Board (EDB) supports companies in advancing cross-sector decarbonization pathways through policy tools, infrastructure, and ecosystem support. For instance, in 2014, the EDB and the Housing & Development Board (HDB) for example rolled out the SolarNova program for the roll out of solar photovoltaic (PV) systems throughout Singapore.

The movement towards carbon neutrality has also been advanced by Singapore with a plan which seeks to transform the power, industrial as well as the transportation system. Concrete goals include "zero private vehicle growth," achieving "90% of peak-hour trips using 'walk, cycle, ride' modes," and ensuring that "all vehicles use clean energy by 2040."

The government mobilizes the population at the school level, community, and media channels through sustainable development programs so that the population become aware and knowledgeable of sustainability. Various measures promote green commuting and promote energy saving household appliances under the program "Climate-Friendly Households."

Furthermore, Singapore remains open to engage with other countries on its smart nation projects. It has entered into digital economy agreements with countries including China, New Zealand, Chile, and Australia and promotes green economy agreements. Singapore has also entered into agreements with several countries for carbon credit trading.

Overall, Singapore has made remarkable achievements in both smart city and digital finance development, and its low-carbon transition efforts are particularly noteworthy. Through its "Smart Nation 2025" initiative, Singapore intends to leverage on technology to improve its economy and the people's participation through better and wiser planning and management. The comprehensiveness of the vision also underlies the inclusion of smart nation elements in the language of healthcare, transport, urban governance, finance, and education, bringing efficiency in public service delivery and enabling change through technology at all scales.

In digital finance, the focus of the Singaporean government is to develop a public sector that is effective and able to withstand the test of time with digital information at its center. E-governance focuses on services that are citizen centric, the unification of various policies and the promotion of change through improving digital and data platforms, developing operational systems, and enhancing the digital capacity.

As for the low-carbon transition, the Singaporean government also implemented the “Singapore Green Plan 2030,” which lays out systematic planning for sustainable development over the next decade, demonstrating its commitment to the UN 2030 Sustainable Development Agenda and the Paris Agreement. Singapore aims to reduce emissions to around 60 million tons of CO_2 equivalent by 2030 and achieve net-zero emissions by 2050. The country is still working on enhancing its sustainability ecosystem and is still willing to honor its international obligation of partnering with nations such as China on renewable energy, green architecture, and green finance.

These comprehensive measures have shaped Singapore into a resource-efficient, environmentally friendly modern metropolis, offering valuable insights and experiences for global smart city and sustainable development initiatives.

17.2 Korea’s Digital Payment Ecosystem Development and Transformation

17.2.1 Accelerating the Cashless Era: An Overview of South Korea’s Mobile Payment Landscape

The shift towards a cashless society in South Korea has rapidly gained momentum. Following the trend seen in regions like the U.S., Europe, and China, South Korea has formally entered the financial technology (fintech) era. A growing number of South Koreans are turning to e-wallets and mobile payments. By 2021, only 14.6% of payments in South Korea were made with cash. This highlights a significant transformation in consumer behavior and payment systems.

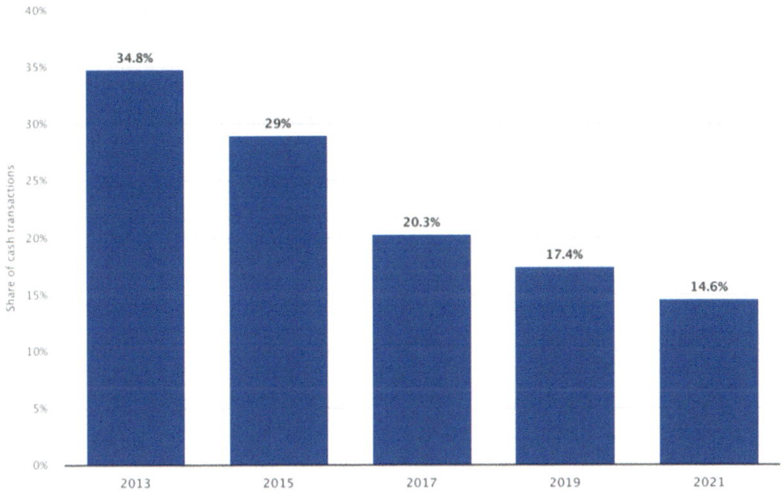

Share of cash in total payments in South Korea from 2013 to 2021 (*Source* Statista)

The COVID-19 pandemic has accelerated the adoption of digital payments, with contactless payments via mobile wallets or e-wallets becoming the norm. The pattern and trend continued after the end and state of the pandemic. Quite simply, people and businesses have become accustomed to the concept of the digital payment's ecosystem.

When it comes to payment trends of the population, the market is divided into proximity and remote payments, with a particular focus on QR code payments. The increase of mobile phone penetration that already comes with the NFC built in, plus the POS systems that accept mobile payments, are expected to help mobile payments to grow.

The expansion of mobile payment services is also a function of the e-commerce development of the country, where companies like Coupang provide us with same-day delivery and online grocery shopping services.

17.2.1.1 Kakao Pay: A Digital Financial Strategy in Line with Global Trends

Ready to blossom is the South Korean mobile payment market considering the level of technology and the changing trends in consumer behaviors and the number of players within the market operation including both local and international. The emergence of the market will remain to complement the overall digital payments revolution in the future.

The South Korean mobile payment market is characterized by a highly competitive landscape with several major players dominating the scene. In 2023, the market size was USD 36.96 billion and this is projected to hit USD 40.67 billion in the year 2024 and grow to USD 62.96 billion by the year 2029, representing a compound annual growth rate of 9.13% during the period 2024 to 2029.[18] The market is fueled by growth in the e-commerce industry, the increase in mobile payments via QR codes and more businesses are incorporating mobile payment systems. This growth is encouraged by existing key players in the market coupled with increased foreign competitors entering the market.

In the South Korean mobile payment industry, key players include Kakao Pay, Naver Pay, and Toss. Certainly, such platforms have helped ease the transition to a cashless economy where Naver Pay and Kakao Pay are considered frequent practices. The rise of the fintech sectors has also posed a threat to the traditional banks and card companies, who have responded by broadening their taxable services to include payment options.

The Korean messenger app leader, Kakao Group, created a payment service called Kakao Pay in 2014. Kakao Pay was developed within the much-adored social application KakaoTalk with oodles of people. The application allows for online and mobile commerce through the use of NFC technology and QR codes. In the course of time, Kakao has morphed from a payment service into a full-fledged financial service provider that offers not only payments but remittances, bills, and more.

Kakao Pay successfully takes advantage of its recent user acquisition wave through Kakao Talk to position itself as a major player in the developing Fintech ecosystem in South Korea. Emerging as one of the mobile payment leaders in South Korea, Kakao Pay has indeed transformed the general landscape regarding the adoption of digital payments in Korea. As of 2022, the company has 36.5 million registered users and around 20 million of them are monthly active users which is a large chunk of the market. For the twelve months ended June 2021, the platform had a total reported transaction volume of $72.4 billion which placed them

[18] Mordor Intelligence, 2024, https://www.researchandmarkets.com/reports/5624552/south-korea-mobile-payment-market-share#product–toc.

as the leaders in mobile payments. In the same year, the payment platform, recorded Kakao Pay revenues of 2840 billion KRW (about $2.45 billion) representing an impressive growth rate of over 102% year-on-year change.[19]

Kakao Pay's market size in 2023 was around 26 trillion KRW, trailing Naver Pay's 43.5 trillion KRW, as stated by Statista.[20] This situation allows Kakao Pay to remain as one of the dominant forces within the mobile payments market in South Korea and its significance informs the changing trends regarding consumer payments and technologies surrounding them.

The rapid growth of Kakao Pay has also attracted significant attention from investors. In 2017, Ant Financial, a subsidiary of Alibaba, invested $200 million in Kakao Pay, acquiring a 45% stake in the company. In 2017, the directors of Kakao decided that it would be best for the company to form a separate organization for its financial services. This structure aims to have a broader base, which includes incorporating Kakao Pay, which is a facilitating platform for multiple services like facilitating payments physically, P2P transfers, paying bills, online banking, and much more, to make it into a broader platform that includes lending and capital assistance.

Ant Financial has been involved in these areas in China and knows the ropes, so the cooperation works for both sides. While this effort is expected to hasten the expansion of Kakao, the parent firm, Alibaba Group, will also gain considerably. This partnership could permit Alibaba to actively pursue its e-commerce sites in South Korea, and make it easier for Chinese visitors to Alipay when in the country.

In November 2022, Kakao Pay came to an arrangement with Alipay+ and managed to expand its payment features to more areas, including e-commerce, digital entertainment, dining, travel, duty-free, convenience stores, shopping centers, and transport. Due to integration with Alipay+, KPP reported earnings, making a profit of around 27.5 billion KRW (nearly $19 million) due to the operational activities of that quarter.[21]

[19] Kakao Pay, 2024, https://www.kakaocorp.com/page/.

[20] Statista, 2024, https://www.statista.com/statistics/1255503/south-korea-mobile-payment-market-size-by-provider/.

[21] Jiemian Global, 2023, https://www.jiemian.com/article/8913211.html.

This working integration has enabled Kakao Pay in collaboration with Alipay+ to ensure that their services offer global remnants and loyalty opportunities. Alipay+ will allow you to make purchases through Kakao Pay at various physical and online retail locations in China. This partnership not only aims to assist South Korean tourists and consumers in China but also contributes to the development of today's connected payment technologies in the region.

Kakao Pay also provides users with transaction services as well as works in the financial sector by developing and providing a variety of finance activities, including investment, loans, and insurance, giving it a wide consumer framework. The presence of this platform in the ecosystem of other subsidiaries such as Kakao Talk and Kakao Bank provides great possibilities for market strengthening and interaction with users.

Kakao Pay has established itself in the mobile payment sector in South Korea by providing its millions of users with secure transactions while consistently modernizing operations in line with new technologies in the region. The agreement with Alipay+ complements the cross-border payment experience, thereby maintaining its market position in the international digital payment market.

17.2.1.2 Naver Pay: Inclusive Growth Through Partnership with Alipay+

Naver Pay is another practitioner in the mobile payment services in South Korea. It is a service operated by NAVER, Korea's leading internet service provider, which had its launch in 2015. Its usage is comparable to PayPal since it allows users to link their designated bank accounts and perform payments online or offline. In June 2021, Naver Pay boasted about 28 million registered users with over 10 million active users.[22] This service plays a crucial role in South Korea's mobile payment market due to its widespread usage in both online and offline transactions.

The service is harder to avoid, considering the major age 20 to 49 representative consumption categories own the age group statistics and dominate the user base of the platform. Naver Pay is thus a potent weapon in the arsenal of the digital economy.

[22] Silkpay, 2023, https://www.silkpay.cn.com/blog/accept-naver-pay-through-alipay-to-unlock-global-markets.

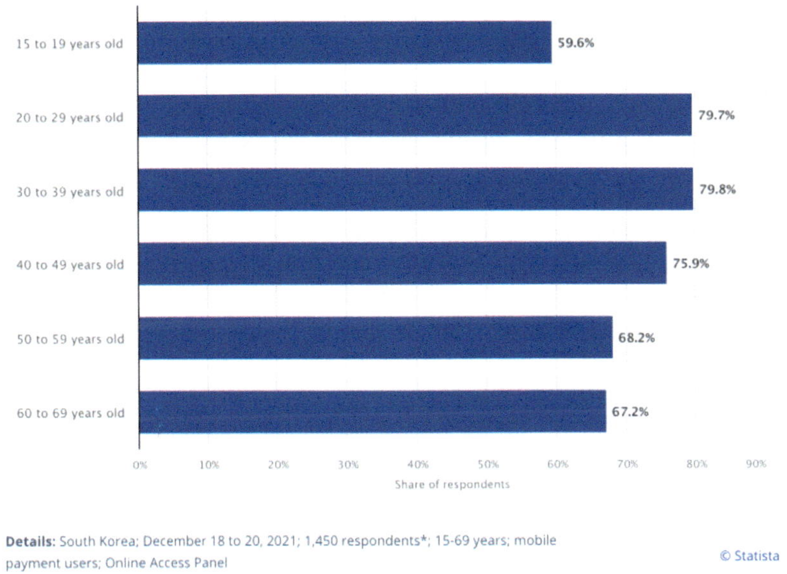

The primary consumer demographic for Naver Pay consists of users aged 20 to 49 (*Source* Statista)

Post-pandemic, in 2023, there was a sharp increase in offline shopping through Naver Pay. Between March and April, the average spending per person in stores grew by 123%, with 72% of users making two or more offline payments in April alone.[23] In the same month, Naver Pay's monthly transaction volume hit 900 billion KRW (around $798 million USD), highlighting its strong influence in the mobile payment market.

Naver Pay is now accelerating its digital financial transformation, aiming to broaden its financial services and become a comprehensive platform. In September 2023, Naver Pay announced that it would begin offering Naver Pay Money and Point QR-based payment services at all Alipay merchants across China, including at the 19th Asian Games in Hangzhou.

Alipay+ was an official sponsor of the 2023 Asian Games in Hangzhou, providing exclusive payment services at key venues such as stadiums and

[23] The Korea Herald, 2023, https://www.koreaherald.com/view.php?ud=20230519000524.

the athletes' village. With this partnership, Naver Pay's QR code payment system could be used anywhere in China where Alipay is accepted. A Naver Pay representative commented, "Starting with the Hangzhou Asian Games, we will expand global payment collaborations, offering more unique experiences and benefits exclusive to Naver Pay in additional countries."[24]

As a result of the collaboration between Naver Pay and Alipay+ Naver Pay, users are able to use their platform to make payments directly in China, thus streamlining the payment process. This not only promotes cross-border accessibility for Naver Pay users but also assists Alipay+ in further developing its international markets.

Naver Pay's global expansion through Alipay+ has provided ease of use to its users and established advantageous dependability on both parties.

For Naver Pay:

1. Hassle-Free Payment Satisfactions: With the aid of Naver Pay, App users no longer have to worry about carrying cash or the hassle of exchanging currency when traveling to China as they are connected to Alipay's wide range of advertisers. This allows payment to be done directly from mobile devices, which most people do habitually back home in South Korea.
2. Transparent Exchange Rates: The partnership avails transparent and market-competitive exchange rates for users, which is important for transacting internationally and also assists in cost-cutting for the users during currency conversion.
3. Promotion Deals and Rewards: Some promotional deals and discounts are also available through Alipay+ Rewards for Naver Pay users, increasing the benefits of shopping in China for the users.
4. Customer Support: While traveling abroad, Naver Pay's team that cooperates with Alipay+ provides users with expected customer support. This guarantees that consumers have adequate assistance during transactions which is backed by Alipay's technology in payment facilities.

For Alipay+:

[24] The Korea Economic Daily Global Edition, 2023, https://www.kedglobal.com/fintech/newsView/ked202309190018.

1. Expanded User Base: For Alipay, Naver Pay being integrated in its system provides an opportunity to acquire additional user base of South Korean tourists and businesses that will enhance the volume of transactions and revenue.
2. Market Expansion: The partnership provides opportunities for Alibaba, the parent company of Naver Pay to penetrate the South Korean market and step-up its marketing of the e-commerce services. It also allows Chinese users to use Alipay in South Korea making the payment system more integrated.
3. Global Merchant Network: The latest addition to Alipay's cross-border payment service is the integration of Naver Pay which expands its network to more international users.

With Naver Pay's expansion to 42 countries and regions through Alipay+, users can enjoy a more convenient travel experience, making payments with their familiar payment app without the need to exchange currencies. The Alipay+ and Naver Pay partnership has not only made international transactions more accessible and user-friendly but has also created strategic opportunities for market expansion and service enhancement for both platforms. This collaboration exemplifies the growing trend of cross-border digital commerce and the importance of mobile payment solutions in facilitating global travel and trade.

17.2.1.3 Toss: Expanding User Base and Entering the Tourism Market

Toss is a leading mobile payment platform in South Korea, launched by fintech company Viva Republica. It supports a wide range of financial services, including money transfers, loans, insurance, credit score management, and cross-border payments. "Toss Pay" allows users to link Visa, MasterCard, and nearly all Korean domestic bank accounts, enabling bank transfers and online payments.[25] The platform is highly functional, offering real-time transactions without chargebacks, making it particularly beneficial for merchants in industries such as e-commerce, live streaming, gaming, and content subscriptions. Its high-security features have also

[25] SilkPay, 2023, https://internationalbanker.com/technology/why-singapore-is-consistently-regarded-as-asias-smartest-city/.

contributed to its reputation as a reliable e-wallet, with no security issues reported since its launch.

Toss boasts a substantial user base in South Korea, with over 20 million registered users as of 2023, covering 98% of the country's banking customers.[26]

Ant Group in the year 2023 made announcements about forming a strategic partnership with Toss Payments. The two companies will assist international merchants in expanding into the South Korean market to enhance their services to local customers.[27] This made Ant Group the second-largest shareholder in Toss Payments after it bought a 72% stake in Blitz Fast, a property funded by Viva Republica, Toss Payments, LB Private Equity, and Premier Partners. Consequently, to this acquisition, Ant Group also secured 36% shareholding in Toss Payments after taking participation in a new shares issue which was massively worth billions of Koreans won. This tactical partnership enables Toss to widen its scope in South Korea and promises a solid platform for the company to enter foreign markets particularly China. Through this partnership, Toss Pay became the first Korean payment operator which provided its customers with the ability to make payments at specific offline shops in China.

The relationship with Ant Group saw Toss increasing its global competitiveness and its market reach. Toss Pay's overly rudimentary payment provider has now been extended to 42 countries including the United States, Japan, United Kingdom, Singapore, and Australia, where offline payments systems are applicable.[28]

Toss Pay users can make easy payments at "Alipay+" partner merchants by scanning barcodes or QR codes with the Toss app. The amount is deducted in Korean won from the linked account. A Toss representative explained, "It's convenient because there's no need to exchange into USD or EUR, and users can instantly verify the amount deducted."

Also, the cooperation of the partners Toss Pay and Alipay+ makes life easier and creates more opportunities for the Chinese tourists in South Korea. For one, Chinese tourists can use Alipay at more than 1.7 million

[26] PPRO, 2024, https://www.ppro.com/payment-methods/tosspay-wallet/.

[27] SilkPay, 2023, https://internationalbanker.com/technology/why-singapore-is-consistently-regarded-as-asias-smartest-city/.

[28] The Chosun Daily, 2023, https://www.chosun.com/economy/tech_it/2023/11/20/SLNA5FTRM5GWZABYKRH2YPPB5E/.

merchants, which means paying is a lot easier. And, because of the partnership, there are special offers for customers such as up to 20% off when purchasing Alipay+ in areas like Myeongdong, Seoul.

This convenience in payment applies across different travel scenarios. With Alipay+, the tourists from China can use Alipay in taxis, duty-free stores, convenience stores, and restaurant chains and enjoy a payment system like in their hometown. This unified system further eliminates the need for exchanging currency and carrying change and this enhances the travel experience of the visitors significantly.

Since Alipay+ is operational in South Korea, it gives a chance for more Korean merchants to showcase their business to the Chinese tourists which allows the South Korean tourism sector to venture to other countries. However, Chinese tourists have the advantage of easier mobile payment and purchasing goods at a cheaper price through discounts and promotions available on Alipay+, enabling them to enjoy their trips more. The partnership between Toss Pay and Alipay+ facilitates the globalization and the digitization of the payment system in South Korean market. It also highlights the need of building corporate alliances for the digital financial system's development in the era of globalized economy.

17.2.2 Digital Payment Integration Boosts Tourism Recovery: Alipay+ in Korea's Payment Ecosystem

On August 10, 2023, China's Ministry of Culture and Tourism announced the full resumption of group tours to South Korea. This marked the end of a six-year and five-month hiatus, initiated in March 2017 when the government halted group visa issuance due to the deployment of THAAD (Terminal High Altitude Area Defense). In response, the South Korean Ministry of Culture, Sports, and Tourism released a plan to attract Chinese tourists, aiming to bring in 2 million visitors by the end of the year.

The preferences and consumption habits of Chinese tourists have changed drastically due to the COVID-19 pandemic in the Korean tourism market. Chinese tourists have been seen to showcase different traits in their travel habits post the pandemic. Firstly, the current overseas travel population mainly consists of millennials born after the 1980s who are more inclined to customized travel experiences and expect a certain level of service quality. As far as South Koreans are concerned, tourism expenditures are made predominantly on the beauty and hair, medical services, dermatology, and fashion. Although the demand for group tours with an objective of shopping and visiting famous tourist attractions has somewhat decreased, food tourism and local tourism with the participation of local people and appropriate tourism are gaining popularity.

The recent surge of Chinese tourists has heavily influenced the tourism industry of South Korea which has led to the Korean government implementing the use of mobile payments in the country. One of the main objectives is to increase the number of merchants who accept mobile payment systems such as Alipay in the most popular tourist areas visited by Chinese tourists. Small businesses are also welcomed to adjust to the consumption changes to tap into the economic gain that comes with the Chinese tourists.

On Jeju Island, which offers visa-free entry to Chinese tourists, this initiative has drawn a large number of visitors. According to data from the Jeju Tourism Organization, Alipay+ transactions in Jeju's Dongmun Traditional Market and Seogwipo Daily Olle Market skyrocketed, with the payment amount rising 15 times from 17 million KRW in March to 250 million KRW in May. Payments through Alipay+ also increased from about 700 in March to 8000 transactions by May.[29]

Jeju Island is also offering its tourists various discounts and effective bargains via Alipay+. To begin with, tourists visiting Jeju can use Alipay to pay for taxis, duty-free shops, convenience stores, fast food chains, and many other merchants throughout the island. Approximately 120,000 merchants throughout the island's territories have provided Alipay+ services, allowing a more convenient payment process, similar to China, while on the island.[30]

Additionally, Alipay+ has partnered with the Korea Easy Payment Foundation (KEPF) and the Jeju Tourism Organization (JTO) to enhance international visitors' experiences by offering convenient digital payment services and exclusive deals. Tourists from various regions—such as mainland China using Alipay, Hong Kong with AlipayHK, Macau's MPay, Thailand's TrueMoney, and other e-wallets like Malaysia's Touch 'n Go e-Wallet, Singapore's OCBC Digital and Changi Pay, Mongolia's

[29] The Chosun Daily, 2024, https://www.chosun.com/national/regional/2024/06/10/GEESCPXCRNFKXO3VKQYQW6CG6U/.

[30] Ant Group, 2022, https://www.antgroup.com/news-media/press-releases/1664262883000.

HiPay, the Philippines' GCash, and Italy's Tinaba—can all scan ZeroPay QR codes to access exclusive offers at ZeroPay merchants across Jeju.[31] This localized collaboration ensures that tourists using Alipay+ enjoy tailored services and discounts throughout the island.

Data from Jeju Tourism Organization states that 377,734 foreign visitors traveled to the island between January and March 2024, a 532.5% increase compared to 58,609 visitors during the same period the previous year. This marked the first time since February 2020 that Jeju saw more than 100,000 international visitors each month for three consecutive months. Around 80% of these tourists were from China, followed by Taiwan, Singapore, and Japan.[32]

Given the significant increase in international visitors to Jeju, "we hope this partnership will help revive inbound tourism and support local businesses," said Danny Chung, general manager of Ant Group in Korea, Australia, New Zealand, and Mongolia.

The three major payment platforms in South Korea are listed among Alipay+ partners (*Source* Alipay+)

[31] The Digital Banker, 2024, https://thedigitalbanker.com/alipay-elevates-seamless-travel-experience-for-overseas-travelers-visiting-jeju-under-partnership-with-korea-easy-payment-foundation-and-jeju-tourism-organization/.

[32] Jeju Tourism Organization, 2024, https://www.ijto.or.kr/english/.

With the help of Ant Group's advanced technology and Alipay's collaboration with APIs and Korean local payment service providers, the outbound travel spending of Chinese travelers to South Korea is expected to continue to increase. Chung stated at a press conference, "As travel demand from Chinese and Southeast Asian tourists to South Korea has returned to 50% to 70% of pre-pandemic levels, coupled with improvements in payment infrastructure, offline payments in South Korea through Alipay+ have increased by approximately 700% from January to October 2023."[33]

At the same time, through Alipay+, Chinese visitors can also take advantage of a wide range of discounts and promotional campaigns. Alipay has previously entered into partnerships with many destination tourism organizations and a million global merchants to reach out to Chinese travelers who are planning to travel and offer them incentives such as purchasing the Global Travel Card and the Destination Savings Pass before or during their trips which give discounts, travel benefits and free data. Alipay+ has partnered with Starbucks in Korea to offer special deals and promotions for foreign guests to improve their brand experience.

Under the framework of the marketing agreement signed between the two parties, Starbucks in Korea has pledged to offer exclusive benefits to users of Alipay+ partner e-wallets for the entire year of 2024. The Alipay+ program enables users to make sales via their local e-wallets at all 1916 Starbucks branches in Korea.[34]

The practice and influence of Alipay+ in the Korean market manifests as a positive boost to the tourism and lifestyle industries, particularly benefiting merchants such as retail stores, taxi services, hotels, and tourist attractions. Even through Alipay+, local merchants are able to effectively expand their international clientele while providing tourists with a payment solution similar to what they are used to. These initiatives by Alipay+ not only boost its competitiveness in the international payments market but also foster development and dynamism to the digital payment landscape in Korea.

[33] The Korea Times, 2023, https://www.koreatimes.co.kr/www/nation/2023/12/126_364868.html.

[34] Electronic Payments International, 2024, https://www.electronicpaymentsinternational.com/news/alipay-partners-with-starbucks-korea/.

17.3 Japan's Mobile Payment Expansion and Sustainable Tourism Ecosystem

17.3.1 Japan's Efforts in Digital Payment Development: PayPay's Local Services

PayPay is one of the most used mobile payment methods in Japan established in 2018 as a partnership between SoftBank Group and Yahoo Japan Incorporation and Paytm from India. It was launched at a time when Japan was actively transitioning to a "cashless society" in order to enhance the payment activities' efficiency and ease along with improving the economy's operational transparency.

PayPay has rapidly developed within Japan's mobile payment market, with registered users surpassing 60 million by October 2023, accounting for about half of Japan's population and roughly two-thirds of smartphone users.[35] In 2024, the estimated size of Japan's mobile payment market is projected to reach $211.13 billion, with expectations to grow to $815.79 billion by 2029, reflecting a compound annual growth rate of 31.04%.[36]

Since its inception, PayPay has quickly established a significant presence in the Japanese market. It offers an e-wallet where the users can link their bank accounts and credit cards to the PayPay app, so they can charge money to the app for payments and payments directly within that app. With PayPay, once a user pays through QR code, they can complete the transaction by either displaying their QR code to merchants or allowing the merchants to display their QR code. Payments are very common, including convenience stores, drugstores, television electronics shops, restaurants, and the izakayas. A PayPay survey was conducted in 2024 by the Mobile Marketing Data Laboratory, making it the only mobile payment system where the QR code was used by 64% of Japanese users.[37]

[35] PayPay, 2023, https://about.paypay.ne.jp/pr/20231025/01/.

[36] Mordor Intelligence, 2024, https://www.mordorintelligence.com/zh-CN/industry-reports/japan-mobile-payments-market.

[37] MMD Laboratory, 2024, https://mmdlabo.jp/investigation/detail_2306.html.

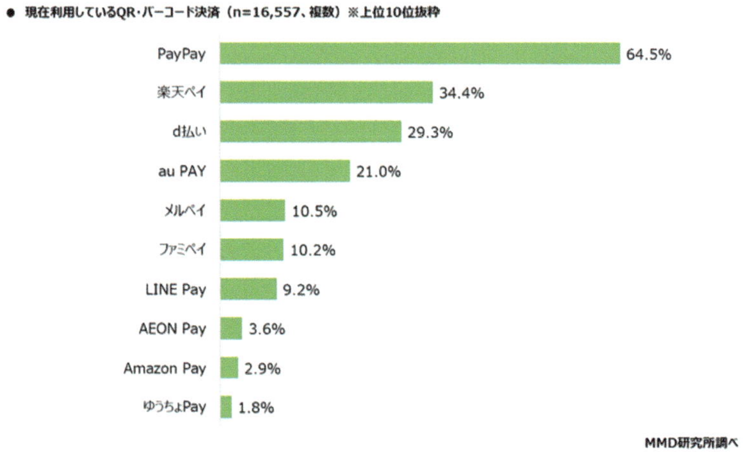

More than 64% of people use PayPay for mobile payments (*Source* MMD Labo)

Users of PayPay need to create an account for the service to function in Japan, requiring the user to have a Japanese smartphone number, making it less convenient for tourists. In 2018, Alipay began providing services through a partnership with PayPay, which means that people who have Alipay accounts will not need to register for the PayPay account and can easily make payments at local merchants in Japan by QR code scanning only.

17.3.2 Promoting Digital Service Solutions in Japan: Ant Group's Collaboration with PayPay

The partnership between PayPay and Ant Group dates back to 2018, when PayPay began collaborating with Alipay to enable payments through Alipay by scanning QR codes at merchants that support PayPay. It is not only the improvement of the convenience for the mainland Chinese tourists who came to Japan, but this opened up opportunities for the mobile payment industry in Japan as well.

PayPal, branded Alipay+, is a cross-border mobile payment and marketing solution offered by Ant Group and aims to integrate international e-wallets providers and help the users to spend the wallets they would like on their travels while cutting cross-border trading cost for small enterprises. The recent launch of Alipay+ shows the shift of Ant

Group in its international approach from merely a focus on payment services but now seeks collaboration in the provision of different digital technologies and services.

Alipay+ has already launched some of its projects in Japan. Alipay+ has maintained the partnership with PayPay, which is now the official acquiring service provider of Alipay+. With this improvement, more users of overseas e-wallets are able to spend lavishly in Japan while paying for their purchases using the same familiar pay service they are used to. On June 15, 2023, PayPay partnered with Thailand's TrueMoney, Malaysia's Touch 'n Go e-Wallet, Philippines' GCash and HelloMoney by AUB, which expanded its service scope further.[38] This suggests that PayPay users can also use the QR codes of these services while on holiday and therefore boost their appeal towards the inbound demand. Therefore, over 1 billion customers in Asia including the ones using Alipay, AlipayHK, and Kakao Pay are able to use PayPay for all their leisure, purchase accommodation, dining, and souvenir requirements while in Japan through PayPay's merchant installations.[39]

PayPay's overseas E-Wallet partners (*Source* Alibaba Japan)

[38] PayPay, 2023, https://about.paypay.ne.jp/pr/20230615/01/.

[39] Alibaba Japan, 2023, https://www.alibaba.co.jp/news/2023/06/alipay150.html.

.

Through the cross-border electronic payment and digital technology solution Alipay+, Japan now supports 16 e-wallets and banking apps, including Alipay (China), Alipay HK (Hong Kong), Mpay (Macau), Korea's KakaoPay, Naver Pay, and Toss, Singapore's OCBC, Changi Pay, and EZ-Link, Malaysia's Touch'n Go e-Wallet and MyPB by Public Bank, GCash and HelloMoney from the Philippines, Thailand's TrueMoney, Mongolia's Hipay, and Italy's Tinaba. The merger of PayPay and Alipay+ has both reinforced the relationship between Japan and China in terms of mobile payment systems and offers more effective payment and marketing services to the international consumers and merchants.

On the whole, the mobile payment market in Japan remains predominantly untapped. With PayPay as one of the major players, the Japanese society seamlessly moves to be drug free despite constant cash injection into the society. Japan's fragmented market for mobile payments gained characteristics five years ago. Visitors had payment solutions that targeted them. Alipay was targeting Chinese tourists, PayPal was for Western tourists, and PayPay was more local users. However, after the joint campaign of Alipay+'s market entry into the Japanese market and its growth, the country's mobile payments market was enriched by the international perspective.

As of March 29, 2024, Alipay+ integrated more than two million local merchants in Japan, allowing businesses to reach out to a vast market of international travelers who can make hassle-free payments through their domestic e-money wallets. And for the first three weeks of March, on a year-to-year basis, Alipay plus e-wallet and its banking apps made in Japan, got more than 200% growth looking back 2023.[40]Ant Group's international president, Douglas Feagin, stated at a press conference in Tokyo that these transactions were projected to account for 10% of travelers' total expenditures in Japan.

Feagin expressed gratitude to more than 40 local acquiring partners in Japan, ranging from large institutions to small-scale businesses, for their support in offering a seamless, transparent, and efficient experience to both local businesses and international tourists. By working with these

[40] Shanghai Business, 2024, https://shanghai.com.my/%e5%aa%92%e4%bd%93%e5%8f%91%e5%b8%83/alipay%e8%bf%9e%e6%8e%a5%e6%97%a5%e6%9c%ac%e7%9a%84200%e4%b8%87%e5%95%86%e6%88%b7-%e4%b8%ba%e5%85%a8%e7%90%83%e6%97%85%e5%ae%a2%e8%b5%b4%e6%97%a5%e8%b5%8f%e6%a8%b1%e8%8a%b1%e5%ad%a3/.

partners, Alipay+ helps its users to have the same secured payment experience as if they are in their home countries. This frustration of worrying about exchanging currency or if they understand the language is removed since all vendors accept QR codes from the tourists.

In the past, Japan was a popular place for over nine million Hong Kongers to visit and was also where AlipayHK sent its services out into the world. According to Venetia Lee, CEO of AlipayHK, with the help of Alipay+, Hong Kong residents can shop in Japanese stores and pay for taxis in Hong Kong dollars with highly competitive exchange rates. In the first quarter of 2024, AlipayHK saw a 3.5-fold increase in average daily transaction value and a fivefold surge in transaction volume, signaling a strong growth trajectory that instills confidence in future developments.[41]

The rise in mobile payment use across various tourism contexts in Japan makes the experience even more appealing, particularly for the Chinese tourists who are known to spend major amounts. For example, Asakusa, which is culturally appealing and is an interesting region in Tokyo, has been greatly receiving visitors. Even in the many small souvenir shops and food stalls, QR code technology is becoming common which makes mobile payments easy for both insiders and visitors. This is especially good for foreigners since the number of these transactions in the region has more than quintupled in a month, demonstrating a pronounced trend towards payment via smartphones.

The Alipay+ partnered e-wallets and banking apps and their volume of transactions at Daimaru Matsuzakaya Department Store's websites increased more than ten times year-on-year within the first two months of the year 2024.[42]While covering customer issues, especially inbound ones, Akito Kimura, the department head at Daimaru Matsuzakaya commented that having Alipay within payment options assists in making Chinese clients more welcoming. He expressed excitement about further expanding the store's relationship with Ant Group and integrating Alipay+ services, allowing international customers to use their preferred payment methods seamlessly. Kimura emphasized that this collaboration enhances the store's ability to provide a memorable and worry-free shopping experience for all customers.

[41] Ibid.

[42] Ibid.

Overseas payment methods supported by Daimaru Matsuzakaya department store (*Source* Daimaru Matsuzakaya Department Store)

The partnership of Alipay and PayPay from Japan in developing a digital payment system is another milestone achieved in the global integration of finance. This partnership has not only diversified payment methods available for the consumers but it has also promoted a more progressive and sustainable financial growth model in Japan. The inclusion of Alipay in popular Japanese services such as PayPay facilitated a dramatic expansion in the volume of cashless transactions, mainly thanks to Chinese tourists who are used to mobile payment. This collaboration has also witnessed a major change in the pattern of payment systems in Japan, especially for the Chinese tourists. This tendency has also urged local businesses as well as the local attractions to embrace such digital payment systems and platforms thus improving their service provision as well as competitiveness.

Even in the absence of the tourism-related activities, the effects of this collaboration have spread to the other spheres of the Japanese economy, paving the way for a more sweeping digital transformation and promoting a wider-consumption-oriented culture. The use of PayPay to introduce Alipay in Japan's system of payment depicts the turning point in the cashless dream of Japan and also proves the power of international collaboration in the growth of digital finance.

17.3.3 Chinese Tourists' Payment Needs Drive Japan's Inclusive Partnership and Sustainable Tourism

The attitude of Chinese citizens towards spending while being on a holiday in Japan has changed from prioritizing "novelties and experiences" to "mass consumption and utilities." Mobile payments used in the mainland China market have already reached the Japanese tourism sector. The widespread adoption of mobile payments in mainland China has extended into the Japanese tourism market. In April 2023, during Japan's cherry blossom season, Japan relaxed its COVID-19 entry requirements for Chinese tourists, leading to a sharp increase in the number of Chinese visitors to Japan. This large cohort of Chinese tourists is very important for Japan's regular spending and domestic economic growth.

At the same time, the Japanese yen continued to lose ground as the US Federal Reserve's interest rate cut expectations were pushed further in the future, which resulted in an influx of tourists to Japan. As per figures from the Japan National Tourism Organization (JNTO), apart from Korea, China turned out to be the fastest expanding clientele market on international tourists. Due to the appeals of summer travel high season July 2024 saw many more Chinese visitors than Koreans, with Chinese tourists exhibiting a growth of 148%. That month, the total number of foreign tourists to Japan also reached a new record of over 3.29 million, surpassing the previous high.[43]

[43] Japan National Tourism Organization (JNTO), 2024, https://www.jnto.go.jp/statistics/data/visitors-statistics/.

2024年7月　訪日外客数（JNTO推計値）（対2023年比）

Visitor Arrivals for Jul. 2024 (Preliminary figures by JNTO)(Compared to 2023)

国・地域	Country/Area	総数 Total			総数 Total		
		2023年 7月	2024年 7月	伸率(%)	2023年 1月～7月	2024年 1月～7月	伸率(%)
総数	Grand Total	2,320,694	3,292,500	41.9	13,033,090	21,069,900	61.7
韓国	South Korea	626,830	757,700	20.9	3,755,306	5,199,800	38.5
中国	China	313,328	776,500	147.8	908,096	3,844,600	323.4
台湾	Taiwan	422,356	571,700	35.4	2,192,974	3,550,900	61.9
香港	Hong Kong	216,413	279,100	29.0	1,126,174	1,555,200	38.1
タイ	Thailand	49,613	53,500	7.8	547,412	671,800	22.7
シンガポール	Singapore	25,320	28,100	11.0	278,023	328,100	18.0
マレーシア	Malaysia	17,357	19,300	11.2	211,609	261,100	23.4
インドネシア	Indonesia	28,489	29,600	3.9	230,145	293,400	27.5
フィリピン	Philippines	51,711	55,500	7.3	328,853	457,300	39.1
ベトナム	Vietnam	44,773	49,700	11.0	345,671	381,500	10.4
インド	India	13,150	14,900	13.3	92,588	136,800	47.8
豪州	Australia	35,650	48,600	36.3	310,435	510,600	64.5
米国	U.S.A.	198,845	251,200	26.3	1,171,053	1,594,100	36.1
カナダ	Canada	38,786	48,200	24.3	222,052	320,500	44.3
メキシコ	Mexico	9,986	16,000	60.2	44,892	80,100	78.4
英国	United Kingdom	26,965	32,900	22.0	177,126	251,400	41.9
フランス	France	31,132	37,400	20.1	153,266	221,700	44.7
ドイツ	Germany	19,081	22,700	19.0	125,796	180,300	43.3
イタリア	Italy	13,969	18,400	31.7	73,333	114,500	56.1
スペイン	Spain	12,985	18,700	44.0	52,515	84,600	61.1
ロシア	Russia	3,239	5,700	76.0	19,421	46,800	141.0
北欧地域	Nordic Countries	12,511	15,900	27.1	65,577	89,600	36.6
中東地域	Middle East	9,937	12,800	28.8	61,894	88,100	42.3
その他	Others	98,268	128,400	30.7	538,879	807,100	49.8

◆注1：本資料を引用される際は、出典名を「日本政府観光局(JNTO)」と明示してください。
◆注2：上記の2023年の数値は確定値、2024年の数値は推計値である。
◆注3：訪日外客数とは、法務省集計による出入国管理統計に基づき、算出したものである。訪日外客は、外国人正規入国者から、日本を主たる居住国とする永住者等の外国人を除き、これに外国人一時上陸客等を加えた入国外国人旅行者のことである。駐在員やその家族、留学生等の入国者・再入国者は訪日外客に含まれる。
なお、上記の訪日外客には、乗員は含まれない。

◆注4：北欧地域はスウェーデン、デンマーク、ノルウェー、フィンランドを指す。
◆注5：中東地域はイスラエル、トルコ、GCC6か国（サウジアラビア、アラブ首長国連邦(UAE)、バーレーン、オマーン、カタール、クウェート）を指す。

◆Note 1. If reproduced, your credit line to JAPAN NATIONAL TOURISM ORGANIZATION is mandatory.
◆Note 2. Above figures for 2023 are definitive, while figures for 2024 are the preliminary ones estimated by JNTO.
◆Note 3. Visitor arrivals exclude permanent residents whose primary place of residence is in Japan and include travelers entering Japan for transit.
Foreigners entering or re-entering Japan, such as expatriates and their families, and international students are included in visitor arrivals to Japan. Crew members are excluded.
◆Note 4. Nordic Countries refer to Sweden, Denmark, Norway, and Finland.
◆Note 5. Middle East refers to Israel, Turkey, and the Gulf Cooperation Council countries (Saudi Arabia, UAE, Bahrain, Oman, Qatar, Kuwait).

The number of visitors to Japan in July 2024 (*Source* JNTO)

New Chinese tourist waves are, however, changing the patterns of consumption of Japanese tourists, who used to be the primary focus of tourism strategies, and Japan's tourism industry is gradually working out approaches to satisfy the needs of the new target market. Chinese language personnel were able to fill in sales counters, and various locations at airports and popular shopping areas of Japan that have already been implemented.[44] Nevertheless, young Chinese tourists when visiting traditional tourist venues look forward to more than basic Chinese cuisine

[44] Universal Studios Japan, 2024, https://www.usj.co.jp/web/zh/cn.

or services, but instead wish to find Chinese styled services in more commonplace areas throughout Japan.[45]

With this shift in tourist demographics, Japan's tourism industry has been rethinking how to meet the needs of these young Chinese travelers, who have new consumer demands and travel preferences. Japan has already increased the number of Chinese-speaking staff in airports, popular shopping districts, and sales counters. However, young Chinese tourists now expect more than just simple Chinese services at traditional tourist spots; they wish to experience the same Chinese-style services they are accustomed to in more everyday settings throughout Japan.

Japan, a country with a high prevalence of vending machines, has posed some challenges for Chinese tourists who are accustomed to mobile payments, as many vending machines still require cash. Recently, however, more vending machines that accept Alipay have emerged. Popular tourist areas such as Shinsaibashi, Umeda, and Universal Studios Japan now feature Alipay integration. Currently, 99% of purchases within Universal Studios Japan accept Alipay, making it one of the few places in Japan where Chinese tourists can enjoy the convenience of "one phone for the entire park."[46]

Visitors at Universal Studios Japan using Alipay+ for payments (*Source* ChinaDaily)

[45] Thepaper, 2024, https://www.thepaper.cn/newsDetail_forward_28336149.

[46] Ibid.

Local Osaka-based Chinese tourism professionals have reported that many large Japanese consumer and public facility companies have not yet fully adopted or perfected QR code payment systems. This presents challenges for Chinese tourism companies seeking to implement their advanced technologies and systems in Japan.[47] Osaka Metro, for example, did not introduce QR code ticket purchasing until March 2020, when it began accepting Alipay and WeChat Pay as well. Nowadays, through new tourism concepts and demands, mobile payment habits from China are slowly reshaping the payment ecosystem in Japan. The tourism landscape in Japan has witnessed a significant shift in Chinese tourists' preferences. The increasing prevalence of mobile payment options such as Alipay in various consumption settings is a reflection of this trend.

[47] Thepaper, 2024, https://www.thepaper.cn/newsDetail_forward_28336149.

PART VII

Conclusion

CHAPTER 18

Measurable Social Impacts of the Alipay+ for SDGs

18.1 Project Overview and Environmental and Social Achievements

The Alipay+ for SDG Project, implemented since November 2022, is led by Ant International with support from the Department of Sociology, The Chinese University of Hong Kong. Participants include over 30 digital payment platforms and 300 partners from nine regions including Thailand, Malaysia, Philippines, Indonesia, Singapore, South Korea, Japan, Hong Kong, and Macau. Alipay+, as a leading cross-border digital payment technology and marketing solution, has covered over 66 markets globally with more than 90 million merchants, seamlessly connecting with over 1.5 billion Asian consumers. This enables Asian consumers to use their home country digital wallets for overseas payments without opening local digital wallets or bank accounts while traveling abroad. The Alipay+ for SDG Project aims to leverage the Alipay+ cross-border digital payment infrastructure in supporting the SDGs, drawing from China's experience in digital payments for SDG initiatives such as Ant Forest, digital poverty alleviation, and digital economy-empowered job opportunity programs. By serving global consumers and small and micro enterprises, the project aims to lead in nine Asian markets in using digital payments to support the SDGs.

Location: Thailand, Malaysia, Philippines, Indonesia, Singapore, South Korea, Japan, Hong Kong, Macau

J. Sun, *Sustainable Financialization*, Sustainable Development Goals Series, https://doi.org/10.1007/978-981-96-3580-1_18

Project Implementation Period: November 2022 to present

Project Participants: Ant International; Department of Sociology, The Chinese University of Hong Kong; Thailand's Line Pay, True Money, and K Plus; Malaysia's Touch 'n Go, and MyPB; Philippines' GCash and HelloMoney; Indonesia's Dana; Singapore's Changi Pay, EZ-Link, OCBC Digital, and GrabPay; South Korea's Kakao Pay, Naver Pay, and Toss Pay; Japan's PayPay; Hong Kong's AlipayHK; and, Macau's Alipay (Macau)

Global Compact Ten Principles: all

Sustainable Development Goals (SDGs): Goal 8: Decent Work and Economic Growth; Goal 9: Industry, Innovation, and Infrastructure; Goal 10: Reduced Inequality; Goal 11: Sustainable Cities and Communities; Goal 12: Responsible Consumption and Production; Goal 13: Climate Action; Goal 15: Life on land; Goal 17: Partnerships for the Goals

Challenges and potential difficulties faced during the initial pilot phase in nine Asian markets outside mainland China include:

1. Technical integration challenge: Ensuring system compatibility and stability among over 30 digital payment platforms and 300 acquiring partners in the initial trial regions.
2. Legal and regulatory challenge: Addressing regulatory requirements on payment systems and data privacy across multiple countries and regions in cross-border payments.
3. User acceptance: Promoting cross-border digital payments and encouraging user participation in supporting SDGs require market acceptance, effective advocacy, environmental education, and public communication.
4. Security and privacy concerns: Ensuring the security of payment systems and user privacy in cross-border payments to prevent data breaches and security vulnerabilities.
5. Market competition: Facing fierce competition in the digital payment sector in Asian markets from local and international competitors, necessitating continuous innovation in SDG actions to enhance user recognition of digital payments in supporting the SDGs. Overcoming these challenges, the Alipay+ digital payment supports the SDG project to better achieve the strategic goal of inclusive finance, promote the development of digital payments in

Asian markets, and leverage leading global digital financial technology to support the achievement of sustainable development goals.

Key Environmental and Social Achievements:

1. Enhancing cross-border digital payment infrastructure and state capacity building in developing countries in the global South (see Governance System).
2. Promoting FinTech-enabled public forest models, encouraging green travel and sustainable tourism (see Environmental Performance).
3. Driving decent work and economic growth through the digital economy, reducing inequality, and building sustainable cities (see Social Performance).

18.2 Governance System

In popular tourist destinations such as Thailand and Malaysia, the Alipay+ for SDGs Project supports the digital capacity building of central banks to provide convenient cross-border digital payment facilities for international tourists, helping to achieve the SDGs. Thailand, with a population of 70 million, attracts approximately 28 million international tourists annually, with Chinese tourists being the largest source. Despite significant progress in digital transformation by the Bank of Thailand, its digital payment system primarily serves domestic citizens and does not cover the 40% of the population who are international tourists. In response, Alipay+ offers an all-in-one payment solution, integrating over 30 Asian digital wallets, quickly becoming a key cross-border payment platform in Thailand to promote sustainable tourism.

In 2023, the Tourism Authority of Thailand collaborated with Alipay+ to enhance tourists' mobile payment experience using Alipay+ cross-border infrastructure. In July 2023, they jointly organized the "Alipay Takes You on an Authentic Tour of Thailand" event, with the Deputy Director of the Tourism Authority of Thailand attending to show support, emphasizing the importance of establishing a strategic partnership with Alipay+. This partnership not only provides a better payment experience for tourists but also helps small and medium-sized enterprises in

Thailand acquire digital operational skills. Statistics show that in the first half of 2023, Thailand welcomed over 1.4 million Chinese tourists, significantly boosting the recovery and development of the Thai tourism industry. To further enhance the tourist experience, Alipay introduced electronic visa services, simplifying the visa application process for Chinese tourists. Additionally, during travel in Thailand, Alipay+ ensures transparency in consumption exchange rates and offers special Black Diamond exchange rate discounts for the summer. Alipay+ also partnered with Grab to launch a one-click Chinese car-hailing service with multiple vehicle types, allowing users to enjoy four travel discounts for 9.9 RMB, saving up to 50 RMB. As of now, Alipay+ has connected with over 550,000 local merchants in Thailand, covering 77 provinces nationwide, significantly improving the digitalization and service quality of the Thai tourism industry.

Competitive FinTech Services for Government Governance of Risk Management

In 2019, the Ministry of Finance of Malaysia launched the "ePemula" project to distribute relief funds specifically to young people, with an estimated 2 million Malaysians aged 18 to 20 expected to receive 150 Ringgit (Malaysia's currency).

The lack of e-KYC (online identity verification technology) in Malaysian market created risks of misusing of funds and other dangerous situations in this funding program. Therefore, Malaysian government cooperated with Ant Group to introduce ZOLOZ, a security technology brand of the latter to help with potential technical risks. Once the partnership was sealed, a whopping 75% of TNG e-Wallet users breezed through registration using ZOLOZ's e-KYC online identity verification technology. This move not only turbocharged the sign-up process but also slashed the need for manual checks by a staggering 80%. Besides, this technology assisted to reduce the proportion of account theft by 98% within a few months. With the help of ZOLOZ's technology, Touch 'n Go e-Wallet took the lead, winning almost 70% of the market share and being recognized for its accurate pinpointing and excellent user experience.

Competitive FinTech Services for Government Governance of Risk Management

In March 2024, China and Thailand implemented mutual visa exemptions, marking a new chapter in bilateral tourism. In the first four months of 2024, the number of Chinese tourists visiting Thailand reached 2,351,909, solidifying Thailand's position as the top overseas travel destination for Chinese tourists. According to Alipay+ consumption data, the total spending on Alipay during the Labor Day period increased by 161% compared to the same period in 2023. Beyond the growth in total spending, Chinese tourists' expectations for their travel experiences have also evolved. Today's outbound Chinese travelers are placing increasing importance on their overall experience, seeking a variety of high-quality, seamless, and convenient consumption options. In response to these evolving needs, Alipay+ has partnered with banks in Southeast Asia to promote digital payment infrastructure, offering international tourists—particularly Chinese travelers—a seamless cross-border payment experience.

Kasikornbank launched Alipay+ in Thailand in partnership with Alipay in 2017, which first promoted the introduction of QR code payment systems to facilitate the growth of cashless transactions in the country. As of now, Kasikornbank has expanded this technology to 200,000 merchants operating in the central region including Central Department Stores, Chatuchak Weekend Market, Suvarnabhumi International Airport, and 7 Eleven stores located around Bangkok. Data shows that since the collaboration began, spending by Chinese tourists in shopping malls has increased by around 30%, providing a significant boost to Thailand's digital transformation in tourism.

In Malaysia, Alipay+ has deepened its cooperation with Maybank to promote the widespread use of QR code payments, enabling Malaysian merchants to seamlessly accept payments from multiple e-wallets. Through the unified DuitNow QR payment platform, merchants can now receive payments across different banks, making the payment process more efficient and convenient. As the leader in the Malaysian market, Maybank holds a 37% market share. By 2023, more than 1.8 million merchants had joined the Alipay+ network, offering tourists from eight different markets across Asia a diverse array of payment options.

In an effort to advance the integration of cross-border digital payments in the region, Alipay+ has developed cross-border payment interlinkages by incorporating 22 Southeast Asian and other regions' e-wallets. In this digital payment network, the e-wallets of China, Hong Kong, Macau, Malaysia, Philippines, Thailand, Singapore, South Korea, and Indonesia are linked, enabling local businesses to receive payments from diverse e-wallets. This has greatly enhanced the efficiency and convenience of transactions. As of the year 2023, Alipay+ has acquired 25 key payment service companies worldwide enabling it to reach over 500 million consumer accounts and over 8 million merchants. This cross-border payment network not only enables contact between international customers and sellers but also enables the countries' economies to enhance kanilang strategies in the age of digital expansion.

Through in-depth cooperation with banks and payment platforms in various countries, Alipay+ is building a global digital payment infrastructure. This infrastructure not only provides tourists with a convenient cross-border payment experience, but also opens up channels for global merchants to reach a wider consumer group, helping to achieve the digital upgrade of the tourism industry, thereby achieving the deep integration and sustainable development of the digital economy.

Connecting local e-wallets to provide seamless cross-border payment solutions

The Philippines is currently making a shift towards a digital financial system. The country regards the provision of digital payment services as important largely because of the youth population. Although there has been some progress in areas such as digital payments, issues like poor infrastructure and low financial inclusion still remain. In this context, Alipay has partnered with local players like Akulaku PayLater and BillEase to help enhance the digital ecosystem. In May 2022, the executive order signed by former President Duterte mandated that all public institutions must use digital channels for payments and collections, which aligns with the government's goal of creating an inclusive digital financial system. The growth of Alipay+ in the Philippines has played a key role in the development of local e-wallets and mobile payment systems. It successfully provides Filipinos with a smooth transition to digital transactions. With e-wallets like Gcash HelloMoney and Maya, Alipay allows users to make payments both domestically and internationally, enabling secure cross-border transactions.

Gcash not only plays an important role in domestic payments but also extends to the cross-border payment field through cooperation with Alipay+. Gcash and Ant Financial partnered early on to offer users multi-functional services, including bill payment, merchant payment, P2P lending, and payroll. By 2023, Gcash had reached 66 million users, nearly half the total population of the Philippines. Through Alipay+'s cross-border payment function, Gcash users can make payments abroad without needing to exchange currency. The fees are transparent, and the exchange rate updates in real time.

HelloMoney, a digital wallet launched by Asia United Bank, focuses on contactless financial services. Users can complete common financial services such as account opening, transfers, and bill payments without visiting physical locations. This feature proved especially useful during the pandemic. HelloMoney also supports virtual card payments for online shopping, which has helped cover the e-commerce market in the Philippines. Starting in 2024, HelloMoney's partnership with Alipay+ will enable users to make seamless payments in destinations like South Korea, Hong Kong, and Malaysia, improving the cross-border payment experience for Filipinos.

In the Philippines context, the highest number of users is registered in the digital payment platform Maya, which allows performing a number of payment transactions like, bill payment, cross-border transactions, online shopping, remittance payment, and transfers both domestic and international currencies. It actively promotes the QR code functionality for making payments throughout the Philippines to merchants, restaurants, and public service entities helping tremendously the adoption of cashless payment schemes. In 2023, Maya was able to offer instant loans and added 2.6 million new users that it has gained thanks to its digital banking services along with AI-powered credit scoring feature. The collaboration with Alipay+ has also been critical in facilitating cross-border payments as it gave the ability to the Maya clients to scan QR codes and make payments overseas thus enhancing the cross-border payments experience.

By connecting local e-wallet platforms in the Philippines, Alipay+ has created a cross-border payment network across Southeast Asia. This initiative not only provides a more convenient payment method for Filipino users but also creates opportunities for Filipino merchants to serve international tourists. It helps upgrade the country's digital payment infrastructure and promotes the development of the digital economy.

Providing flexible payment options, BNPL benefits more young consumers

Alipay+ provides convenient and inclusive financial services to young people and ordinary people in the Philippines by empowering the Philippines' "buy now, pay later" (BNPL) platforms Akulaku PayLater and

BillEase. Such cooperation has effectively promoted the popularization of financial inclusion and digital payments, provided reliable credit options for a large number of underserved groups in the Philippines, enabled local economic growth, and created a wider digital life experience.

As one of the important BNPL platforms in Southeast Asia, Akulaku PayLater reached a cooperation with Alipay+ in September 2022 to expand its services to the Philippine market, helping local consumers to make installment payments without having a credit card. The cooperation currently covers more than 4,000 merchants and provides instant credit to a large number of unbanked Filipinos. Akulaku also enables Filipino consumers to enjoy the convenience of installment payments in Southeast Asia and around the world through Alipay+ 's cross-border payment network, improving the consumption experience of ordinary users. Akulaku PayLater's installment payment option helps users without credit cards to achieve a reasonable payment plan with the support of e-wallets or bank transfers, allowing them to manage their consumption within their financial means.

BillEase is an important BNPL platform in the Philippines. Since its establishment in 2017, it has expanded rapidly and has more than 2,000 partner merchants, covering multiple scenarios from air ticket purchases to daily bill payments. BillEase provides consumers with flexible installment plans, allowing small purchases to be paid for 10 or 20 days later, or choose installments for up to 12 months, providing users who cannot obtain traditional credit with a low interest rate option of 0%-3.49%. In addition, BillEase has extended to global payment scenarios through cooperation with Alipay+, supporting users to use installment payments on global shopping websites and travel platforms (such as Agoda), improving the convenience of cross-border consumption for Filipinos and expanding the accessibility of financial services.

Offering Chinese tourists smooth travel experience in South Korea

In August 2023, China fully resumed group tours to South Korea, and the South Korean government quickly proposed a plan to attract Chinese tourists, with the goal of receiving 2 million tourists throughout the year. The consumption habits of Chinese tourists have undergone significant changes, becoming more personalized and high-quality service-oriented, paying more attention to food and local cultural experiences, and showing strong interest in beauty, health, fashion shopping, and other fields. This trend has prompted the South Korean government to optimize tourism supporting services with digital payments as the core, especially by expanding the scope of merchants that can access the Alipay+ payment service that Chinese tourists are familiar with.

Currently, more than 120,000 merchants in South Korea support Alipay+ payment services, covering duty-free shops, convenience stores, restaurant chains, and other locations frequently visited by tourists. The construction of a digital payment ecosystem has made South Korea's tourism market more appealing to international visitors. Driven by the visa-free policy, Chinese tourists' spending power on Jeju Island has been particularly noticeable. In the first quarter of 2024, Jeju Island received more than 370,000 foreign tourists, a year-on-year increase of 532.5%, of which 80% were Chinese tourists. At the same time, Alipay+ has established strategic partnerships with brands such as Starbucks in South Korea to provide tourists with customized preferential activities and enhance the overall attractiveness of the payment ecosystem. From January to October 2023, offline payment transactions completed through Alipay+ increased by approximately 700%. Through advanced technology and a global cooperation model, Alipay+ not only provides Korean merchants with channels to reach the international market, but also brings a familiar payment experience to Chinese tourists.

Alipay+ also further enhances the convenience and attractiveness of the payment ecosystem through cooperation with the Korea Easy Payment Promotion Agency (KEPPA) and the Jeju Tourism Organization (JTO). Users can use multinational e-wallets including Alipay in Mainland China, AlipayHK in Hong Kong, etc., to pay at ZeroPay merchants covering Jeju Island and enjoy exclusive discounts. This cross-border cooperation helps local merchants reach a wider international customer base. From March to May 2023, the total transaction value processed through Alipay+ in Dongmun Market and Seogwipo Daily Olle Market in Jeju Island soared from 17 million won to 250 million won, and the number of transactions also surged from 700 to 8,000. The payment convenience and promotional offers provided by Alipay+ provide a large number of tourists with convenient payment methods and also promote the development of local tourism.

18.3 Environmental Performance

Supporting More Digital Wallets to Set up FinTech-Enabled Public Forest Programs

Ant Forest is Alipay's flagship environmental project in China. The public participates in tree planting through low-carbon activities. According to Ant Group's 2023 Sustainable Development Report, Ant Forest has mobilized over 690 million Chinese people to participate in carbon reduction efforts, accumulating over 31 million tons of carbon emission reductions through daily low-carbon actions. Ant Forest users have planted 475 million trees, covering 5.08 million mu in 11 provinces of China, with a significant portion located in the key places under the "Three-North Shelterbelt" project. Ant Forest has established 31 public welfare protection areas in 13 provinces and cities, protecting over 4,800 square kilometers of land. Its sister project, Magical Ocean, has planted 66 million seaweeds in Weihai, Shandong, and 2.72 million mangroves in Ningde, Fujian, promoting the recovery of seagrass beds and mangrove.

The Philippine archipelago is one of the most biodiverse regions on Earth. However, due to improper human activities such as deforestation, land development, and agricultural practices, the ecosystem of the Philippine archipelago has been severely affected. Alipay supports its Philippine partner GCash in launching the GForest project. GForest collaborates with local environmental partners to plant seedlings and monitor progress. GForest has attracted 12 million local users and planted over 2.7 million trees through local partners, becoming one of the largest public tree planting initiatives in the Philippines. The trees planted by these individual users are enough to cover approximately 5,600 hectares of land in the Philippines, reducing over 30.5 billion grams of carbon emissions. GForest has made a significant contribution to the environmental cause in the Philippines, engaging the Filipino public in green actions and greatly enhancing public awareness of sustainable development.

Encouraging Chinese Tourists Outbound to Public Transportation and Sustainable Tourism

Drawing on China's experience in green travel in Macau, Hong Kong, Japan, and South Korea, enriching the outbound green energy scenes for mainland tourists, encouraging sustainable tourism through public transportation. Taking Hong Kong as an example, in December 2023, the Hong Kong Tourism Board partnered with Alipay to offer a series of travel discounts for mainland tourists. MTR introduced a 50% discount for adult tourists using Alipay for the first time. Just like on the mainland, you can use Alipay to take MTR, busses, ferries, trams, taxis, and other modes of transportation. In the second half of 2023, the overall transaction volume using Alipay by mainland tourists in Hong Kong increased by 40% compared to the first half of the year, with a surge of 80% in transactions for scanning codes to take public transportation. The MTR adult tourist full-day pass half-price promotion officially launched on December 18, 2023, and will last for three weeks, giving away over 100,000 passes in total. Tourists can purchase the MTR adult tourist full-day pass for half price, HK$32.5, instead of the usual HK$65, through Alipay by promoting banners or searching and entering the "Good Reward Buy" mini-program. After purchase, they can exchange the electronic voucher for physical tickets at the customer service center of MTR stations and enjoy unlimited rides on MTR within 24 hours during the promotion period. By diverting traffic through public transport scenes, Alipay and AlipayHK, together with the Hong Kong Retail Management Association, launched the "Hong Kong Festival of Rewards" after the "Hong Kong Happy Shopping Festival," offering benefits to users from various merchants to boost tourism during the peak season at the end of the year and the beginning of the new year in Hong Kong. Currently, over 90% of merchants in Hong Kong support Alipay's QR code payment, enabling convenient cross-border digital payments through Alipay+. In addition to consumption, retail discount activities have landed on Alipay again, covering over 3000 merchants from hundreds of brands. Tourists can access various merchant discounts by entering the "Hong Kong Festival of Rewards" page in "Outbound Benefits."

Advancing smart cities, digital finance inclusion, and low-carbon living in Singapore

Singapore has achieved low-carbon transformation through digital financial inclusion and promoted environmental sustainability through innovative smart city construction and broad green economy initiatives. Singapore has been implementing the "National Computerization Programme" since the 1980s and has subsequently actively promoted information technology and data analysis through a series of smart nation plans, including the latest "Smart Nation 2025" strategy, to enhance the smart capabilities of urban governance and services.

Inclusive finance has accelerated the low-carbon initiatives of Singapore's smart city. In 2018, the Monetary Authority of Singapore (MAS) launched the unified QR code standard (SGQR), which simplified electronic payments and integrated multiple payment methods, allowing consumers to conveniently choose contactless payments. At the same time, MAS collaborated with Ant Group to introduce Alipay+ to Singapore to promote the interconnection of payment systems in Southeast Asia. The SGQR system was subsequently upgraded to SGQR + , which became compatible with 23 local and cross-border payment solutions, reduced technical barriers between merchants and financial institutions, and enhanced the sustainable development potential of local small businesses.

At the same time, Singapore has optimized traffic flow through intelligent transportation systems and sensor networks, reducing traffic congestion and carbon emissions. Singapore has promoted low-carbon transformation through smart city construction and digital financial inclusion. Its innovative measures have improved urban infrastructure and residents' quality of life and also highlighted the potential of inclusive finance in advancing low-carbon urban development.

18.4 Social Performance

Promoting World Heritage and Sustainable Tourism for Decent Work and Economic Growth

Since the end of the epidemic, Alipay+ has closely collaborated with Fliggy and the Macao Government Tourism Office to launch a tourism promotion plan. This initiative has attracted more than 440,000 participants, offering 103 tourist routes, organizing nearly 7,000 tour groups, selling approximately 75,000 hotel packages, and receiving 180,000 visitors. This effort has directly created an economic impact of 190 million Macao Patacas, contributing to the diversified development of Macao's "1 + 4" economy.

In May 2023, Ant International's Macao Pass received approval and partnered with Alipay+ partners electronic wallets GCash, Touch 'n Go e-Wallet, and TrueMoney to conduct cross-border payment business, thereby facilitating tourism recovery in Macao. In May 2024, Macao Pass, along with Alipay+ and other members of Ant International, made an appearance at the International Science and Technology Innovation Expo. Chief Executive Ho Iat Seng encouraged them to introduce new industrial models to Macao, promote digital transformation, and leverage Macao's strengths.

In response to the government's initiative, Macao Pass optimized electronic payment services, expanded Alipay+ payment coverage, met the needs of residents and merchants, and introduced additional overseas payment tools. Since September 2023, MPay's cross-border payment scope has expanded to mainland China, Hong Kong, and over 40 countries and regions overseas through the Alipay+ network. This expansion has enabled global consumption capabilities and enhanced cross-border payment experiences for local residents and international tourists.

MPay collaborates globally with Alipay, WeChat Pay, and Alipay+ to facilitate efficient transactions at over 30,000 payment points in Macao, covering various sectors such as public transport, public utility fees, ticketing, and government services. This collaboration has driven the development of industries like catering, retail, and e-commerce. In early 2024, MPay made Macao's public transport QR code accessible to mainland China, Hong Kong, and Alipay+ international tourists.

With over 5 million digital cards issued, Macao Pass has received high recognition in the local market. Together with local merchants, Macao Pass is building a "Tourism+" ecosystem, participating in numerous promotional activities, benefiting tens of thousands of small and medium-sized businesses, and fostering the digital transformation of Macao's tourism economy.

Building Sustainable Cities, Promoting Local Businesses and Responsible Consumption.

Huggs Coffee, Singapore's largest local coffee chain brand with 22 chain stores across the country, has embraced Alipay+'s cutting-edge business digital solution, D-store, to undergo a digital operational transformation. This strategic move has enabled the brand to enhance its digital presence and operations. During the Singapore FinTech Summit, attendees, including Singapore Prime Minister Lawrence Wong, had the opportunity to witness the allure of "Internet Coffee" firsthand at Huggs Coffee's showcase. Lee Haoming, the Managing Director of Huggs Coffee, emphasized that coffee is more than just a beverage; it serves as a catalyst for fostering social connections. He highlighted the potential for creating unique experiences beyond coffee and underscored the pivotal role of digitalization in elevating customer experiences.

Through Alipay+ D-store, merchants like Huggs Coffee can swiftly establish online stores across various digital wallets and apps. This enables them to streamline their digital operations, including payment processing, service delivery, and marketing efforts through mini-programs or H5 pages. By leveraging multiple channels for customer engagement and transactions, businesses can diversify their revenue streams and enhance operational efficiency. Notably, prominent platforms like Chope, a leading online ordering platform in Southeast Asia, and the global fast food chain Burger King have already adopted Alipay+ 's digital solution, showcasing its effectiveness in driving digital transformation within the industry.

A broad vision of sustainable digital tourism
In response to Alipay's call for green and sustainable travel, InterContinental Hotels Group and Ant Group have launched the "Green Stay" plan to encourage guests to reduce the use of disposable items such as toothpaste and toothbrushes.

Advanced technological infrastructure and support behind these efforts empower this green-digital transition. Green Cloud, Alipay's digital ecosystem partner, has grown into China's leading digital ecosystem service provider in the large accommodation industry. The Green Cloud team has more than 30 years of hotel informatization experience and advanced technical background. Based on the four core businesses of Green Cloud PMS, Oracle Hospitality, data platform, and e-commerce platform, it has formed a digital solution for four major sectors, including hotel accommodation, hotel catering, hotel groups, and cultural tourism complexes with different services. Green Cloud, together with Alipay and upstream and downstream partners in the hotel industry, interconnects and empowers the digital transformation of the industry.

Currently, Green Cloud has served more than half of the top 50 hotel groups in China, with nearly 30,000 signed hotels. Its embedded digital ecosystem in Alipay has brought significant value to the Chinese hotel industry: first, the efficiency of the hotel front desk has been improved, achieving one-click check-in completion in 30 seconds, which is 800% more efficient than the average manual processing time of 3 minutes. Second, it increases customer stickiness and enhances hotel revenue. After customers check-in, the hotel's mini-program in the Alipay APP automatically links to the customers, and the opening rate of the mini-program is higher. Third, it quickly develops members. Through the smart stay facial recognition to join the membership, the hotel group's membership level is matched with the Alipay membership level, which not only helps the hotel acquire more members but also enhances the hotel group's benefits for Alipay's level members.

Sustainable fintech support for small businesses and supply chain management

Since kicking off in 2015, MY Bank has had its sights set on helping small businesses and rural areas, pushing for financial inclusion and growing tiny loans. Using payment and shopping insights from Alibaba, MY Bank turns customer actions into credit scores with big data magic, letting small clients get loans without the hassle of guarantees or collateral. Plus, MY Bank's loan process is super speedy with the "310" approach, namely, apply in 3 minutes, get cash in 1 second, and no human intervention needed—it is all done with a few clicks.

By the end of 2019, MY Bank had handed out a whopping 70.03 billion yuan in loans to 20.87 million small businesses, with each account averaging a modest 31,000 yuan. There were 121.319 million small businesses in China that year, meaning MY Bank reached over 17% of them. Plus, MY Bank's annual report shows that in 2019, it offered credit to more than half of the small shops across the country, with those small shops making less than 30,000 yuan a day seeing a 35% jump in transactions on average.

For efficient supply chain management, MY Bank has developed the "Wild Goose System," a digital credit and fund management tool designed to address the pain points in various scenarios within the supply chains of brand enterprises. The system has developed a range of digital financial products. After three years of practice, the Wild Goose System has linked more than 500 brand cooperation's, starting from the fast-moving consumer goods industry and gradually covering 18 industries; the loan availability rate of upstream and downstream supply chain customers has been greatly improved, reaching 80%, without the need for core enterprise guarantees; dealers who have used loan services have seen an average increase of 20% in their purchase volume, improving the financial blood supply of the brand supply chain "capillaries" and truly realizing financial assistance for business development. MY Bank's 2021 annual report shows that about 70% of micro users' average loan duration is within three months, and the interest expenditure of more than 70% of single loans is not higher than 100 yuan, and the overall interest rate of inclusive micro loans has continued to decline for four years.

Inclusive financial supports for the underbanked/unbanked social groups
At the end of 2016, Ant Group launched a rural finance strategy, relying on the Rural Finance Department of MY Bank to officially enter the agricultural and rural fields, attempting to use existing resources to solve the problems of difficult and expensive rural financing support. MY Bank's solution to this problem is the village and town government. Currently, MY Bank has reached strategic cooperation with more than 850 counties and districts involved in agriculture in China, covering 28 provinces, autonomous regions, and municipalities directly under the Central Government.

The integration of satellite remote sensing, financial services, and agricultural development represents a unique innovation of the "Daurian Starling" project, initiated by MyBank. In July 2019, the MY Bank officially launched the "Daurian Starling" project to facilitate satellite remote sensing technology to obtain effective, sustainable, and dynamically updated data on farmers' core assets with the authorization of farmers. It can help to superimpose algorithmic technology capabilities, quickly and uniformly identify the asset situation of farmers across the country and expand the scale of credit.

By the end of 2022, MyBank had signed cooperation covering 1,230 counties nationwide, involving about 50% of the county-level jurisdictions involved in agriculture in the country. Currently, the "Daurian Starling" has been applied in 31 provinces, autonomous regions, and municipalities in China, covering 15 agricultural product industries nationwide, providing pure credit, mortgage-free financial services for more than 1.5 million planters. This inclusive fintech support has significantly improved the digital and intelligent level of various business links and is recognized as a leading practice in the field of remote unbanked areas and rural finance.

Innovative financial services boost cross-border trade for SMEs in Asia
Anext Bank, a fully owned subsidiary of Ant Group, is a digital wholesale bank based in Singapore and regulated by the Monetary Authority of Singapore. Given the great demand for financial support for SMEs in cross-border e-commerce market in the Asia–Pacific region, Anext Digital Bank rolled out a new financing product tailored for SMEs called Anext Business Loan in 2022. This product allows these businesses to apply for

loans up to $30,000 without any paperwork. It caters to customers with smaller loan needs by offering a digital process and credit assessment, along with competitive annual interest rates and no management fees.

By the end of 2023, 68% of Anext Bank's customer base were micro enterprises, and 31% were foreign entrepreneurs registered in Singapore. As for 2024, Anext Bank's regional and global expansion has also achieved notable success. Anext Bank reported that its customer base grew three-fold over the past 12 months, with cross-border transaction volumes increasing sixfold. This indicates that Anext Bank effectively supports the internationalization efforts of MSMEs. Additionally, cross-border transactions by Anext Bank's customers in sectors such as wholesale and retail trade, professional services, and information and communication technology have also seen significant increases. In terms of customer composition, as of May 31, 2024, 31% of Anext Bank's customers were foreign entrepreneurs from 78 countries.

Digital Infrastructure Development Capabilities Enable the Scalable Upgrading of Digital Services

DANA, also known as "Indonesia's Alipay," provides a range of digital services to Indonesian consumers, including social security contributions, bill and utility payments, cellphone recharges, and more. In Indonesia, DANA boasts more than 150 million users and 5 million transactions every day. With more than 273 million people dispersed among thousands of Indonesian islands, its digital operations are under increasing pressure and obstacles as its transaction volume keeps growing. For example, how can local infrastructure expenses be effectively addressed and how can local digital payment compliance needs be met? For this reason, DANA wants to build a strong, dependable, and flexible cloud-based infrastructure for its mobile app. Within its structure, DANA has implemented a range of Alibaba Cloud Platform-as-a-Service (PaaS) and Infrastructure as a Service (IaaS) goods and services. The DataWorks big data platform from Alibaba Cloud, the MaxCompute large-scale data warehousing solution, and the Elasticsearch analytics solution have improved DANA's data analysis capabilities and helped it comprehend changes in user behavior patterns and market trends. In order to manage databases in many environments, monitor data changes in a visual data management platform, and increase database development efficiency, DANA also makes considerable use of Alibaba Cloud Data Management Services (DMS).

Akulaku Group has evolved into a prominent tech-powered mobile finance platform since its inception in Jakarta in 2014, making waves in Southeast Asia. Entering the market in 2016, the Akulaku PayLater App was designed to cater to the credit cardless demographic in the region by offering virtual credit card installment loans. The Akulaku PayLater ecosystem currently hosts over 4,000 merchants, and the company has set its sights on expanding to serve 10,000 online merchants and over 100,000 brick-and-mortar stores within the next two years through its BNPL solution. Alipay+ has integrated Indonesian Akulaku's BNPL solution into its suite of digital payment methods, allowing Akulaku PayLater users to access their BNPL benefits across all Alipay+ merchants worldwide.

Kredivo, which launched in 2016 and is based in Jakarta, Indonesia, is on a mission to offer local consumers low interest, instant credit financing services. Its parent company, FinAccel, is a Singapore-based fintech firm. Kredivo leverages its strong foothold in Indonesia's BNPL market and Alipay+'s extensive reach and technical backing for digital infrastructures, aims to offer a more seamless payment experience for local users in Indonesia and to foster the growth of cross-border transactions. By October 2023, through its collaboration with Alipay+, Kredivo had already surpassed one million transactions, with a total transaction value of at least 300 billion Indonesian rupiah. Kredivo's BNPL service has served over 6 million users, and its customer base has doubled in the last ten months. Looking ahead, Kredivo plans to extend its services to regional markets such as Vietnam and Thailand, further bolstered by its partnership with Alipay+.

Facilitating the Dual-City Lifestyle

Alipay's Hong Kong version, AlipayHK, was launched in 2017 and has become one of the most popular e-wallets in Hong Kong, with a market penetration rate of 33%. The digital payment scenarios provided by Alipay have brought convenience to Hong Kong people's consumption in the Mainland. In the "2024 Policy Address of the Chief Executive of the Hong Kong Special Administrative Region" in October 2024, Li Jiachao pointed out the need to continue to promote innovative financial services such as central bank digital currency, mobile payments, and virtual banks. According to Alipay data, in 2023, the consumption amount and the number of transactions of Hong Kong residents in the Mainland increased by more than three times compared to the pre-epidemic period, with consumption recovering to 140% of the same period in 2019. At the same time, the Hong Kong Monetary Authority and the People's Bank of China are promoting the interconnection of the two places' payment systems, simplifying the process for cross-border small-value payments. AlipayHK also promoted the recovery of the retail industry through activities such as the "Hong Kong Happy Shopping Festival," injecting new vitality into the Hong Kong economy.

AlipayHK has opened up the payment channel between Hong Kong and the Mainland through cross-border payment functions, greatly facilitating cross-border consumption and daily payments for residents of the two regions. By connecting to AlipayHK, merchants in Shenzhen, such as Hema Fresh Supermarket, have achieved a 183% increase in transaction volume. The convenience of cross-border payments has stimulated consumers' desire to shop and also driven the flow of people between Shenzhen and Hong Kong. In 2023, the number of vehicles passing through the ports between Shenzhen and Hong Kong increased by 64% year-on-year, showing that the interaction between the two places is becoming increasingly frequent.

In Hong Kong, public transportation such as subways and busses have implemented QR code payments, and Hong Kong residents can conveniently use AlipayHK for payment. In addition, AlipayHK has cooperated with the Hong Kong tourism industry to launch cross-border tourism services, providing functions such as Mainland train tickets and hotel reservations, further facilitating the travel of Hong Kong citizens. In terms of localized services, AlipayHK has launched a series of financial services, including cross-border remittances, bill payments, and credit card binding, through cooperation with local banks in Hong Kong. Especially for ethnic minority groups in Hong Kong, AlipayHK has launched a Philippine remittance service to help a large number of Filipino foreign workers reduce remittance costs. In summary, AlipayHK has brought a more convenient cross-border life experience to residents of both places through one-stop cross-border payments and localized services, and promoted the economic integration of the Guangdong-Hong Kong-Macao Greater Bay Area and the realization of two-way convenient living.

Promoting Diverse Tourism Experiences in Macau

Macao's tourism industry is undergoing all-round diversified development, gradually moving away from the traditional gambling tourism model and transforming towards a more comprehensive and creative direction. By providing convenient cross-border payment solutions, Alipay+ has brought convenient payment experience to Macao for global tourists, especially mainland tourists, and promoted the recovery of Macao's tourism market in the post-epidemic era. According to data from the China Tourism Academy, Macao became the preferred destination for mainland outbound tourism in the first half of 2023. In the first seven months of 2024, Macao received nearly 19.75 million tourists, an increase of more than 37% year-on-year. In terms of tourism promotion and discounts, in collaboration with e-commerce platforms and travel agencies, Macao has launched many discount activities and customized tourism products, attracting more domestic and foreign tourists.

As Macao has become a popular concert and exhibition destination, more and more tourists choose to stay overnight here. In the first half of 2024, the hotel occupancy rate in Macao reached 88.5%, an increase of nearly 11% from 2019 before the epidemic. After entering the summer vacation, the average occupancy rate of hotels exceeded 90%, and the cost-effectiveness of three-star and four-star hotels also attracted a large number of tourists. Macao's cultural and creative tourism activities are also gradually enriched. For example, in the summer of 2024, the Macao Government Tourism Office (MGTO) launched the "Cool Macao" campaign to encourage tourists to redeem local food and souvenirs for free by spending at least MOP 100 and checking in on social media. Alipay+ further promotes the payment and consumption convenience of tourists through the convenience of cross-border payment, allowing them to easily enjoy various entertainment activities.

Fostering Japan's economic growth and facilitates cross-border consumption.

The introduction and popularization of Alipay+ in Japan has enhanced the internationalization of the Japanese mobile payment market and promoted the market's diversification. Alipay+ integrates merchants from different regions and sizes into a seamless and unified ecosystem through international payment solutions. As of March 29, 2024, Alipay+ has connected 2 million local merchants in Japan, allowing businesses of all sizes to better serve global travelers and benefit from a seamless payment experience facilitated by their local e-wallets. The transaction volume of Alipay+ 's electronic wallet and banking applications in Japan surged by more than 200% year-on-year in the first three weeks of March compared with 2023.

The establishment of an international payment network helps local Japanese businesses optimize services and expand markets. Daimaru Matsuzakaya Department Store provides Chinese tourists with familiar and convenient payment options by connecting to the Alipay+ payment system. In the first two months of 2024, the store's transaction volume through Alipay+ partners grew by over tenfold year-on-year, enhancing both customer experience and competitiveness. Akito Kimura, manager of the inbound customer experience improvement department, said that the popularity of digital payments allows them to better meet the needs of international customers while simplifying operational processes. Many small merchants in Asakusa, a tourist hotspot, have also experienced a surge in cross-border transactions by accessing QR code payment technology. At the beginning of 2024, the region's cross-border payment transaction volume increased by more than five times on average every month. This transformation not only provides tourists with a hassle-free payment experience, but also accelerates the digital transformation of local merchants.

Alipay+ also actively supports the payment needs of tourists from Hong Kong and other regions. For example, Hong Kong residents use AlipayHK to make payments in Hong Kong dollars in Japanese stores and taxis, benefiting from efficient and competitive exchange rates. In the first quarter of 2024, AlipayHK's transaction volume and count increased 3.5-fold and fivefold quarter-on-quarter, highlighting the strong demand for cross-border payments. The in-depth cooperation with PayPay not only simplifies the payment experience for tourists, but also enables more small and medium-sized merchants to leverage the global digital payment ecosystem.

18.5 Experience and Insights

In establishing a robust governance system, particularly in popular tourist destinations like Thailand and Malaysia, the Alipay+ for SDGs initiative collaborates with central banks to offer cross-border digital payment services for international tourists and bolster digital national capacity building. This endeavor aims to provide a seamless cross-border digital payment infrastructure for international tourists, aligning with Sustainable Development Goal 9. The "ePemula" project in Malaysia exemplifies local governance by leveraging technology to enhance the efficiency and security of government-funded relief programs. This initiative aligns with SDG 16: Peace, Justice, and Strong Institutions, specifically by ensuring public funds are distributed transparently and without corruption. The introduction of ZOLOZ's e-KYC technology not only streamlined the registration process but also reduced the risk of fraud, thereby supporting SDG 5: Gender Equality and SDG 10: Reduced Inequalities by ensuring that funds reach the intended beneficiaries without discrimination.

Regarding environmental conservation, the Alipay+ for SDGs project upholds the principles of the global compact in human rights, labor, environmental, and anti-corruption practices. The project promotes the adoption of Ant Forest in locations such as the Philippines, supporting local partners under Sustainable Development Goal 17 to establish FinTech-enabled public forest projects known as G Forest. Drawing insights from China's eco-friendly travel practices in destinations like Macao and Hong Kong, the initiative enhances green travel experiences for mainland tourists. It advocates for sustainable tourism through promoting public transportation to Chinese tourists outbound, contributing to Sustainable Development Goal 13. To enhance the development of green and sustainable tourism regarding environmental protection, the "Green Stay" plan, a collaboration between InterContinental Hotels Group and Ant Group, promotes sustainable tourism by reducing the use of disposable items, directly supporting SDG 12: Responsible Consumption and Production. This initiative encourages environmental consciousness among guests and leverages Green Cloud's digital ecosystem to enhance the efficiency of the hotel industry, which in turn supports SDG 8. Besides, Singapore optimizes urban operations through intelligent transportation systems and digital payment infrastructure, reducing carbon emissions and promoting a low-carbon transformation, contributing to Sustainable Development Goal 13.

In the realm of social development, Alipay+ plays a pivotal role in fostering sustainable tourism and driving job creation and economic growth in regions like Thailand, Malaysia, Macao, and Hong Kong. The initiative supports the local digital economy, addresses employment disparities under Sustainable Development Goal 10, and contributes to building sustainable cities and communities as outlined in Sustainable Development Goal 11. By guiding tourists to explore world heritage sites and local businesses, Alipay+ promotes responsible consumption in line with Sustainable Development Goal 12. For social partnership and inclusive collaboration, MY Bank's focus on small businesses and rural areas through financial inclusion and microloans aligns with SDG 1: No Poverty and SDG 2: Zero Hunger by providing capital to businesses that can stimulate local economies and agricultural development. Ant Group's rural finance strategy and the "Daurian Starling" project aim to address the financial needs of the underbanked in rural areas by providing access to financial services that can lead to economic empowerment, also supporting these goals. The "310" loan process and the "Wild Goose System" for supply chain management support SDG 9: Industry, Innovation, and Infrastructure by facilitating access to credit and improving the efficiency of supply chains, which are crucial for business growth and economic development. Moreover, these practices also align with SDG 10: Reduced Inequalities by bridging the gap between urban and rural financial services. Anext Bank's Anext Business Loan product caters to the needs of SMEs in cross-border e-commerce, supporting SDG 9: Industry, Innovation, and Infrastructure by providing accessible financing options that can foster international trade and economic growth. This initiative also contributes to SDG 17: Partnerships for the Goals by promoting regional and global economic cooperation. In Indonesia, DANA's adoption of Alibaba Cloud services to enhance its digital operations supports SDG 9: Industry, Innovation, and Infrastructure by improving the resilience and efficiency of digital services, which are essential for economic development and social inclusion. Akulaku Group, PayLater, BillEase, and Kredivo's expansion of their digital payment services, particularly in the BNPL market, also aligns with SDG 9 by promoting financial innovation and accessibility. Alipay+ integrates local e-wallets in the Philippines, such as Gcash, Maya, and HelloMoney, to offer convenient cross-border payment solutions, support the development of the digital payment ecosystem, and promote the coverage of inclusive finance, in line with SDG 8. By collaborating with local and international payment platforms, Alipay+ supports the digital transformation of the Philippines and the collaborative development of regional payment networks, contributing to SDG 17. In advancing digital payment infrastructure, Alipay+ collaborates with merchants and institutions in South Korea and Japan in adopting digital payments, enhancing payment convenience for international tourists, and facilitating the digital transformation of the tourism market in East Asia, contributing to SDG 9. In 2023, Alipay International expanded pilots to Malaysia, Thailand, Singapore, South Korea, and Japan through the Sustainable Travel Program

(STP) with 13 international e-wallets. In 2024, Alipay+ connects 90 million local merchants with 1.5 billion users across 32 popular digital wallets in Asia and beyond. In 2025, Ant International launched the Alipay+ Global Travel Gala, covering ten popular tourist destinations favored by mainland Chinese tourists, including Japan, South Korea, Hong Kong, Macau, Singapore, Malaysia, Australia, Italy, France, and the United States, to raise their awareness of SDGs. Partners in Japan include the Osaka Nagai Botanical Garden, Universal Studios Japan, Legoland Discovery Center Osaka, Osaka Solaniwa Onsen, Tokyo Tower, Tokyo Skytree, Shinjuku Gyoen National Garden, and Sanrio Puroland. Partners in South Korea include N Seoul Tower, Lotte World, Gyeongbokgung Palace, Myeongdong Theater, Busan Songdo Marine Cable Car, Busan X the Sky, Busan Spa Land, and Busan SEA LIFE Aquarium. Partners in Hong Kong include Ocean Park, Hong Kong Disneyland, Ngong Ping 360, and the Central Ferris Wheel. Partners in Macau include Studio City Water Park, Macau Tower, teamLab SuperNature Macao, and The Parisian Macao's Eiffel Tower. Partners in Singapore include Universal Studios Singapore, Gardens by the Bay, S.E.A. Aquarium, and Night Safari. Partners in Malaysia include the National Zoo of Malaysia, Sunway Lagoon, KL Hop-On Hop-Off Bus, Aquaria KLCC, Langkawi Crocodile Adventureland, Langkawi Wildlife Park, Langkawi Underwater World, and Langkawi SkyCab. Partners in Australia include Taronga Zoo, Scenic World Blue Mountains, Featherdale Wildlife Park, Luna Park Sydney, Melbourne Skydeck, Puffing Billy Railway, Melbourne Museum, and Werribee Open Range Zoo. Partners in Italy include the Milan Cathedral, Teatro alla Scala, Pinacoteca di Brera, Uffizi Gallery, Galleria dell'Accademia, Palazzo Pitti, and Basilica of Santa Croce. Partners in France include the Arc de Triomphe, Eiffel Tower, Louvre Museum, Palais Garnier, Château de Chenonceau, Château de Chambord, Château d'Amboise, and Puy du Fou. Partners in the United States include Disneyland California, Universal Studios Hollywood, Griffith Observatory, Natural History Museum of Los Angeles County, Museum of Modern Art New York, Statue of Liberty, Rockefeller Center, and Empire State Building. For the reputation from Alipay+ for SDGs, I have been invited to participate in sustainable finance standards across various sectors like real estate and supply chain. The Alipay+ for SDGs receives accolades from the United Nations and leaders in Singapore (Prime Minister, Mr. Lawrence Wong), Macau (Chief Executive Officer, Mr. Ho Iat-seng), Malaysia (YBrs. Tuan Samuel Lee Thai Hung, Deputy Director-General,

Tourism Malaysia), and Thailand (Tanes Petsuwan, Deputy Governor, Tourism Authority of Thailand) for promoting sustainable tourism, digital transformation, and economic development.

18.6 Conclusion

This chapter evaluates the measurable social impacts of these efforts on achieving the SDGs, providing a critical analysis of the effectiveness of FinTech in driving sustainable change. These efforts have been recognized by governments, the industry, and international organizations. For example, at the International Science and Technology Innovation Expo in 2024, Macau's Chief Executive Region, Mr. Ho Iat Seng, visited MPay, a member company of Ant International, saying: "Many thanks to MPay, and other member companies of Ant Group and Ant International for bringing new industries, new business models, and new opportunities to Macau, enabling the digital transformation of traditional industries. This will help promote Macau's six advantages and, serving as a 'golden business card' of Macau, make this international metropolis more dazzling."

At the Singapore FinTech Festival (SFF), renowned as one of the largest FinTech summits globally, Singapore's Prime Minister, Lawrence Wong, paid a visit to the booth of Ant Group. He personally engaged with the Alipay+ D-store solution, using his phone to scan and order a cup of Huggs Coffee. After seeing that Huggs Coffee, Singapore's largest local coffee chain brand with 22 chain stores across the country, has embraced Alipay+ 's D-store to undergo a digital operational transformation, Mr. Wong confirmed the contribution of this project as follows: "I am thrilled to witness the local physical coffee shops in Singapore embracing digital transformation!" Indeed, prominent platforms like Chope, and the global fast food chain Burger King have already adopted Alipay+ 's digital solution.

Sanda Ojiambo, Assistant Secretary-General of the United Nations, and Director-General of the UN Global Compact (UNGC), said, "We sincerely hope that these cases can provide ideas and experience for global companies, especially those in developing countries." On November 29, 2024, the UNGC issued the Best Sustainable Development Cases in the 20th anniversary of the ESG principle since it has been proposed by the UNGC. Ant Group's "Green Computing and Public Participation in Low-Carbon Innovation" was featured, along with other prominent companies like China Development Bank. As examined by this book, Ant Group's sustainable development strategy is acknowledged as it "integrates commercial and social value through digital technological innovation, green inclusion initiatives, and open ecology for SMEs. Ant plans to continue investing in AI and green computing to support China's national decarbonization strategy and global SDGs."

It shows that Sustainable Financialization is essential reading for policymakers, business leaders, academics, and anyone interested in the intersection of finance, technology, and sustainability. It demonstrates that FinTech is not just a tool for financial transactions but a powerful agent for positive social change and sustainable development.

References

Alipay+ for HK users in shopping at Shenzhen: https://www.alipayplus.com/news/detail/Alipayplus_shenzhen

Alipay+ for Asian E-Wallet users in Japan: https://www.alipayplus.com/news/detail/alipayplus-transactions-triple-in-2024

Alipay+ in collaboration with Macau Gov in consumption coupon: https://www.alipayplus.com/news/detail/Macao_AlipayHKoffers

Alipay+ in partnership with Singapore Tourism Board: https://www.alipayplus.com/news/detail/stb_partnership

Alipay+ in partnership with over 10 Asian E-Wallets in Mainland China: https://www.alipayplus.com/news/detail/ant_introduces_10_asian_e wallets_to_chinese_mainland

Alipay+ transactions triple in 2024 as travel industry embraces cross-border digital payment and marketing technology: https://www.morningstar.com/news/business-wire/20241009199791/alipay-transactions-triple-in-2024-as-travel-industry-embraces-cross-border-digital-payment-and-marketing-techno logy

Ant Group doubles down on global expansion with cross-border payments offering Alipay+: https://www.cnbc.com/2024/05/06/chinese-fintech-ant-group-doubles-down-on-global-expansion-with-alipay.html

Index

J. Sun, *Sustainable Financialization*, Sustainable Development Goals Series, https://doi.org/10.1007/978-981-96-3580-1

E

F

G

GPSR Compliance

The European Union's (EU) General Product Safety Regulation (GPSR) is a set of rules that requires consumer products to be safe and our obligations to ensure this.

If you have any concerns about our products, you can contact us on ProductSafety@springernature.com

In case Publisher is established outside the EU, the EU authorized representative is:

Springer Nature Customer Service Center GmbH
Europaplatz 3
69115 Heidelberg, Germany

Batch number: 10703843

Printed by Printforce, the Netherlands